Stata for the Behavioral Sciences

Stata for the Behavioral Sciences

Michael N. Mitchell

A Stata Press Publication
STATA CORPORATION
College Station, Texas

Published by Stata Press, 4905 Lakeway Drive, College Station, Texas 77845
Typeset in LaTeX 2_ε
Printed in the United States of America

10 9 8 7 6 5 4 3 2 1

ISBN-10: 1-59718-173-0
ISBN-13: 978-1-59718-173-0

Library of Congress Control Number: 2015947163

Acknowledgments

This book was made possible by the help and input of many people. First and foremost, I want to thank my technical editor, Kristin MacDonald. I felt as if she read the book with X-ray vision, seeing not just the words on the page but the underlying skeletal structure of the book. With this insight, she provided many helpful thoughts and suggestions that helped me to revise, restructure, and grow this book into the version that you see. I also want to thank Chuck Huber for creating so many fantastic Stata tutorial videos and permitting me to reference them from within the book. I am very grateful for the help from the Stata Press production team, namely, Stephanie White and Patricia Branton, as well as David Culwell for providing copy-editing that excises the blemishes in my writing while accentuating my voice of expression. As always, I am grateful to Vince Wiggins for his support in navigating some of the choppy waters of writing a book and to Annette Fett for the creative cover design of the book you are holding. Finally, I want to thank the entire StataCorp and Stata Press team—I wish all of you could know, as I do, how their intelligence and creativity are exceeded only by their genuine warmth of spirit.

I also want to thank the National Opinion Research Center for granting me permission to use the General Social Survey dataset for this book. My thanks to Tom W. Smith and Jibum Kim for facilitating this process and helping to create a unique dataset reflecting opinions and attitudes spanning 40 years and 5 decades.

Contents

Tables

Figures

Preface

I worked as a statistical consultant at the University of California, Los Angeles, ATS statistical consulting group for over 12 years. Before the start of every walk-in consulting session, I would wonder about the questions that would walk into our office. Every session was different, with people bringing questions from all parts of the campus. When a new client walked in the door, one of my first questions was, What department are you from? It might have seemed like a polite question as we got to know each other before diving into the heart of his or her problem. But this was an essential question for me: the answer would guide the way I handled the entire visit.

In working as a consultant who served so many schools and departments, I discovered that there are many regional dialects of statistics. Depending on your school or department, there are certain types of statistical models you emphasize, customs you embrace, and types of terminology you favor. This is why there are statistics books that are written specifically for certain disciplines—to address the statistical customs and traditions within that discipline.

My home discipline is psychology, in which I received my bachelor's, master's, and doctorate. As I was taught statistics, my professors focused on forming specific hypothesis tests for the exact predicted pattern of results. They emphasized taking a laserlike focus on specific contrasts that would directly test the hypothesis of interest. This was especially the case when I was taught about factorial analysis of variance (ANOVA), where a finding of an overall interaction effect was nothing to get excited about because the significant interaction could be consistent with a variety of patterns of results, some of which could be contrary to our hypotheses. We were taught to graph the interactions and probe and dissect the interaction using planned contrasts to test for the exact pattern of results that we hypothesized.

I brought this training to my statistical consulting and my use of statistical packages. However, I frequently found that my training regarding ways of dissecting interactions was not so easily supported by statistical packages—that is, until the most recent versions of Stata. Unlike any other statistical package that I have ever used, Stata provides a suite of tools that allows us to probe, interpret, understand, and graph the results of ANOVA models. These tools are incredibly powerful; they are also very simple and intuitive. In this book, I show how you can use this suite of ANOVA tools to easily form contrasts among groups and dissect interactions with surgical precision. This allows you to present tests of hypotheses regarding the specific pattern of your results, establishing not only that your results are significant but also that they are in the pattern predicted. The suite of tools also integrates graphing tools so that you can use graphics as a means of interpreting your results and for presenting results to others.

The heart and spirit of this book is about showcasing this suite of ANOVA tools that Stata offers, but that does not mean this book is limited to just the presentation of ANOVA. This is because this suite of ANOVA tools can be applied to a wide variety of designs, including analysis of covariance, analysis of covariance with interactions, repeated measures designs, longitudinal designs, and the analysis of survey data to name just a few. This suite of tools can also be used in the context of a wide variety of regression modeling methods, including ordinary regression, robust regression, multilevel models, logistic regression, and Poisson regression.

As I see it, one of the strengths of learning statistics from a behavioral science perspective is seeing how factorial designs can help us understand how the effect of one variable is moderated by another variable through testing of interactions. With most statistical packages, you are handcuffed to using these tools only in the context of a traditional ANOVA. Once you extend your reach outside that realm, these tools are taken away from you. In Stata, you carry this suite of tools with you as you run a multilevel model, a robust regression, a logistic regression, or even a regression based on complex survey sampling. In this book, my aim is to show how you can use these familiar tools and to enable you to apply them across a wide variety of designs and modeling methods.

While this book draws upon my statistical training from the perspective of psychology, it is written for anyone in the behavioral sciences and anyone who would like to learn how to apply ANOVA (and ANOVA-like tools) to a variety of designs and modeling techniques using Stata. Regardless of your home discipline, I hope this book shows how you can use Stata to understand your results so that you can interpret and present them with clarity and confidence.

Valencia, California   Michael N. Mitchell
July 2015

Part I

Warming up

As implied by the title, this part of the book warms us up for the more substantial parts of the book.

The first chapter provides information to help you get the most out of the book. It includes information about how to download the associated datasets, reasons why I think a behavioral scientist would want to use Stata, an overview of the book, and recommended supplemental resources.

Chapter 2 covers descriptive statistics such as tabulations (frequency distributions), summary statistics, cross tabulations, and summary statistics for specific subgroups.

This part concludes with chapter 3, which introduces basic inferential statistics such as two-sample t tests, one-sample t tests, and one- and two-sample tests of proportions.

1 Introduction

1.1   Read me first!

This book is filled with examples I hope you will try on your own. This section shows how you can download the example datasets and supplemental programs.

1.1.1   Downloading the example datasets and programs

All the example datasets and programs can be downloaded from within Stata using the following commands:

```
. net from http://www.stata-press.com/data/sbs
. net install sbs
. net get sbs
```

The `net install` command downloads the programs, I wrote for the book. These programs are listed below with a brief description and the chapters in which they are used:

- The `showcoding` command shows the coding of an original variable as compared with the recoded version. This program is used in chapter 13.
- The `power multreg` command computes power for simple and multiple regression analyses. It is used in chapter 19.
- The `power nestreg` command computes power for a nested regression. It is used in chapter 19.

The `net get` command downloads the example datasets. I encourage you to download these so that you can reproduce and extend the examples illustrated in this book.

1.1.2   Other user-written programs

The book also uses a number of user-written programs. All of these user-written programs are stored at the Statistical Software Components (SSC) repository and can be downloaded using the `ssc` command. The programs are described below along with the `ssc` command that you can use to download the program.

The fre command

This program shows frequencies of a variable with the values of the variable and the value labels. It is an alternative to the `tabulate` command for one-way tabulations. You can download it by typing

```
. ssc install fre, replace
```

The `fre` command, written by Ben Jann (2007a), is used in many chapters throughout the book.

The esttab command

This program creates formatted estimation tables. It is an extended version of the Stata `estimates table` command. You can download it by typing[1]

    . ssc install estout, replace

The `esttab` program is extensively used in chapter 16, showing how to create presentation-quality regression tables. This program was also written by Jann (2007b).

The extremes command

The `extremes` command displays extreme values for a variable. You can download it from the SSC repository using the `ssc` command as shown below:

    . ssc install extremes, replace

This program, written by Nicholas J. Cox (2003), is used in chapter 18 on regression diagnostics.

I am grateful to Jann and Cox for their kind permission to use their programs in my book. These programs not only helped to make my book better but also showed the kinds of user-written programs that have been created by the skilled and generous members of the Stata community. I touch on this point in section 1.2.7 later in this chapter.

1.2   Why use Stata?

I have extensively used and supported many statistical packages, both general and specialized. As someone whose background is in psychology and who feels strong connections to the behavioral sciences, I think there are many reasons for behavioral scientists to choose Stata as their first statistical package or to switch from their current package over to Stata, as described below.

1.2.1   ANOVA

Analysis of variance (ANOVA) is a cornerstone statistical technique of the behavioral sciences, especially factorial ANOVA with its ability to dissect the interactions to answer very meaningful substantive hypotheses. Stata offers exceptionally powerful (yet easy-to-use) tools that allow you to analyze and dissect results from ANOVA. Let me illustrate with an example.

---

1. The command `esttab` is part of the `estout` package of programs, so it is downloaded by typing `ssc install estout`.

Imagine a study in which subjects either are assigned to a control group or will receive a kind of therapy called optimism therapy, which is geared toward increasing optimism. All subjects in the study were screened for depression and only nondepressed subjects were included. At the conclusion of the study, the optimism of the participants was measured, and optimism was found to be greater in the optimism therapy group than in the control group. As we can see in figure 1.1, the optimism of those who received optimism therapy is 10 units greater than the optimism of the control group. The effect of optimism therapy is that it boosted optimism by 10 units.

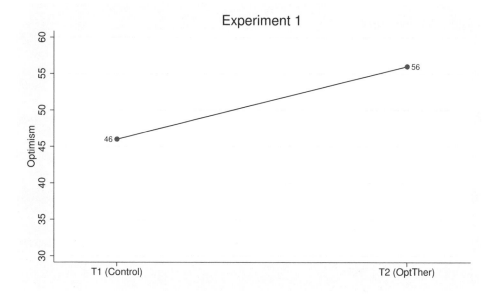

Figure 1.1: Results of experiment 1

The experimenter decides to replicate the previous study but with a twist: in addition to recruiting nondepressed participants, the experimenter will also recruit participants who are depressed. The experimenter expects that among those who are nondepressed, he or she will find the same kind of pattern as was found in the first study (that is, shown in figure 1.1). The experimenter wonders how those who are depressed will respond to this therapy. Will they equally benefit from the therapy? Will they benefit from the therapy but not as much as those who are nondepressed? Will they not benefit from the therapy at all? These three possible patterns are described further below and also illustrated in figure 1.2.

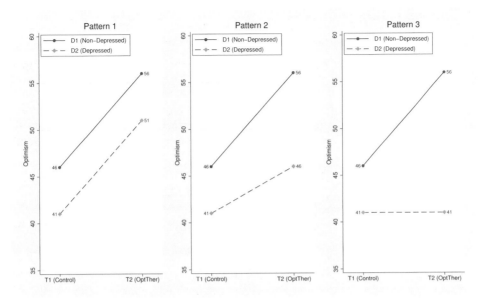

Figure 1.2: Three possible patterns of results from a 2 by 2 factorial design

Pattern 1. Optimism therapy benefits those who are depressed and nondepressed equally alike. Among those who are nondepressed, the effect of optimism therapy is 10 units (56 versus 46), the same gain as for those who are depressed (51 versus 41).

Pattern 2. Optimism therapy boosts optimism by 5 points for those who are depressed but does not yield the 10-point boost experienced by nondepressed participants. Optimism therapy is effective for those who are depressed but just not to the same degree as for those who are not depressed.

Pattern 3. Optimism therapy has no effect for those who are depressed (yielding no change from the control group, 41 versus 41), while optimism therapy delivered a 10-point boost in optimism for nondepressed participants.

The comparisons to distinguish patterns 1, 2, and 3 involve tests of interactions and tests to dissect the specific pattern of the interaction. As illustrated in chapter 7, Stata has exceptionally powerful, yet easy-to-use, tools for dissecting interactions with surgical precision.

But this is not the only reason I think Stata is an outstanding statistical package for behavioral scientists. Here are some other reasons.

1.2.2   Supercharging your ANOVA

Suppose the outcome of the previous study was changed to a binary outcome, necessitating the use of a logistic regression. Not to worry, because if you are using Stata, you can perform a two-by-two factorial logistic regression drawing upon the same concepts and tools as you could in performing a two-by-two ANOVA. The same can be said if you were using a count outcome and wanted to run a two-by-two Poisson regression model. In fact, Stata has brought ANOVA-like analysis technology to virtually all of its analysis commands. As a behavioral scientist, you can apply these familiar designs to an exceptionally wide variety of statistical models.

1.2.3   Stata is economical

Stata has very attractive academic pricing, including student pricing, that can put Stata into your hands (at the time this book was written) for \$125 a year (or \$54 a year if you use very small datasets). In addition, there is no extra price for extra modules. Some statistical software packages charge extra for modules that address missing data, the analysis of complex survey data, structural equation modeling, or bootstrapping. In fact, there are some specialized statistical packages that people buy to add these features to their existing statistical package. With Stata, all of these features are included at no extra cost.

Note: Stat/Transfer

If you want the ability to translate datasets from a variety of formats (for example, from Stata to SAS, from SPSS to Stata), then I suggest you consider purchasing Stat/Transfer, which you can obtain directly from StataCorp. Stat/Transfer can convert virtually any kind of dataset from one format to another format (for example, from Stata to SAS, from SPSS to Stata, from SAS to SPSS, and from Access to Excel). I have used Stat/Transfer for many years and been consistently impressed with its ease of use and how well it works. Speaking for myself, I cannot imagine doing my daily work without Stat/Transfer.

1.2.4   Statistical powerhouse

Stata is a statistical powerhouse in terms of its statistical features. Rather than listing all the features, I will point you to http://www.stata.com/features/ so that you can look at them yourself.

Tip: Stata/MP for multiprocessing

If you want to really unleash the power of Stata, consider Stata/MP. Taking advantage of multicore and multiprocessor computers, Stata/MP speeds up computations by using a divide and conquer strategy. Using 8 cores, the median time to complete an estimation command is 4.1 times faster. The *Stata/MP Performance Report* (available at http://www.stata.com/statamp/statamp.pdf) describes the concepts involved in parallel processing and details the performance gains achieved across Stata commands. Consider these performance gains that can be achieved on a 16-core machine—a linear regression using the `regress` command runs 16.5 times faster, a 2-way ANOVA using the `anova` command runs 12.7 times faster, and a logistic regression using the `logistic` command runs 11.1 times faster.

1.2.5 Easy to learn

Looking at the list of statistical features, you might feel overwhelmed wondering how you could learn all of these different commands. The amazing thing is how similarly the commands work. To run a regression predicting y from x1, x2, and x3, you type

```
. regress y x1 x2 x3
```

To run a robust regression predicting y from x1, x2, and x3, you type

```
. rreg y x1 x2 x3
```

To run a logistic regression predicting y from x1, x2, and x3, you type

```
. logistic y x1 x2 x3
```

To run a Poisson regression predicting y from x1, x2, and x3, you type

```
. poisson y x1 x2 x3
```

If Stata had a regression style command called `xyzreg`, you would likely be able to use it to predict y from x1, x2, and x3 by typing

```
. xyzreg y x1 x2 x3
```

What could be simpler?

1.2.6   Simple and powerful data management

With its extensive statistical capabilities, I think many overlook the power of Stata for data management. Stata features many specialized commands that are like data management shortcuts, directly handling commonly difficult data management tasks. Example commands include `reshape long`, `reshape wide`, `egen`, `collapse`, and `merge`.

1.2.7   Access to user-written programs

One of the greatest virtues of Stata is the way it dovetails the ease of developing add-on programs with a great support structure for finding and downloading these programs. This virtue has led to a rich and diverse network of user-written Stata programs that extend the capabilities of Stata. As a result, the power of Stata is greatly extended and enhanced by these user contributions. Such programs are easily found, downloaded, and installed with the `search` command.

The `search` command connects to Stata's own search engine, which indexes user-written Stata programs from all around the world. Typing, for example, `search regression` searches for and displays Stata resources associated with the keyword `regression`. The resources searched include the Stata online help, Stata frequently asked questions (FAQs), the *Stata Journal* and its predecessor, the *Stata Technical Bulletin*, and programs posted on the websites of Stata users from around the world. All of these results are culled together and displayed in the Stata Viewer window. You can then point to and click on the programs you want to download and install.

Video tutorial: Downloading user-written programs

See a video demonstration of how to find and download user-written programs at http://www.stata.com/sbs/user-written.

Many of these programs are hosted at the SSC archive. This repository makes it easy for people to contribute programs to the Stata community and makes it easy for end users like you and me to easily download such programs.

You can see the newest additions to this archive by typing

```
. ssc new
```

You can also see the most popular downloads by typing

```
. ssc hot
```

Even more information: Downloading user-written programs

The *Getting Started* manual has more information about finding and downloading user-written programs. Just type `help gs`, and see the chapter titled "Updating and extending Stata—Internet functionality"

1.2.8   Point and click or commands: Your choice

You can use Stata with a point-and-click interface or by typing in commands. In this book, I focus exclusively on showing commands, but at any time, you can explore Stata via the drop-down menus, which give you point-and-click access to data management commands (via the **Data** drop-down menu), graphics commands (via the **Graphics** drop-down menu), and statistics commands (via the **Statistics** drop-down menu). Whenever you execute commands via the menus, Stata will display the command equivalent in the output window, teaching you the commands even as you use the point-and-click interface.

1.2.9   Powerful yet simple

In life, you often face the dilemma of choosing power or simplicity. With Stata, you can have both. It offers the simplicity of a point-and-click interface and commands that are simple to use; it also offers the power of being able to write your own programs using its Mata programming language.

1.2.10   Access to Stata source code

Stata is not just a statistical software program; it is a statistical programming environment. You can write your own Stata commands and programs, and you can view the source code for nearly all Stata commands. For example, would you like to see the source code for the `ttest` command? If so, just type

```
. viewsource ttest.ado
```

You can view the source code for nearly all Stata commands in this way. This allows you to see how every command works. Moreover, it means that you could even make your own version of any such Stata command with your own personal customizations.

1.2.11   Online resources for learning Stata

Another reason to use Stata is that there are so many terrific online resources to help you learn Stata.

The Stata resources and support page provides a comprehensive list of online resources available for Stata. It lists official resources available from StataCorp as well as from the Stata community. See http://www.stata.com/support/.

The Stata Resource links page provides a list of resources created by the Stata community to help you learn and use Stata; see http://www.stata.com/links/. Among the links included there, I highly recommend the UCLA Institute for Digital Research and Education (IDRE) Stata web resources at http://www.ats.ucla.edu/stat/stata/, which include FAQs, annotated Stata output, textbook examples solved in Stata, and online classes and seminars about Stata.

The Video tutorials on using Stata page contains links to numerous videos illustrating a wide variety of topics about Stata. These videos uniquely exploit the ability to show you about the use of Stata in a way that a written explanation cannot convey. Further, the videos are brief and to the point (usually lasting between two and five minutes). While you can find the videos on the StataCorp YouTube channel at http://www.youtube.com/user/statacorp, the "Video tutorials on using Stata" page (at http://www.stata.com/links/video-tutorials/) shows the videos organized by topic.

The Stata Frequently Asked Questions page is special because it not only contains many frequently asked questions but also includes answers! The FAQs cover common questions (for example, How do I export tables from Stata?) as well as esoteric questions (for example, How are estimates of rho outside the bounds $[-1, 1]$ handled in the two-step Heckman estimator?). You can search the FAQs using keywords, or you can browse the FAQs by topic. See http://www.stata.com/support/faqs/.

Statalist is an independently run web forum that connects Stata users from all over the world. It began in 1994 as a listserv (hence the name "Statalist") and was relaunched in March 2014 as a web forum. The community is both extremely knowledgeable and friendly, welcoming questions from newbies and experts alike. Even if you never post a question, you can learn quite a bit by reading the questions and answers posted by others. You can visit the web forum at http://www.statalist.org/. And if you wish to read questions and answers that predate the web forum (going all the way back to 2002), you can visit the "Statalist archives" page at http://www.stata.com/statalist/archive/.

The Stata Blog covers many interesting and technical aspects of Stata. Entries are written by Stata's developers and technical support team; see http://blog.stata.com/.

The Stata Journal is published quarterly with articles that integrate various aspects of statistical practice with Stata. Although current issues and articles are available by subscription, articles over three years old are available for free online as PDF files. See http://www.stata-journal.com/.

The Stata Technical Bulletin is the predecessor of the *Stata Journal*. All of these issues are available for free online. Although many articles may be out of date, there are many gems that contain timeless information. For more information, see http://www.stata.com/bookstore/individual-stata-technical-bulletin-issues/.

Note: Help menu

It is easy to overlook or forget that the Stata **Help** drop-down menu is a central hub for directing you to many helpful resources, including the Stata documentation in PDF, the help files organized by content, and information about what's new in Stata. My favorite is the **Resources** item, which brings up *Resources for learning more about Stata*, which provides a concise and comprehensive list of resources to help you learn and use Stata. You can also access this help file by typing `help resources`.

1.2.12 And yet there is more!

For even more information and more reasons why you would enjoy using Stata, see the Stata webpage titled "Why use Stata" at http://www.stata.com/why-use-stata/.

1.3 Overview of the book

The book is divided into five parts, described below.

1.3.1 Part I: Warming up

As implied by the title, this part warms us up for the more substantial parts. This includes the current chapter you are reading. The next chapter, chapter 2, covers descriptive statistics such as tabulations (frequency distributions), summary statistics, cross-tabulations, and summary statistics for specific subgroups. This part concludes with chapter 3, which introduces basic inferential statistics such as two-sample t tests, one-sample t tests, and one- and two-sample tests of proportions.

1.3.2 Part II: Between-subjects ANOVA models

This part covers between-subjects ANOVA models, beginning with chapter 4, which covers one-way between-subjects ANOVA. This is followed by chapter 5, which illustrates contrasts that you can use for making comparisons among groups in a one-way ANOVA. Next, chapter 6 covers analysis of covariance (ANCOVA), illustrating its use in experimental designs (to increase power) and in nonexperimental designs (to attempt to statistically control for confounding variables). Chapter 7 introduces factorial designs, covering two-by-two designs, two-by-three designs, and three-by-three designs. This chapter emphasizes how to visualize and interpret the interactions. It also illustrates how to dissect two-way interactions using simple effects, simple contrasts, partial interactions, and interaction contrasts. The `contrast` command is illustrated for dissecting the interactions, and the `margins` and `marginsplot` commands are used to display and graph the means associated with the interactions. Chapter 8 illustrates ANCOVA-type analyses with the focus on interactions of the independent variable (IV) and the covariate (in other words, categorical by continuous variable interactions). Chapter 9 covers factorial models with three IVs. This chapter, like chapter 7, emphasizes the visualization and interpretation of the interactions, in this case focusing on the three-way interactions. Figures are used to visually understand the three-way interactions, and a variety of analytic and graphical methods is illustrated to also help understand them. Chapter 10 shows how you can extend the power of ANOVA by blending the ANOVA designs with regression commands. This chapter illustrates how you can extend your ANOVA design to analyze data that come from complex surveys, data that violate the homogeneity of variance assumption, or data with influential observations (via robust regression or quantile regression). This part concludes with chapter 11, which illustrates power analysis for ANOVA and ANCOVA.

1.3.3 Part III: Repeated measures and longitudinal models

This part covers two different strategies for analyzing designs with multiple observations on the same subject.

Chapter 12 covers repeated measures ANOVA designs. In such designs, participants are observed at more than one time point. All participants are observed according to the same time schedule. This chapter shows three examples illustrating the analysis of repeated measures designs. The first example includes a single repeated measures IV (see section 12.2). The second example illustrates a two-by-three between-within design where the between-subjects IV has two levels and the repeated measures IV has three levels (see section 12.3). The third example illustrates a three-by-three between-within design where the between-subjects IV has three levels and the repeated measures IV also has three levels (see section 12.4).

Chapter 13 covers longitudinal models. These models, in contrast to repeated measures designs, typically have a larger number of observations per subject, and the time gaps between the repeated measures can vary between people. This chapter includes four examples, all of which use multilevel modeling as the main analysis strategy. The first example models the dependent variable (DV) as a linear function of time (see section 13.2). The second example adds a between-subjects IV, which allows us to model the linear effect of time and explore IV by time interaction (see section 13.3). This example is similar to an ANCOVA with a treatment by covariate interaction (for instance, like the examples in chapter 8). The third example includes time as the only predictor but uses a piecewise modeling strategy for the effect of time (see section 13.4). The fourth example adds a between-subjects IV to the third example, modeling the interaction of the IV with the piecewise effects of time (see section 13.5).

1.3.4   Part IV: Regression models

This part illustrates how to use Stata commands to fit regression models. The chapters are ordered like a meal in which you decide to eat dessert first. The sweet and delicious chapters are presented first, deferring nutritional topics such as regression diagnostics and power analysis to the end. This part begins with chapter 14, which shows you how to perform multiple regression using Stata (showing you how to fit a simple linear regression model and multiple regression models) and how to test multiple coefficients within a multiple regression model. Chapter 15 covers more details about using the `regress` command, showing options for customizing output and how to create summary statistics based on the sample of observations included in the most recent regression analysis. This chapter also shows you how to store results of regression models for use later in your Stata session. This feature is used in chapter 16, which shows tools that you can use to create formatted regression tables. The chapter illustrates how to create such tables for display on the screen and how to create customized formatted output that can be used within a word processor like Word to create presentation-quality regression tables. Chapter 17 illustrates model-building tools, showing you how to fit multiple models using the same sample of observations. The chapter also shows you how to fit nested regression models and perform stepwise regression models. Chapter 18 illustrates commands for performing regression diagnostics, demonstrating analytic and graphical methods for identifying outliers. The chapter also illustrates analytic and graphical methods that you can use for testing for nonlinearity and how you can detect multicollinearity, assess the homoskedasticity assumption, and evaluate the normality of the residuals. This part concludes with chapter 19, which illustrates how to perform power analysis for a simple regression model and a multiple regression model and how to compute power for a nested multiple regression model.

1.3.5   Part V: Stata overview

The previous parts have focused on different statistical techniques, providing examples of how to perform analyses using those techniques in Stata. This part provides command-centric information, offering an overview of the use of Stata.

The first chapter (chapter 20) shows common features of estimation commands. Even though Stata has a very large number of estimation commands, they share a number of common features. This is by design, not by accident. Because estimation commands work similarly, what you learn about the behavior of one estimation command translates over to the use of other estimation commands. This chapter is about the features (behaviors) that estimation commands share.

Chapter 21 discusses a special set of commands called postestimation commands. They are called this because they are used after an estimation command (for example, after the `anova` or `regress` command). In particular, this chapter provides additional details about the `contrast`, `margins`, `marginsplot`, and `pwcompare` commands.

Chapter 22 provides brief information about basic data management commands in Stata. It illustrates reading data into Stata, keeping and dropping variables and observations, labeling data, creating variables, appending datasets, merging datasets, and reshaping datasets wide to long and long to wide.

The final chapter, chapter 23, recognizes that many readers of this book might be familiar with IBM® SPSS®. If you are such a reader, you might find yourself asking, for a given SPSS command, What is the equivalent Stata command? To answer such questions, this chapter lists commonly used SPSS commands (in alphabetical order) and shows the equivalent (or near equivalent) Stata command along with a brief example of the Stata command.

1.3.6 The GSS dataset

One of the commonly used datasets in this book is based on the General Social Survey (GSS). The GSS dataset was created and is collected by the National Opinion Research Center (NORC). To learn more, see http://www.norc.uchicago.edu/GSS+Website/. The GSS is a unique survey and dataset. It contains numerous variables measuring demographics and societal trends from 1972 to 2012. This is a cross-sectional dataset; thus the data for each year represent different respondents. (Note that the GSS does have a panel dataset for 2006, 2008, and 2010, but this is not used here.) In some years, certain demographic groups were oversampled. For simplicity, I will overlook this and treat the sample as though simple random sampling was used.

The version of the dataset we will use for the book is based on the GSS from 2012. This dataset was accessed by visiting http://www3.norc.org/GSS+Website/Download/ STATA+v8.0+Format/ and looking under the heading "Download Individual Year Data Sets (cross-section only)" and the subheading "GSS 1972–2012 Release 6". Clicking on the link *2012*, I downloaded a file named 2012_stata.zip, and unzipping that file yielded the dataset named GSS2012.dta. I created a Stata do-file that subsets and recodes the variables to create the analytic data file we will use, named gss2012_sbs.dta. This dataset is used below.

```
. use gss2012_sbs
```

The `describe` command shows that the dataset contains 1,974 observations and 42 variables.

```
. describe, short
Contains data from gss2012_sbs.dta
  obs:         1,974
  vars:            42                    17 Jul 2015 09:41
  size:      157,920
Sorted by:
```

Note that http://www3.norc.org/GSS+Website/Download/STATA+v8.0+Format/ provides some key information about missing value codes.

There are four missing values in the data:

- `.c`: Cannot choose.
- `.i`: Inapplicable. Respondents who are not asked to answer a specific question are assigned to IAP.
- `.d`: Don't know.
- `.n`: No answer.

This suggests that the special missing value code of `.c` indicates that the value is missing because the respondent could not choose a rating that reflected his or her happiness. The missing value code of `.i` indicates that the response is missing because the question was not asked of the respondent. (Note that some groups of questions are asked of only some randomly chosen respondents.) The missing value code of `.d` means that the answer is missing because the respondent did not know. Finally, a missing value code of `.n` means that there is no answer (for example, the respondent preferred not to respond). For more documentation about the GSS, you can visit http://www3.norc.org/GSS+Website/Documentation/. You can also learn more about missing value codes in Stata by typing `help missing`.

1.3.7 Language used in the book

I would like to comment on the language used in this book. I use language from the tradition of experimental design and the behavioral sciences. Here are some of the terms that I will be using and my intended meaning.

Independent and dependent variables. In ANOVA, it is traditional to call the categorical predictor an IV and the outcome the DV. This usage reflects the tradition that ANOVA was most commonly used for the analysis of designed experiments. I will continue to describe the categorical predictor as the IV and the outcome as the DV—even if the study is not a designed experiment and even if the design is not an ANOVA design. I choose this terminology to emphasize that the role these variables play (in a statistical sense) are the same, even if they do not arise from a designed experiment and even if they are not used in a traditional ANOVA analysis.

Note: Factor variables

Sometimes, when referring to an IV or a categorical variable, I will also use the term "factor variable". This is a Stata-specific term that refers to a categorical variable that has been entered into an ANOVA model or into a regression model by using the `i.` prefix (for example, `i.race`). Stata treats factor variables differently, knowing that they are categorical variables, and most Stata commands understand how to treat such variables differently from continuous variables (for example, age).

Covariate. In an ANCOVA design, a covariate is a continuous predictor, in addition to the IV in the prediction of the DV.

Effect. It can be very parsimonious to talk about a "treatment effect" when talking about the difference in the means for a treatment versus control group. Or in the context of regression, it can be useful to call the regression coefficient for a variable (such as age) the "effect of age". Whenever I use this term, I am not using it in the context of cause and effect but to describe an observed statistical relationship.

Note! Internal validity

A key question in many studies is whether statistically significant associations reflect underlying causal relationships. The issue of causal inference concerns the scientific integrity study design, and the ability to draw such causal conclusions is often described as the "internal validity" of the study. In this book, I will sidestep such issues but instead refer you to your favorite book in experimental methods for more information about the conditions that are necessary for drawing causal conclusions regarding statistically significant findings.

1.3.8   Online resources for this book

The online resources for this book can be found at the book's website:

http://www.stata-press.com/books/sbs.html

Resources you will find there include the following:

- All the datasets used in the book. I encourage you to download the datasets, reproduce the examples, and try variations on your own. You can download all the datasets into your current working directory from within Stata by typing

```
. net from http://www.stata-press.com/data/sbs
. net get sbs
```

- Errata (which I hope will be short or blank). Although I have tried hard to make this book error free, I know that some errors will be found, and they will be listed in the errata.
- Other resources that may be placed on the site after this book goes to press. Be sure to visit the site to see what else may appear there.

1.4   Recommended resources and books

This book focuses on how to analyze data from a behavioral science perspective and is not a general purpose book about the overall use of Stata. It omits topics such as an overall introduction to Stata and general principles of using Stata and provides very little details about data management or graphics. I made this deliberate choice because there are so many other resources that cover these topics and because the coverage of these topics is not specific to a behavioral scientist. Thus here I provide recommendations for resources to help you acquire this information.

1.4.1   Getting started

If you are new to Stata, I highly recommend the Stata *Getting Started* manual. There is a unique version of the *Getting Started* manual that shows what Stata will look like and how it works on your platform. (The manual comes in three separate versions written for Windows, Mac, and Unix.) You can access the *Getting Started* manuals by typing

```
. help gs
```

To get you started, I suggest you read the chapter titled "Introducing Stata—sample session". In addition, the chapters titled "The Stata user interface", "Using the Viewer", and "Getting help" should help you quickly feel comfortable in the Stata environment.

Video tutorial: The Stata interface

Take a video tour of the Stata interface at http://www.stata.com/sbs/interface.

1.4.2   Data management in Stata

The *Getting Started* manual shows how to get data into Stata in the chapters titled "Opening and saving Stata datasets" and "Importing data". It also covers "Creating new variables" and "Deleting variables and observations".

For more information about general topics in data management, I recommend the IDRE (formerly Academic Technology Services [ATS]) UCLA website. There is a special page devoted to the topic of data management at http://www.ats.ucla.edu/stat/stata/topics/data_management.htm.

For comprehensive coverage of the topic of data management (from basic tasks such as labeling variables and recoding variables to advanced tasks such as merging or reshaping datasets), I recommend my book titled *Data Management Using Stata: A Practical Handbook*, which is available at http://www.stata.com/bookstore/data-management-using-stata/.

Video tutorial: Getting help in Stata

See a video demonstration of how to get help in Stata at http://www.stata.com/sbs/help.

1.4.3   Reproducing your results

One topic that I have not addressed but is extremely important concerns how you can reproduce your results. This book illustrates commands that you can use for performing your analyses, but it does not show you how to create a procedure for saving these commands so you can easily execute them again. A related topic is how to save the results of your commands so that you can refer to the results in the future. The *Getting Started* manual has an excellent introduction to those topics. The chapter "Using the Do-file Editor—automating Stata" shows you how to save a sequence of Stata commands in a file called a do-file, which you can execute at a later time.[2] The chapter "Saving and printing results by using logs" shows you how to can save your results in a log file, which provides you a transcript of your commands and output from previous analyses. These two features can be combined so that your do-files automatically generate log files.

---

2. For IBM® SPSS® users, this is the equivalent of an SPSS syntax file.

The topic of creating and using do-files is also covered in my book *Data Management Using Stata: A Practical Handbook.*

Video tutorial: PDF documentation in Stata

Did you know that Stata has well over 12,000 pages of documentation that are just one click away? From the **Help** menu, click on **PDF documentation**. You can also see a video demonstration of how you can access PDF documentation within Stata at http://www.stata.com/sbs/pdf-documentation.

1.4.4   Recommended Stata Press books

StataCorp has a publishing arm called Stata Press, which issues books like this one. You can see a list of all the books at http://www.stata-press.com/catalog/. At that site, you can see a description of each book, including a detailed table of contents, comments from the Stata technical group, and a sample chapter. Every Stata Press book I have read has excelled in providing useful information about the use of Stata for researchers. Among these books, I particularly recommend the following as books that build upon what is presented in this book:

- *Discovering Structural Equation Modeling Using Stata, Revised Edition* by Alan C. Acock (2013). This book provides an excellent introduction to the use of structural equation modeling using Stata.
- *An Introduction to Survival Analysis Using Stata, Third Edition* by Mario Cleves, William Gould, Roberto G. Gutierrez, and Yulia V. Marchenko (2010). This book provides excellent and detailed information about how to perform survival analysis using Stata.
- *Multilevel and Longitudinal Modeling Using Stata, Third Edition (Volumes I and II)* by Sophia Rabe-Hesketh and Anders Skrondal (2012b). This two-volume set provides extensive and very detailed information about fitting multilevel and longitudinal models using Stata.
- *An Introduction to Stata Programming* by Christopher F. Baum (2009). This is an excellent book to help you learn and explore the power of Stata programming.

I would be remiss if I did not mention my other books published by Stata Press, listed below.

- *Data Management Using Stata: A Practical Handbook* by Michael N. Mitchell (2010). This book provides data management information to complement the statistical examples shown in the book you are holding.
- *A Visual Guide to Stata Graphics, Third Edition* by Michael N. Mitchell (2012b). This book visually illustrates the use of Stata graphics.

- *Interpreting and Visualizing Regression Models Using Stata* by Michael N. Mitchell (2012a). This book overlaps with the book you are holding, covering many of the same topics but discussing them in a way that would appeal to a more general Stata audience. It discusses modeling of categorical, continuous, and categorical and continuous interactions in a much more general fashion than the book you are holding.

2 Descriptive statistics

2.1 Chapter overview

This chapter introduces how to perform descriptive statistics. It begins with section 2.2, which introduces the General Social Survey (GSS) dataset as well as some of the Stata commands you can use to become familiar with a new dataset. This is followed by a series of sections that illustrate how to perform different kinds of descriptive statistics, namely, one-way tabulations (section 2.3), summary statistics (section 2.4), summary statistics by one group (section 2.5), two-way tabulations (section 2.6), and summary statistics by two groups (section 2.7).

2.2 Using and describing the GSS dataset

The examples from this chapter are based on analyses of the GSS from the year 2012 using a dataset named `gss2012_sbs.dta`. Once you have downloaded the datasets for the book (as described in section 1.1), you can load this dataset into Stata by typing

```
. use gss2012_sbs
```

We can use the `describe` command to obtain information about the dataset, including the number of observations, the number of variables, and a listing of all the variables and labels. The command is shown below, but the output is very long, so I have omitted it to save space. I suggest you try the command so that you can see the output for yourself.

```
. describe
(output omitted)
```

We can use the `describe` command with the `short` option to obtain a short description of the dataset. Note how the `short` option appears after the comma. This is how options are specified in Stata.

```
. describe, short
Contains data from gss2012_sbs.dta
  obs:         1,974
  vars:             42                       17 Jul 2015 09:41
  size:       157,920
Sorted by:
```

The output tells us that the dataset has 1,974 observations and 42 variables. It also tells us the size of the dataset and the date it was created.

Note: Viewing the dataset

If you want to view the dataset in memory, you can use the `edit` command. This invokes the Data Editor, which shows the dataset much like a spreadsheet, with one row per observation and one column per variable. You can type `help edit` for more information.

If we want a very space-saving listing of all the variables, we can type the `describe` command with the `simple` option. This provides a simple listing of all the variables in the dataset.

```
. describe, simple
id          female      marital3    educ4       realrinc    fepol       socbar
wtss        race        married     educ3       incomet     health      satfam7
vpsu        age         separated   hsgrad      faminc      weekswrk    rank
vstrat      agedec      nevmarry    cograd      happy3      socrel      polviews
sampcode    class       children    paeduc      vhappy      socommun    webhh
yrborn      marital     educ        maeduc      happy7      socfrend    emailhh
```

As you can see, many of the variable names are self-explanatory, such as `yrborn` for the year the respondent was born, `educ` for the education level of the respondent, and `marital` for the marital status of the respondent.

Stata allows us to store documentation within the dataset, adding documentation either about the overall dataset or about specific variables. Stata calls each documentation entry a note. You can see all the notes that I stored in the dataset by using the `notes` command. The output of this command is omitted to save space.

```
. notes
  (output omitted)
```

If you type the `notes` command yourself, the first set of notes (labeled with the heading `dta`) are notes that I created to describe the overall dataset. The output is then followed by notes associated with each of the variables.

In the example below, I have used the `notes` command followed by the variable name `happy7`. This command shows the notes associated with the variable `happy7`. This note documents the fact that I reverse coded this variable (compared with the original GSS dataset from the website). I did this recoding so that higher values on `happy7` indicate greater levels of happiness.

```
. notes happy7
happy7:
   1.  Reverse coded the GSS variable "happy7" so 1=Completely unhappy and
       7=Completely happy
```

The above note could be very useful if you were to download the GSS dataset yourself because you might be confused as to why their version of `happy7` differs from my version of `happy7`.

Note: Make a note of it

You can add notes to this dataset (or any dataset) by using the `note` command. I use the `note` command frequently as part of my daily work to add documentation to my datasets. When I use such datasets in the future, the notes help me remember key pieces of information that help me understand and use the dataset. You can type `help notes` to learn more about how to create and view notes.

Now that we know some of the basics of this dataset, let's use it as an example for illustrating descriptive statistics in this chapter and for illustrating inferential statistics in the following chapter (chapter 3). Because of the large sample size of the entire dataset ($N = 1974$), many of the inferential statistics (in the next chapter) would not be very interesting: even tiny differences would be significant with such a large sample size.

Let's say that our research interest motivating the use of this dataset for this chapter and the following chapter is on people under the age of 40 with a focus on topics such as their marital status, gender, happiness, level of education, and number of children. Using the `keep if` command below, we will keep only observations where people are less than 40 years old. The `count` command counts up the number of observations in the dataset, showing that we have 720 observations for people who are less than 40 years old.

```
. keep if age < 40
(1,254 observations deleted)
. count
  720
```

2.3   One-way tabulations

For all the examples in this section, we will use the GSS dataset, keeping just the people
who are under age 40. The `use` command below reads this dataset into memory. Then,
the `keep if` command selects only people who are under age 40. The `count` command
counts the number of observations, showing that we have 720 observations remaining
after using the `keep if` command.

```
. use gss2012_sbs, clear
. keep if age < 40
(1,254 observations deleted)
. count
  720
```

Tip: Clearing things up

Sometimes, when you go to use a dataset, Stata might refuse to obey your request,
as illustrated below.

```
. use gss2012_sbs
no; data in memory would be lost
r(4);
```

The error message is saying that you have unsaved changes in the dataset currently
in memory and that you will lose those changes if the `use` command reads in the
new dataset. If you like the changes you have made to the current dataset, then
you can `save` the dataset. Otherwise, if you do not care about the changes to the
current dataset in memory, you can then use the `clear` command to clear that
dataset from memory before issuing the `use` command, as shown below.

```
. clear
. use gss2012_sbs
```

Or you can specify the `clear` option on the `use` command, as shown below.

```
. use gss2012_sbs, clear
```

Let's consider how to perform one-way tabulations to describe the marital status of the respondents. The marital status of the respondent is stored in the variable called `marital`. We can use the `tabulate` command to show the frequency distribution of marital status, as shown below.

```
. tabulate marital
      marital |
       status |      Freq.      Percent        Cum.
--------------+-----------------------------------
      married |        267        37.08       37.08
      widowed |          4         0.56       37.64
     divorced |         47         6.53       44.17
    separated |         20         2.78       46.94
never married |        382        53.06      100.00
--------------+-----------------------------------
        Total |        720       100.00
```

This shows us that 267 of the people were married, and this composed 37.08% of the responses. This also shows the frequency and percentage of people who were widowed, divorced, separated, and never married. The output also shows the total number of valid responses was 720, corresponding to the number of observations shown by the `count` command above.

Let's make a tabulation for the variable `happy7`, which is the respondent's rating of his or her happiness on a seven-point scale.

```
. tabulate happy7
  how happy R is (recoded) |      Freq.      Percent        Cum.
---------------------------+-----------------------------------
       Completely unhappy |          1         0.21        0.21
             Very unhappy |          6         1.23        1.44
           Fairly unhappy |         10         2.06        3.50
Neither happy nor unhappy |         28         5.76        9.26
             Fairly happy |        166        34.16       43.42
               Very happy |        221        45.47       88.89
         Completely happy |         54        11.11      100.00
---------------------------+-----------------------------------
                    Total |        486       100.00
```

We see that 54 people described themselves as being `Completely happy`, and this is 11.11% of the valid responses. There are a total of 486 valid responses. By default, the `tabulate` command only includes valid responses. If we want the tabulation to also include missing values, we can specify the `missing` option, as shown below.

```
. tabulate happy7, missing
```

how happy R is (recoded)	Freq.	Percent	Cum.
Completely unhappy	1	0.14	0.14
Very unhappy	6	0.83	0.97
Fairly unhappy	10	1.39	2.36
Neither happy nor unhappy	28	3.89	6.25
Fairly happy	166	23.06	29.31
Very happy	221	30.69	60.00
Completely happy	54	7.50	67.50
Cannot choose	5	0.69	68.19
Inapplicable	229	31.81	100.00
Total	720	100.00	

This output shows all 720 of the responses, including the responses that were missing. We now see two additional response categories: `Cannot choose` and `Inapplicable`. The `Inapplicable` category contains 229 people and is composed of 31.81% of all responses. This missing value category is due to about a third of the respondents (by design) not being asked this question. 5 people could not choose an answer, and their response was coded as `Cannot choose`. Note that the percentages reported in this table are out of a total of 720 observations.

Video tutorial: Tabulations and cross-tabulations

See a video illustrating tabulations and cross-tabulations at http://www.stata.com/sbs/tabulate.

We see that the `tabulate` command without the `missing` option provides percentages out of the total number of valid responses, while the inclusion of the `missing` option provides percentages out of all the observations. I wish there was a way to display both kinds of these percentages in a single table. Even though Stata does not offer a command that provides this functionality, Ben Jann (a Stata user from the University of Bern) wrote such a program and contributed it to the Stata community. The command is called `fre` (think frequencies) and can be downloaded over the Internet using the following Stata command:[1]

```
. ssc install fre
```

After downloading this command, you can execute it like any other Stata command, as shown below.[2]

---

1. See section 1.1 for information about how to download all the add-on programs used in this book.
2. You can also obtain help using this command in the same way you would obtain help using any Stata command by typing `help fre`.

```
. fre happy7
```

happy7 — how happy R is (recoded)

			Freq.	Percent	Valid	Cum.
Valid	1	Completely unhappy	1	0.14	0.21	0.21
	2	Very unhappy	6	0.83	1.23	1.44
	3	Fairly unhappy	10	1.39	2.06	3.50
	4	Neither happy nor unhappy	28	3.89	5.76	9.26
	5	Fairly happy	166	23.06	34.16	43.42
	6	Very happy	221	30.69	45.47	88.89
	7	Completely happy	54	7.50	11.11	100.00
	Total		486	67.50	100.00	
Missing	.c	Cannot choose	5	0.69		
	.i	Inapplicable	229	31.81		
	Total		234	32.50		
Total			720	100.00		

There are a few things I particularly like about the fre output. Note that it explicitly shows the overall sample size ($N = 720$) and the number of valid responses ($N = 486$). It also computes the percentages for each category expressed in two ways: 1) as a percentage of all responses in the column titled Percent; and 2) as a percentage of the valid responses in the column titled Valid. For example, we see that 54 respondents indicated that they are Completely happy. This represents 7.50% of all responses (with respect to $N = 720$) and 11.11% of the valid responses (with respect to $N = 486$).

What I especially like about the fre command is that it displays both the values and labels associated with each response. We can see that the response for Completely unhappy was coded as 1 and the response for Completely happy was coded as 7. (Section 22.4 shows how you can create and assign value labels for numeric variables.) Further, we can see the codes for the missing values as well. A value of .c is used for Cannot choose and a value of .i for Inapplicable. You can see help missing for more information about missing value codes in Stata.

We can use the fre command with multiple variables. The fre command below is used to show tabulations for the variables marital and female, providing one-way tabulations of each variable.

```
. fre marital female
```

marital ─ marital status

		Freq.	Percent	Valid	Cum.
Valid	1 married	267	37.08	37.08	37.08
	2 widowed	4	0.56	0.56	37.64
	3 divorced	47	6.53	6.53	44.17
	4 separated	20	2.78	2.78	46.94
	5 never married	382	53.06	53.06	100.00
	Total	720	100.00	100.00	

female ─ Is R female (yes=1 no=0)?

		Freq.	Percent	Valid	Cum.
Valid	0 Male	338	46.94	46.94	46.94
	1 Female	382	53.06	53.06	100.00
	Total	720	100.00	100.00	

We can also use the `tab1` command (which is a command built into Stata to obtain multiple one-way tabulations), as shown below.

```
. tab1 marital female
-> tabulation of marital
```

marital status	Freq.	Percent	Cum.
married	267	37.08	37.08
widowed	4	0.56	37.64
divorced	47	6.53	44.17
separated	20	2.78	46.94
never married	382	53.06	100.00
Total	720	100.00	

```
-> tabulation of female
```

Is R female (yes=1 no=0)?	Freq.	Percent	Cum.
Male	338	46.94	46.94
Female	382	53.06	100.00
Total	720	100.00	

If we used `tabulate` instead of `tab1`, Stata would compute a two-way tabulation (that is, a cross-tabulation), as discussed in section 2.6.

Tip: Exporting one-way tabulations to Excel

The Stata Blog describes a user-written command called `tab2xl`, which you can use to export one-way tabulations to Excel. Such tables can be dynamically linked to Word documents, leading to dynamically updated documents. For information, see http://blog.stata.com/2014/02/04/retaining-an-excel-cells-format-when-using-putexcel/.

2.4 Summary statistics

Let's now turn our attention to the computation of summary statistics. As with the previous section, we will use the GSS dataset for the year 2012, focusing on those who are under the age of 40. I have repeated the commands for using the dataset and keeping just those who are under the age of 40. I then use `count` to show that we now have 720 observations in the dataset in memory.

```
. use gss2012_sbs, clear
. keep if age < 40
(1,254 observations deleted)
. count
  720
```

Let's use the `summarize` command to compute summary statistics for the variable `educ`, which represents the number of years of education for the respondent. The output shows that 720 observations had valid responses for this variable, and the mean is 13.54 with a standard deviation of 2.8. The minimum education was 0 and the maximum was 20.

```
. summarize educ

    Variable |        Obs        Mean    Std. Dev.       Min        Max
-------------+--------------------------------------------------------
        educ |        720    13.54444    2.803876          0         20
```

By including the `detail` option, we can see even more detailed summary statistics. Focusing on the right column of the output, we see the number of valid observations, the sum of the weights,[3] mean, standard deviation, variance, skewness, and kurtosis. The leftmost column shows percentile rankings of education. For example, 5% of the respondents had no more than a 9th grade education, and 90% of the respondents had at most 17 years of education. Finally, the middle column shows that the 4 lowest education scores were 0, 3, 4 and 5, and it also shows the 4 highest education scores were all 20.

---

3. This is applicable only for analyses involving weights, so we will ignore this part of the output.

```
. summarize educ, detail

                 highest year of school completed

            Percentiles      Smallest
    1%           6               0
    5%           9               3
   10%          11               4        Obs                   720
   25%          12               5        Sum of Wgt.           720

   50%          13                        Mean             13.54444
                              Largest     Std. Dev.        2.803876
   75%          16              20
   90%          17              20         Variance         7.861722
   95%          18              20         Skewness        -.2544962
   99%          20              20         Kurtosis         4.100558
```

We can use the `summarize` command for more than one variable at a time. (In fact, if you issue the `summarize` command without specifying any variables, it will show you summary statistics for all variables.) Let's compute summary statistics for the variables `happy7`, `educ`, and `children`. For each of these variables, Stata shows us the number of valid observations, mean, standard deviation, minimum, and maximum. We see that `happy7` ranges from 1 to 7 and has a mean of 5.5. As we saw before, the average number of years of education is 13.5 and ranges from 0 to 20. The average number of children is 1.3 and ranges from 0 to 8.

```
. summarize happy7 educ children
      Variable |       Obs        Mean    Std. Dev.       Min        Max
 --------------+--------------------------------------------------------
        happy7 |       486    5.532922    .9491974         1          7
          educ |       720    13.54444    2.803876         0         20
      children |       720    1.255556    1.468718         0          8
```

2.5   Summary statistics by one group

In this section, let's consider the computation of summary statistics computed for specific groups. As we have done in each of the previous sections, we use the GSS dataset, keeping just the observations for those who are less than 40 years old. Just in case you are jumping into the chapter right here, I have repeated the `use` and `keep` commands below; I also use the `count` command to show that we have subset the dataset to 720 observations.

```
. use gss2012_sbs, clear

. keep if age < 40
(1,254 observations deleted)

. count
  720
```

In the previous section, we considered the variable `happy7`, which is the respondent's rating of happiness on a 7-point scale on which a rating of 1 corresponds to being Completely unhappy and a rating of 7 corresponds to being Completely happy. Let's now compute the average happiness for this sample of people who are under the age of 40.

```
. summarize happy7
    Variable |        Obs        Mean    Std. Dev.        Min        Max
-------------+-------------------------------------------------------------
      happy7 |        486    5.532922    .9491974          1          7
```

We see that the average happiness is about 5.5, which lies near the middle of `fairly happy` and `very happy`. Say that we were interested in obtaining the happiness of those who are married. We can include an `if` specification as shown below to restrict the analysis to just those who are married.[4] We see that the average happiness is somewhat greater, $M = 5.81$, for those who are married.

```
. summarize happy7 if married==1
    Variable |        Obs        Mean    Std. Dev.        Min        Max
-------------+-------------------------------------------------------------
      happy7 |        170    5.805882    .7791443          4          7
```

Note: Double equals

Note that I typed `if married==1` and not `if married=1`. This is because Stata (like many programming languages) considers `married=1` (one equal sign) to signify the assignment of the value 1 to the variable `married`, whereas `married==1` (two equal signs) is a test of whether the variable `married` is equal to 1.

While we are at it, let's compute the average happiness for those who are not married by adding `if married==0` to the `summarize` command; this will yield results that focus solely on the respondents who indicated they were not married. We see that the average happiness for those who are not married is 5.39.

```
. summarize happy7 if married==0
    Variable |        Obs        Mean    Std. Dev.        Min        Max
-------------+-------------------------------------------------------------
      happy7 |        316    5.386076    .9998392          1          7
```

Applying the `if` specification is useful if we want to focus on a certain subgroup. But if our goal was to obtain the summary statistics for those who are married as well

---

4. I created the variable named `married`, and it is coded 1 if the person is married and 0 if the person is not married.

as for those who are not married, this method can be cumbersome. Instead, consider the `tabulate` command below, which includes the `summarize(happy7)` option. The behavior of this command is almost self-explanatory. It performs a tabulation of the variable `married`, but while doing so, it shows us summary statistics for the variable `happy7`. It produces output much like the two previous `summarize` commands (except it does not display the minimum and maximum value).

```
. tabulate married, summarize(happy7)
   marital: |
 married=1, |
 unmarried=0 | Summary of how happy R is (recoded)
  (recoded) |        Mean      Std. Dev.       Freq.
------------+-----------------------------------------
  Unmarried |    5.3860759    .99983925          316
    Married |    5.8058824    .77914426          170
------------+-----------------------------------------
      Total |    5.5329218    .94919744          486
```

The above table more concisely shows us that the average happiness for the 316 people who are unmarried is 5.39, compared with 5.81 for the 170 people who are married. It also shows the overall average of happiness is 5.53 for the 486 people. (In the following chapter, we will see how to perform a *t* test that compares the happiness of those who are married with that of those who are not married, see section 3.2.)

2.6   Two-way tabulations

Let's now look at how we can perform two-way tabulations (cross-tabulations) using Stata. Let's continue to use the GSS dataset, keeping just the observations for those who are less than 40 years old. The commands for using the dataset and keeping just those who are under 40 years of age are repeated below. The `count` command confirms we have subset the dataset to 720 observations.

```
. use gss2012_sbs, clear
. keep if age < 40
(1,254 observations deleted)
. count
  720
```

When we were looking at one-way tabulations in section 2.3, we considered one-way tabulations of the variables `female` and `marital`. I am interested in looking at a two-way tabulation of these variables to compare the distribution of marital status between males and females. Before performing that two-way cross-tabulation, let's first repeat the one-way tabulations of these variables using the `tab1` command.

```
. tab1 marital female

-> tabulation of marital
      marital |
       status |      Freq.      Percent         Cum.
--------------+-----------------------------------------
      married |        267        37.08        37.08
      widowed |          4         0.56        37.64
     divorced |         47         6.53        44.17
    separated |         20         2.78        46.94
never married |        382        53.06       100.00
--------------+-----------------------------------------
        Total |        720       100.00

-> tabulation of female
    Is R female |
        (yes=1  |
         no=0)? |      Freq.      Percent         Cum.
----------------+-----------------------------------------
           Male |        338        46.94        46.94
         Female |        382        53.06       100.00
----------------+-----------------------------------------
          Total |        720       100.00
```

Now, let's perform a two-way cross-tabulation of the variable `marital` by `female`.

```
. tabulate marital female
              | Is R female (yes=1
      marital |      no=0)?
       status |      Male     Female |     Total
--------------+----------------------+----------
      married |       117        150 |       267
      widowed |         2          2 |         4
     divorced |        17         30 |        47
    separated |         4         16 |        20
never married |       198        184 |       382
--------------+----------------------+----------
        Total |       338        382 |       720
```

The first variable (`marital`) forms the rows of the cross-tabulation, and the second variable (`female`) forms the columns of the cross-tabulation. Focusing on the first row of the output, we see that there are 117 men who are married, 150 women who are married, and a total of 267 people who are married. The total of 267 conforms to what we saw previously from the one-way tabulation of `marital`.

Focusing on the first column, we see that there is a total of 338 men, as we saw in the one-way tabulation of `female`. Of the 338 men, 117 of them are married, 2 of them are widowed, 17 are divorced, 4 are separated, and 198 have never been married.

Let's now focus on those who are married. It appears that women are more likely to be married than men, with 150 women being married and 117 men being married. But these raw numbers can be misleading because there are more women in the sample than men (382 women versus 338 men). To make a more accurate comparison, we will compute and compare the percentage who are married between men versus women through the use of column percentages. The `tabulate` command is repeated below; we add the `column` option to request column percentages.

```
. tabulate marital female, column
```

Key
frequency
column percentage

marital status	Is R female (yes=1 no=0)? Male	Female	Total
married	117 34.62	150 39.27	267 37.08
widowed	2 0.59	2 0.52	4 0.56
divorced	17 5.03	30 7.85	47 6.53
separated	4 1.18	16 4.19	20 2.78
never married	198 58.58	184 48.17	382 53.06
Total	338 100.00	382 100.00	720 100.00

The portion of the output labeled `Key` shows us how to read the values contained in the table. The column percentages show that among the 338 males, 34.6% are married, while among the 382 females, 39.3% are married. Even after we account for the differences in the sample sizes, these percentages seem to suggest that women are more likely to be married (before age 40) than men. We cannot say anything more definitive without a significance test. In the following chapter, we will revisit this cross-tabulation to consider whether the percentage of men who are married differs significantly from the percentage of women who are married (see section 3.5). In that chapter, we will also revisit this cross-tabulation when considering whether the distribution of marital status differs by gender (see section 3.7).

Note: Other tabulate options

In addition to the `column` option, you can use the `row` option to display row percentages, and you can use the `cell` option to display cell percentages.

2.7   Cross-tabulations with summary statistics

Say that we were interested in considering the average happiness as a function of both whether one is married and gender. We can form a cross-tabulation of married (yes/no) by gender (male/female) and show the average happiness for each cell. This is illustrated below using the `tabulate` command.

```
. tabulate married female, summarize(happy7)
  Means, Standard Deviations and Frequencies of how happy R is (recoded)

   marital:
   married=1,
   unmarried=  | Is R female (yes=1
           0   |     no=0)?
   (recoded)   |    Male      Female  |    Total
-------------+-----------------------+------------
   Unmarried | 5.3741935   5.3975155 | 5.3860759
             | 1.1174437    .87521071 |  .99983925
             |       155          161 |       316
-------------+-----------------------+------------
     Married | 5.7468354   5.8571429 | 5.8058824
             |  .80810564   .75382622 |  .77914426
             |        79           91 |       170
-------------+-----------------------+------------
       Total |       5.5   5.5634921 | 5.5329218
             | 1.0368738    .86079766 |  .94919744
             |       234          252 |       486
```

By looking at the row totals, we see the average happiness for those who are unmarried and married is 5.39 and 5.81 (respectively). Looking at the column totals, we can see that the average happiness for males and females is 5.5 and 5.56 (respectively). We can then inspect the individual cells to view the average happiness associated with the combinations of gender and marital status. For example, for an unmarried male, the average happiness is 5.37.

To perform hypothesis testing regarding whether happiness varies as a function of gender and whether one is married implies a two-way factorial analysis of variance. Such a test is beyond the scope of this chapter but is considered in detail in chapter 7.

Video tutorial: Cross-tabulations and summary statistics

See a video showing how you can obtain summary statistics as part of a cross-tabulation at http://www.stata.com/sbs/tabulate-summarize.

2.8   Closing thoughts

This chapter has provided a very brief introduction to descriptive statistics using Stata. Some of these examples have begged for a statistical test to determine whether the differences shown are significant. Such tests are covered in the next chapter (chapter 3) on basic inferential statistics.

Video tutorial: Descriptive statistics

See a video illustrating descriptive statistics in Stata at http://www.stata.com/sbs/descriptive-stats.

---

3 Basic inferential statistics

3.1 Chapter overview

This chapter describes how to perform basic inferential statistics using Stata. In other words, this chapter is about hypothesis testing, focusing on tests of some very basic hypotheses. In fact, some of the hypotheses that are covered here probably refer back to tests you have not thought about since your very first statistics class. These kinds of tests are not the meat and potatoes of our daily work as research statisticians, but these basic hypothesis tests can be very useful when the situation arises.

All the examples from this chapter (like the previous chapter) are based on analyses of the General Social Survey (GSS) from the year 2012 using a dataset named gss2012_sbs.dta. As we did in the previous chapter, we keep just the observations for those who are less than 40 years old. The `use` command below uses the GSS dataset and the `keep if` command keeps just the observations for those who are under the age of 40. The `count` command then shows that we have subset the dataset to 720 observations.

```
. use gss2012_sbs, clear
. keep if age < 40
(1,254 observations deleted)
. count
  720
```

All the examples in this chapter will be based on this subset of the GSS dataset.

3.2   Two-sample t tests

Consider the example from section 2.5 of the previous chapter in which we summarized the average happiness separately for those who are married versus those who are unmarried. Say that we wanted to perform a test to determine whether the population average of happiness differs for those who are married versus not married. In other words, say we wanted to test the following null hypothesis,

$$H_0: \mu_{\text{Married}} = \mu_{\text{Unmarried}}$$

where μ_{Married} is the population mean of happiness for those who are married and $\mu_{\text{Unmarried}}$ is the population mean of happiness for those who are unmarried. We can easily test this null hypothesis using a t test via the ttest command.[1]

```
. ttest happy7, by(married)
Two-sample t test with equal variances
```

Group	Obs	Mean	Std. Err.	Std. Dev.	[95% Conf.	Interval]
Unmarrie	316	5.386076	.0562454	.9998392	5.275412	5.49674
Married	170	5.805882	.0597576	.7791443	5.687915	5.92385
combined	486	5.532922	.0430565	.9491974	5.448322	5.617522
diff		-.4198064	.0883389		-.5933815	-.2462313

```
    diff = mean(Unmarrie) - mean(Married)                      t =  -4.7522
Ho: diff = 0                                  degrees of freedom =       484

    Ha: diff < 0              Ha: diff != 0              Ha: diff > 0
 Pr(T < t) = 0.0000      Pr(|T| > |t|) = 0.0000       Pr(T > t) = 1.0000
```

Video tutorial: Two-sample t test

See a video showing how you can perform a two-sample t test at http://www.stata.com/sbs/2-sample-t-test.

---

1. Some might be bothered by using a Likert scale like happy7 as though it were measured on an interval scale. For the sake of example, let's assume that happy7 is measured on an interval scale.

The body of the table includes columns indicating the group membership, number of observations, mean, standard error, standard deviation (SD), and 95% confidence interval (CI). These statistics are shown for the unmarried group ($N = 316$), the married group ($N = 170$), and the two groups combined ($N = 486$). The fourth row of the table (labeled `diff`) shows the difference of the first group minus the second group (that is, unmarried minus married). Focusing on this row shows that the difference in the average happiness of those who are unmarried versus married is -0.42, 95% CI $= [-0.59, -0.25]$.

Below the body of the table, we see the t statistic and the associated degrees of freedom. Finally, the bottom left part of the table shows the p-value associated with a left-tailed test ($p = 0.000$), the bottom right shows the p-value associated with a right-tailed test ($p = 1.000$), and the bottom center shows the p-value associated with a two-tailed test ($p = 0.000$).

Note: More on two-sample tests

For more information about two-sample t tests, you can see section 4.2. There you will find even more details about how to read the output. That is followed by section 4.3, which shows how you can compare the means of two groups using the `anova` command.

We could summarize these findings by saying that the average happiness for the $N = 316$ unmarried respondents was $M = 5.39$ (SD $= 1.00$) compared with $M = 5.81$ (SD $= 0.78$) for the $N = 170$ married respondents. The difference in these means was -0.42 (95% CI $= [-0.59, -0.25]$) and was statistically significant using a two-tailed test $[t(484) = -4.75, p < 0.001]$.

Note! A note about significance

In this book, I will use the term "significant" as a shorthand for "statistically significant using a two-tailed test with an alpha of 0.05". Such references to an effect being significant does not speak to matters such as clinical significance or practical significance.

3.3   Paired sample t tests

The GSS dataset contains a variable representing the years of education of the respondent's father (`paeduc`) and one representing the years of education of the respondent's mother (`maeduc`). Say that we wanted to perform a test asking whether there are differences in the population average years of education for the respondent's father versus mother. In other words, we want to test the following null hypothesis,

$$H_0 : \mu_{\text{father}} = \mu_{\text{mother}}$$

where μ_{father} is the population mean years of education for the fathers and μ_{mother} is the population mean years of education for the mothers. We can test this null hypothesis with a paired sample t test using the `ttest` command.

```
. ttest paeduc=maeduc

Paired t test
```

Variable	Obs	Mean	Std. Err.	Std. Dev.	[95% Conf. Interval]	
paeduc	483	12.69151	.1845263	4.055383	12.32894	13.05409
maeduc	483	12.72464	.1698772	3.733436	12.39085	13.05843
diff	483	-.0331263	.1335055	2.934086	-.295451	.2291984

```
    mean(diff) = mean(paeduc - maeduc)                         t =   -0.2481
Ho: mean(diff) = 0                              degrees of freedom =       482

Ha: mean(diff) < 0            Ha: mean(diff) != 0              Ha: mean(diff) > 0
Pr(T < t) = 0.4021        Pr(|T| > |t|) = 0.8041              Pr(T > t) = 0.5979
```

The output is formatted very much like the independent groups t test, so I will not explain these results in the same level of detail. There are a couple of key differences to focus on. For example, the first row of output represents the education of the father (the first variable mentioned on the `ttest` command), and the second row represents the education of the mother (the second variable mentioned on the `ttest` command). Also, note that the output includes only the $N = 483$ observations in which the respondent provided information about the education of his or her father and mother.

We can summarize these findings by saying that the average years of education of the fathers is not significantly different from the average years of education of the mothers using a two-tailed test $[t(482) = -0.25, p = 0.8041]$. The average difference in years of education (between fathers and mothers) was -0.03 (95% CI $= [-0.30, 0.23]$).

Video tutorial: Paired t test

See a video showing how you can perform a paired t test at
http://www.stata.com/sbs/paired-t-test.

3.4   One-sample t tests

In the previous chapter (see section 2.4), we computed descriptive statistics for the variable `educ`, representing the number of years of education. Say that our interest was to test whether the population average years of education is equal to 13. That is, say we wanted to test the following null hypothesis,

$$H_0 \colon \mu = 13$$

where μ is the average years of education in the population. We can test this hypothesis via a one-sample t test, as illustrated below.

```
. ttest educ=13
One-sample t test
```

Variable	Obs	Mean	Std. Err.	Std. Dev.	[95% Conf. Interval]
educ	720	13.54444	.1044943	2.803876	13.33929 13.74959

```
        mean = mean(educ)                                       t =    5.2103
Ho: mean = 13                                    degrees of freedom =      719

    Ha: mean < 13                 Ha: mean != 13                 Ha: mean > 13
  Pr(T < t) = 1.0000       Pr(|T| > |t|) = 0.0000           Pr(T > t) = 0.0000
```

The `ttest` command shows the mean, standard error, SD, and 95% confidence interval for the mean. The t-value comparing the average years of education to 13 is 5.21 with 719 degrees of freedom. The output shows a p-value for a left-tailed test ($p = 1.000$) at the left, a two-tailed test ($p = 0.000$) in the center, and a right-tailed test at the right ($p = 0.000$). The average years of education significantly differ from 13 ($M = 13.54$, $t(719) = 5.21$, $p < 0.001$, 95% CI $= [13.34, 13.75]$).

> **Video tutorial: One-sample t test**
>
> See a video showing how you can perform a one-sample t test at http://www.stata.com/sbs/one-sample-t-test.

3.5   Two-sample test of proportions

Back in section 2.6, we considered the cross-tabulation of marital status by gender. I have repeated this cross-tabulation below, including the `column` option to show column percentages.

```
. tabulate marital female, column
```

```
┌─────────────────────┐
│ Key                 │
├─────────────────────┤
│           frequency │
│   column percentage │
└─────────────────────┘
```

marital status	Is R female (yes=1 no=0)? Male	Female	Total
married	117	150	267
	34.62	39.27	37.08
widowed	2	2	4
	0.59	0.52	0.56
divorced	17	30	47
	5.03	7.85	6.53
separated	4	16	20
	1.18	4.19	2.78
never married	198	184	382
	58.58	48.17	53.06
Total	338	382	720
	100.00	100.00	100.00

Let's focus on the married category. For males, 34.62% are married compared with 39.27% for females. Say that we are interested in testing the null hypothesis that the population proportion of males who are married is equal to the population proportion of females who are married; that is,

$$H_0 : P_{\text{male}} = P_{\text{female}}$$

where P_{male} is the population proportion of males who are married and P_{female} is the population proportion of females who are married. To form such a test, we need a variable that is coded 1 if the person is married and 0 for everyone else (that is, 0 for those who are widowed, divorced, separated, or never married). In fact, the dataset contains such a variable that I created called `married`. Below I show a cross-tabulation of `married` by `female`.

```
. tabulate married female, column
```

```
┌─────────────────────────────┐
│ Key                         │
├─────────────────────────────┤
│         frequency           │
│    column percentage        │
└─────────────────────────────┘
```

```
 marital:
married=1,
unmarried=    Is R female (yes=1
    0               no=0)?
(recoded)      Male    Female  │    Total
───────────┼──────────────────┼──────────
Unmarried  │     221       232 │      453
           │   65.38     60.73 │    62.92
───────────┼──────────────────┼──────────
  Married  │     117       150 │      267
           │   34.62     39.27 │    37.08
───────────┼──────────────────┼──────────
    Total  │     338       382 │      720
           │  100.00    100.00 │   100.00
```

We can test the above null hypothesis comparing the proportion married between males and females using the prtest command. (Note how the syntax of this command resembles the syntax of the two-sample ttest command.)

```
. prtest married, by(female)
Two-sample test of proportions                  Male: Number of obs =      338
                                              Female: Number of obs =      382
─────────────────────────────────────────────────────────────────────────────
    Variable │    Mean    Std. Err.     z     P>|z|     [95% Conf. Interval]
─────────────┼─────────────────────────────────────────────────────────────────
       Male  │ .3461538   .025877                      .2954358    .3968718
     Female  │ .3926702   .0249859                     .3436987    .4416416
─────────────┼─────────────────────────────────────────────────────────────────
       diff  │ -.0465163  .035971                     -.1170182    .0239856
             │ under Ho:  .0360702   -1.29   0.197
─────────────────────────────────────────────────────────────────────────────
       diff = prop(Male) - prop(Female)                           z =  -1.2896
   Ho: diff = 0

   Ha: diff < 0                 Ha: diff != 0                 Ha: diff > 0
 Pr(Z < z) = 0.0986     Pr(|Z| < |z|) = 0.1972         Pr(Z > z) = 0.9014
```

The very top of the output shows the number of observations for the males ($N = 338$) and females ($N = 382$). The body of the table shows the proportion married for males and for females (0.346 versus 0.393) and the confidence interval for each of these proportions. The next line shows that the difference in the proportion married for males versus females is -0.047 and a 95% confidence interval for difference $= [-0.12, 0.02]$. Looking at the fourth line of the body of the table, we see the z statistic ($z = -1.29$) and p-value for the two-tailed test ($p = 0.1972$). The very bottom of the output shows p-values associated with a left-tailed test ($p = 0.0986$) and a right-tailed test ($p = 0.9014$)

and repeats the two-tailed p-value ($p = 0.1972$). We can say that there is no significant difference in the proportion married for males versus females ($z = -1.29$, $p = 0.197$).

3.6 One-sample test of proportions

In the previous section, we used the variable `married` to compare males and females in terms of the proportion married. Say that our interest was in the overall proportion of people who are married. Using the `fre` command below, we see that 37.08% of the people in this sample are married.

```
. fre married
married — marital: married=1, unmarried=0 (recoded)
```

		Freq.	Percent	Valid	Cum.
Valid	0 Unmarried	453	62.92	62.92	62.92
	1 Married	267	37.08	37.08	100.00
	Total	720	100.00	100.00	

Say that we wanted to test whether the proportion of people who are married was different from 0.4 (that is, 40%). That is, say we wanted to test the following null hypothesis,

$$H_0 : P = 0.40$$

where P is the population proportion of people who are married. We can test this null hypothesis using the `prtest` command and specifying `married=.4`, as shown below.

```
. prtest married=.4
One-sample test of proportion                 married: Number of obs =      720
```

Variable	Mean	Std. Err.	[95% Conf. Interval]	
married	.3708333	.0180014	.3355513	.4061154

```
       p = proportion(married)                              z =   -1.5975
Ho: p = 0.4
      Ha: p < 0.4                Ha: p != 0.4                Ha: p > 0.4
   Pr(Z < z) = 0.0551       Pr(|Z| > |z|) = 0.1101       Pr(Z > z) = 0.9449
```

The output shows that the proportion married is 0.37 and that the 95% confidence interval for this proportion is 0.336 to 0.406. The output reports the z statistic for the test of the above null hypothesis ($z = -1.598$) as well as a left-tailed p-value ($p = 0.0551$), a right-tailed p-value ($p = 0.9449$), and a two-tailed p-value ($p = 0.1101$). Using a two-tailed test, we see that the proportion married is not significantly different from 0.40 ($z = -1.598$, $N = 720$, $p = 0.1101$).

Let's consider the same kind of test but with respect to the proportion of people who are separated, for which there is a variable called `separated`, which is coded 1 if the respondent is separated and 0 otherwise. Using the `fre` command we see that 2.78% are separated. (This matches what we saw from the variable `marital`.)

```
. fre separated
separated — marital: Separated=1, Not Separated=0 (recoded)
```

		Freq.	Percent	Valid	Cum.
Valid	0 Not Separated	700	97.22	97.22	97.22
	1 Separated	20	2.78	2.78	100.00
	Total	720	100.00	100.00	

Say that we want to test the null hypothesis that this proportion is 0.04 (that is, 4%). This null hypothesis is

$$H_0 \colon P = 0.04$$

where P is the population proportion of people who are separated. We can test this null hypothesis using the `prtest` command.

```
. prtest separated=0.04
One-sample test of proportion                  separated: Number of obs =       720
```

Variable	Mean	Std. Err.	[95% Conf. Interval]	
separated	.0277778	.0061244	.0157741	.0397814

```
     p = proportion(separated)                              z =  -1.6736
Ho: p = 0.04
    Ha: p < 0.04               Ha: p != 0.04                Ha: p > 0.04
 Pr(Z < z) = 0.0471      Pr(|Z| > |z|) = 0.0942         Pr(Z > z) = 0.9529
```

Expressed as percentages, the output shows that the percentage who are separated is 2.78% (95% CI = [1.58%, 3.98%]). Using the normal approximation to the binomial distribution, we find that the proportion who are separated is not significantly different from 0.04 ($z = -1.67$, $N = 720$, $p = 0.0942$).

As noted above, the `prtest` command uses the normal approximation method. For large sample sizes, this method is quickly computed and sufficiently accurate. However, it can become inaccurate as the sample size becomes small (say, below 30) or when the number of observed events becomes small (again, say, below 30). The binomial exact method, provided by the `bitest` command, is an alternative to the `bitest` command that is accurate in all situations. The `bitest` command below is used to apply the binomial exact method to test the null hypothesis $H_0 \colon P = 0.04$.

```
. bitest separated=0.04
    Variable │       N   Observed k   Expected k   Assumed p   Observed p
   ──────────┼──────────────────────────────────────────────────────────
   separated │     720          20        28.8      0.04000      0.02778
    Pr(k >= 20)                = 0.967391  (one-sided test)
    Pr(k <= 20)                = 0.051486  (one-sided test)
    Pr(k <= 20 or k >= 38) = 0.105043  (two-sided test)
```

The format of the output of the `bitest` command differs from that of the `prtest` command, reflecting the exact binomial computational methods. The output shows that we have 720 observations, the proportion of whom were separated is 0.02778. It also shows we want to test whether this proportion differs from 0.04. The output shows this same information in terms of counts (which it calls k). In our sample, we observed 20 people who were separated, and if the proportion were 0.04, then 28.8 would have been separated. The bottom three lines of output show p-values for a right-tailed test ($p = 0.9674$), a left-tailed test ($p = 0.0515$), and a two-tailed test ($p = 0.1050$). Using the binomial exact test with a two-sided test, we see that the proportion of people who are separated is not significantly different from 0.04 ($N = 720$, $p = 0.1050$).

3.7   Chi-squared and Fisher's exact test

When performing a cross-tabulation, we can use a chi-squared test to test whether the distribution of the row variable is independent of the column variable. (Of course, we can also frame this as whether the distribution of the column variable is independent of the row variable.) For the example of gender and marital status, we can ask whether the response for marital status differs as a function of gender. The `tabulate` command is used to cross-tabulate `marital` by `female`; we also add the `chi2` option and `exact` option, as shown below.

```
. tabulate marital female, column chi2 exact
```

```
┌─────────────────┐
│ Key             │
├─────────────────┤
│       frequency │
│ column percentage │
└─────────────────┘
```

```
Enumerating sample-space combinations:
stage 5:   enumerations = 1
stage 4:   enumerations = 5
stage 3:   enumerations = 69
stage 2:   enumerations = 1170
stage 1:   enumerations = 0
```

marital status	Is R female (yes=1 no=0)? Male	Female	Total
married	117 34.62	150 39.27	267 37.08
widowed	2 0.59	2 0.52	4 0.56
divorced	17 5.03	30 7.85	47 6.53
separated	4 1.18	16 4.19	20 2.78
never married	198 58.58	184 48.17	382 53.06
Total	338 100.00	382 100.00	720 100.00

```
             Pearson chi2(4) =   12.7462   Pr = 0.013
             Fisher's exact =               0.009
```

The chi-squared test shows that the distribution of marital status differs as a function of gender [$\chi^2(4, N = 720) = 12.75$, $p = 0.013$]. However, because there are some cells with very small sample sizes, I also requested Fisher's exact test via the exact option. Fisher's exact test provides exact p-values even with very small cell sizes or even empty cells. In this case, Fisher's exact test yields a very similar p-value to the chi-squared test ($p = 0.009$). Even though we have found a significant association between marital status and gender, we cannot comment on any specific differences, for example, regarding whether women are more likely to be married than men. Such specific comparisons can be made as described earlier in this chapter in section 3.5.

Video tutorial: chi-squared and Fisher's exact tests

See a video illustrating the applications of the chi-squared test and Fisher's exact test at http://www.stata.com/sbs/chi-squared-test.

3.8   Correlations

As with the previous sections, let's continue using the GSS dataset and focusing on those who are less than 40 years old. (Section 3.1 shows the commands to read and subset the dataset.) Say we wanted to compute the correlations among education, happiness, and health. We can compute these correlations using the `correlate` command.

```
. correlate happy7 educ health
(obs=483)
                 happy7      educ    health

      happy7     1.0000
        educ     0.0761    1.0000
      health     0.2905    0.1672    1.0000
```

The `correlate` command shows the correlations among the variables while using listwise deletion with respect to missing values. This can lead to a needless loss of observations.

Instead, let's use the `pwcorr` command (shown below) to compute correlations using pairwise deletion for missing values. This command includes the `obs` and `sig` options. The `obs` option results in a display of the number of observations for each pairwise correlation. The `sig` option requests a display of a p-value for each correlation that tests the null hypothesis that the population correlation coefficient is equal to zero.

```
. pwcorr happy7 educ health, obs sig
                 happy7      educ    health

      happy7     1.0000

                    486

        educ     0.0740    1.0000
                 0.1031
                    486       720

      health     0.2905    0.1854    1.0000
                 0.0000    0.0000
                    483       488       488
```

The correlation between self-reported health with self-reported happiness is significant ($r = 0.29$, $N = 483$, $p < 0.001$), as is the correlation between self-reported rating of health with education ($r = 0.19$, $N = 488$, $p < 0.001$). The correlation between education and self-reported happiness was not significant ($r = 0.07$, $N = 486$, $p = 0.103$).

> **Video tutorial: Pearson's correlation**
>
> See a video demonstration of Pearson's correlation at
> http://www.stata.com/sbs/correlate.

3.9 Immediate commands

There might be situations where you want to perform one of the hypothesis tests illustrated in this chapter but you have access only to summary statistics and do not have access to the original dataset. In some cases, you are able to perform the hypothesis test based solely on summary data using one of the immediate Stata commands. These are called "immediate" commands because they allow you to perform immediate analyses using summary data. This section illustrates the use of the `ttesti`, `prtesti`, `bitesti`, and `tabi` commands. Note how all of these commands end with the letter `i` to emphasize that they are all immediate commands.

> **Note: More immediate commands**
>
> You can see a list of all the immediate commands offered by Stata by typing `help immed`. In addition to the immediate commands illustrated in this section, you can see immediate commands for CIs (see `help cii`), tables for epidemiologists (see `help cci`), and tests of SDs (see `help sdtesti`).

3.9.1 Immediate test of two means

Consider the two-sample t test illustrated in section 3.2 that tested the null hypothesis

$$H_0 : \mu_{\text{Married}} = \mu_{\text{Unmarried}}$$

where μ_{Married} is the population mean of happiness for those who are married and $\mu_{\text{Unmarried}}$ is the population mean for those who are unmarried. We can use the `ttesti` command to replicate that analysis based solely on knowing the sample size, mean, and SD from each group, as shown below.

Following the `ttesti` command, I supply the N, mean, and SD for group 1 (unmarried) followed by the N, mean, and SD for group 2 (married). You can compare the output below to that from section 3.2 to see that the results are the same (aside from slight differences due to rounding).

```
. ttesti 316 5.3861 .9998 170 5.8059 .7791
Two-sample t test with equal variances
```

	Obs	Mean	Std. Err.	Std. Dev.	[95% Conf.	Interval]
x	316	5.3861	.0562431	.9998	5.27544	5.49676
y	170	5.8059	.0597542	.7791	5.687939	5.923861
combined	486	5.532944	.0430546	.9491573	5.448347	5.61754
diff		-.4198	.088335		-.5933675	-.2462325

```
       diff = mean(x) - mean(y)                                    t =  -4.7524
Ho: diff = 0                                      degrees of freedom =       484

    Ha: diff < 0                 Ha: diff != 0                  Ha: diff > 0
 Pr(T < t) = 0.0000       Pr(|T| > |t|) = 0.0000          Pr(T > t) = 1.0000
```

Video tutorial: Two-sample t test on summary data

See a video showing how you can perform a two-sample t test using summary data via the `ttesti` command at http://www.stata.com/sbs/two-sample-ttesti.

3.9.2 Immediate test of one mean

In section 3.4, we considered whether the population average years of education is equal to 13. In other words, we were interested in the following null hypothesis, are

$$H_0 : \mu = 13$$

where μ is the average years of education in the population. We can test this hypothesis via the `ttesti` command, shown below. After the `ttesti` command, I supply the sample size, the mean of education, the SD of education, and the hypothesized value for comparison. As you can see, `ttesti` yields the same output as the original `ttest` command (aside from slight differences due to rounding).

```
. ttesti 720 13.54444 2.803876 13
One-sample t test
```

	Obs	Mean	Std. Err.	Std. Dev.	[95% Conf.	Interval]
x	720	13.54444	.1044943	2.803876	13.33929	13.74959

```
      mean = mean(x)                                         t =   5.2102
Ho: mean = 13                                degrees of freedom =      719

   Ha: mean < 13               Ha: mean != 13                Ha: mean > 13
Pr(T < t) = 1.0000       Pr(|T| > |t|) = 0.0000          Pr(T > t) = 0.0000
```

> **Video tutorial: One-sample t test on summary data**
>
> See a video showing how you can perform a one-sample *t* test using summary data at http://www.stata.com/sbs/one-sample-ttesti.

3.9.3 Immediate test of two proportions

Let's consider how we can perform a hypothesis test of two proportions using summary data. Consider the example from section 3.5, where we tested the null hypothesis that the population proportion of males who are married is equal to the population proportion of females who are married; that is,

$$H_0: P_{\text{male}} = P_{\text{female}}$$

where P_{male} is the population proportion of males who are married and P_{female} is the population proportion of females who are married. We can test this null hypothesis using the `prtesti` command, as shown below. After the `prtesti` command, I supply the sample size and proportion married for group 1 (males) followed by the sample size and proportion married for group 2 (females). Aside from differences due to rounding, the results match those from section 3.5.

```
. prtesti 338 .3462 382 .3927
Two-sample test of proportions                    x: Number of obs =     338
                                                  y: Number of obs =     382
```

Variable	Mean	Std. Err.	z	P>|z|	[95% Conf. Interval]	
x	.3462	.0258778			.2954804	.3969196
y	.3927	.0249862			.3437279	.4416721
diff	−.0465	.0359718			−.1170035	.0240035
	under Ho:	.0360709	−1.29	0.197		

```
        diff = prop(x) - prop(y)                             z =  -1.2891
    Ho: diff = 0

    Ha: diff < 0                 Ha: diff != 0                 Ha: diff > 0
 Pr(Z < z) = 0.0987      Pr(|Z| < |z|) = 0.1974          Pr(Z > z) = 0.9013
```

3.9.4 Immediate test of one proportion

Consider the one-sample test of a proportion we performed in section 3.6. We tested the null hypothesis that the proportion married equals 0.40; that is,

$$H_0: P = 0.40$$

Using the `prtesti` command shown below, we can perform such a test. Note how I specify that the overall sample size is 720, that the proportion married is 0.3708, and that we want to test the null hypothesis that this proportion equals 0.40. As you can see below, the `prtesti` command yields the same results as those from section 3.6, except for minor differences due to rounding.

```
. prtesti 720 .3708 .4
One-sample test of proportion                    x: Number of obs =       720

     Variable │      Mean   Std. Err.                    [95% Conf. Interval]

            x │     .3708   .0180011                      .3355186    .4060814

        p = proportion(x)                                         z =   -1.5993
    Ho: p = 0.4

        Ha: p < 0.4                Ha: p != 0.4                 Ha: p > 0.4
    Pr(Z < z) = 0.0549        Pr(|Z| > |z|) = 0.1097        Pr(Z > z) = 0.9451
```

In section 3.6, we also considered a one-sample test of a proportion regarding the proportion of people who are separated. We tested the null hypothesis that this proportion equals 0.04 (or 4%); that is,

$$H_0: P = 0.04$$

where P is the population proportion of people who are separated. Because so few people were separated, we used the binomial exact method via the `bitest` command. We can perform an immediate version of this test via the `bitesti` command below. Following the `bitesti` command, I supply the overall sample size (720), the number of people who are separated (20), and the population proportion for comparison, 0.04. The output of this command matches that from section 3.6 exactly.

```
. bitesti 720 20 0.04
           N   Observed k   Expected k   Assumed p   Observed p

         720           20         28.8     0.04000      0.02778
    Pr(k >= 20)              = 0.967391  (one-sided test)
    Pr(k <= 20)              = 0.051486  (one-sided test)
    Pr(k <= 20 or k >= 38) = 0.105043  (two-sided test)
```

3.9.5 Immediate cross-tabulations

There is an immediate version of the `tabulate` command called `tabi` which allows you to replicate cross-tabulations based on knowing the frequencies within each of the cells. This allows you not only to replicate the cross-tabulation table but also to perform chi-squared tests and Fisher's exact tests.

Video tutorial: Cross-tabulations on summary data

See a video showing how you can create cross-tabulations and perform chi-squared tests using summary data at http://www.stata.com/sbs/tabi.

Consider the cross-tabulation of marital status and gender illustrated in section 3.7. We can replicate those results by typing in the cell frequencies for each of the cells. Note that a backslash is used to demarcate one row from the next.

```
. tabi 117 150 \ 2 2 \ 17 30 \ 4 16 \ 198 184, column chi2 exact
```

```
  Key

        frequency
  column percentage
```

```
Enumerating sample-space combinations:
stage 5:   enumerations = 1
stage 4:   enumerations = 5
stage 3:   enumerations = 69
stage 2:   enumerations = 1170
stage 1:   enumerations = 0
```

| | col | | |
	1	2	Total
1	117	150	267
	34.62	39.27	37.08
2	2	2	4
	0.59	0.52	0.56
3	17	30	47
	5.03	7.85	6.53
4	4	16	20
	1.18	4.19	2.78
5	198	184	382
	58.58	48.17	53.06
Total	338	382	720
	100.00	100.00	100.00

```
          Pearson chi2(4) =  12.7462   Pr = 0.013
          Fisher's exact =                 0.009
```

Note: Nonparametric statistics

In this chapter, I omitted coverage of nonparametric statistics, but Stata has a wide array of offerings in this regard. You can type `help contents_stat_nonparametric` to see an extensive listing of the nonparametric statistics that Stata offers.

3.10   Closing thoughts

This chapter has illustrated how to perform basic inferential statistics using Stata. Many of the tests, especially the one-sample tests, may not be part of your daily work. However, I find that every once in a while, I am interested in doing a test of a one-sample proportion, a simple confidence interval for one mean, or a test comparing two proportions. In those instances, the tools described in this chapter can be very useful.

Part II

Between-subjects ANOVA models

This part of the book covers between-subjects analysis of variance (ANOVA) models, beginning with chapter 4, which covers one-way between-subjects ANOVA. Chapter 5 illustrates contrasts that you can use for making comparisons among groups in a one-way ANOVA. Chapter 6 covers analysis of covariance (ANCOVA), illustrating its use in experimental designs (to increase power) and in nonexperimental designs (to attempt to statistically control for confounding variables).

Chapter 7 introduces factorial designs, covering two-by-two designs, two-by-three designs, and three-by-three designs. This chapter emphasizes how to visualize and interpret the interactions. It also illustrates how to dissect two-way interactions using simple effects, simple contrasts, partial interactions, and interaction contrasts. The `contrast` command is illustrated for dissecting the interactions, and the `margins` and `marginsplot` commands are used to display and graph the means associated with the interactions.

Chapter 8 illustrates ANCOVA-type analyses with the focus on interactions of the independent variable and the covariate (in other words, categorical by continuous variable interactions).

Chapter 9 covers factorial models with three independent variables. This chapter, like chapter 7, emphasizes the visualization and interpretation of the interactions, in this case focusing on three-way interactions. Figures are used to visually understand three-way interactions, and a variety of analytic and graphical methods are illustrated for understanding three-way interactions.

Chapter 10 shows how you can extend the power of ANOVA by blending the ANOVA designs with regression commands. This chapter illustrates how you can extend your ANOVA design to analyze data that come from complex surveys or how to analyze data that violate the homogeneity of variance assumption or data with influential observations (via robust regression or quantile regression). This part concludes with chapter 11, which illustrates power analysis for ANOVA and ANCOVA.

4 One-way between-subjects ANOVA

4.1 Chapter overview

This chapter shows you how to perform a one-way analysis of variance (ANOVA) using Stata. It begins by showing you how to perform a two-group t test (section 4.2) and includes a discussion of computing effect sizes for the comparison of two groups. The two-group analysis is repeated using ANOVA (section 4.3). The chapter then shows you how to perform an ANOVA with three groups (section 4.4) and includes a brief discussion of testing specific comparisons among levels of the independent variable (IV) and a discussion of computing effect sizes associated with specific comparisons. This is followed by a discussion of estimation and postestimation commands (section 4.5) and concludes with comments on interpreting confidence intervals (CIs) (section 4.6).

4.2 Comparing two groups using a t test

The field of psychology has a long history of studying depression and trying to find ways of lessening it. Imagine a hypothetical research psychologist (Professor Cheer) who focuses her research on studying optimism and finding ways of increasing it. She developed a validated measure of optimism that, although correlated with conventional measures of depression, is conceptually and operationally distinct from depression. As part of her research, she has created a new kind of therapy, called optimism therapy, that she believes can be effective in increasing optimism. In this first hypothetical study, she seeks to determine the effectiveness of optimism therapy by comparing the optimism of people who have completed optimism therapy treatment with the optimism of a control group who received no treatment.

The dataset for this example is used below, and the first five observations are displayed. (See section 1.1 for information about how to download the example datasets.) The variable `treat` indicates the treatment assignment, coded as follows: 1 = control (`Con`), 2 = optimism therapy (`OT`). The variable `opt` is the optimism score at the end of the study. In this study, the variable `treat` is the IV with two levels, and `opt` is the dependent variable (DV).

```
. use opt-2, clear
. list in 1/5

      id    opt    treat

 1.    1   50.0     Con
 2.    2   28.0      OT
 3.    3   29.0     Con
 4.    4   30.0     Con
 5.    5   42.0     Con
```

The `tabulate` command is used to show the frequencies broken down by `treat`; it shows that there were 100 participants in the control group and 100 participants in the treatment (optimism therapy) group. Each of the 200 people was randomly assigned (in equal numbers) to the treatment or control group.

```
. tabulate treat

 Treatment
     group
assignment        Freq.      Percent        Cum.

       Con          100        50.00       50.00
        OT          100        50.00      100.00

     Total          200       100.00
```

The `fre` command is used below to show the frequency distribution of the variable `treat`. This user-written command (which can be installed as described in section 1.1) is very similar to the `tabulate` command. A key difference is that it shows both the numeric values for `treat` and the value labels. It emphasizes that a value of 1 stands for being in the control group and that a value of 2 represents being in the optimism therapy group

```
. fre treat
```

treat — Treatment group assignment

		Freq.	Percent	Valid	Cum.
Valid	1 Con	100	50.00	50.00	50.00
	2 OT	100	50.00	50.00	100.00
	Total	200	100.00	100.00	

The `summarize` command is used to display summary statistics for optimism. The mean optimism in the sample is 47.1 with a standard deviation of 10.49. The values of optimism range from a minimum of 24 to a maximum of 77.

```
. summarize opt
```

Variable	Obs	Mean	Std. Dev.	Min	Max
opt	200	47.1	10.48953	24	77

We can obtain detailed summary statistics for optimism by using the `summarize` command with the `detail` option. The key part of the output I want to note is that the variance of `opt` is 110.03. We will see this value later when we perform an ANOVA on the optimism scores.

```
. summarize opt, detail
```

Optimism score

	Percentiles	Smallest		
1%	27.5	24		
5%	30	27		
10%	32.5	28	Obs	200
25%	40	28	Sum of Wgt.	200
50%	48		Mean	47.1
		Largest	Std. Dev.	10.48953
75%	55	68		
90%	60.5	69	Variance	110.0302
95%	65.5	71	Skewness	.1183874
99%	70	77	Kurtosis	2.519853

The `tabulate` command combined with the `summarize(opt)` option displays summary statistics for optimism broken down by treatment group assignment. The mean optimism in the control group is 44.3, compared with 49.9 for the treatment group. The results are in the predicted direction, with the average optimism being higher for those who received optimism therapy (as compared with the control group).

```
. tabulate treat, sum(opt)
  Treatment
      group                Summary of Optimism score
  assignment        Mean      Std. Dev.          Freq.

         Con        44.3          10.1             100
          OT        49.9          10.2             100

       Total        47.1          10.5             200
```

Let's use the `ttest` command to perform a *t* test to compare the average optimism of the optimism therapy group with that of the control group. In the `ttest` command below, the DV is `opt` and the IV is `treat`.

```
. ttest opt, by(treat)
Two-sample t test with equal variances

    Group  |    Obs       Mean     Std. Err.    Std. Dev.    [95% Conf. Interval]

      Con  |    100      44.29     1.00567      10.0567      42.29453    46.28547
       OT  |    100      49.91     1.020209     10.20209     47.88568    51.93432

combined  |    200       47.1     .7417215     10.48953     45.63736    48.56264

    diff  |             -5.62      1.43255                  -8.445014   -2.794986

        diff = mean(Con) - mean(OT)                            t =  -3.9231
Ho: diff = 0                                 degrees of freedom =      198

    Ha: diff < 0                 Ha: diff != 0                    Ha: diff > 0
 Pr(T < t) = 0.0001       Pr(|T| > |t|) = 0.0001            Pr(T > t) = 0.9999
```

The output shows the mean optimism for the control group is 44.29, compared with 49.91 for the treatment group. The *t*-value for the comparison of these means is -3.92 with 198 degrees of freedom. The output shows *p*-values for a left-tailed test ($p = 0.0001$), a right-tailed test ($p = 0.9999$), and a two-tailed test ($p = 0.0001$). Using the *p*-value from the two-tailed test indicates that this result is statistically significant. We can conclude that the optimism scores in the treatment group are significantly different from those in the control group. Specifically, we can say that optimism is significantly greater in the optimism therapy group compared with the control group.

Let's go through the output of the `ttest` command more thoroughly. The body of the output shows a table with four rows, described below.

- Con. This row presents the number of observations, mean, standard error, standard deviation, and 95% confidence interval for the DV among those in the first group, the control group. The average optimism in this group is 44.29 with a standard deviation of 10.06. The standard error is computed by dividing the standard deviation by the square root of N (for this group). The standard error is used to form the 95% confidence interval. We can say with 95% confidence that the mean of optimism for the control group is between 42.3 and 46.3.

- OT. This row presents the same statistics but for the second group, the optimism therapy group. The mean for this group is 49.9 (95% CI = [47.9, 51.9]).
- **Combined**. This row presents the same statistics but for the overall sample. The mean optimism in the overall sample is 47.1 (95% CI = [45.6, 48.6]).
- **diff**. This row shows the same statistics (except for the standard deviation) for the difference in the mean of the first group minus the mean of the second group. This is indicated by the legend below the table, which shows that `diff=mean(Con) - mean(OT)`. The difference in the means is −5.6 and the standard error of the difference is 1.4. The 95% confidence interval for the difference in the means is −8.4 to −2.8. We can reverse the signs of this output to frame the difference in terms of the optimism therapy group versus the control group.

The output below the table shows the *t*-value, the degrees of freedom, and the *p*-values for a left-tailed test (left), two-tailed test (center), and right-tailed test (right). Using the two-tailed *p*-value, we can report that the difference in means is statistically significant $[t(198) = -3.92, p = 0.0001]$.

4.3 Comparing two groups using ANOVA

As we know, data from a two-group *t* test can also be analyzed using ANOVA. Let's illustrate how we can use ANOVA to perform this same analysis. Instead of using the `ttest` command, we use the **anova** command, as shown below. For the **anova** command, we list the DV (`opt`) followed by the IV (`treat`). Stata knows that the first variable listed is the DV and that the second (and subsequent) variables are the IVs.

```
. anova opt treat
                          Number of obs =        200    R-squared     =  0.0721
                          Root MSE      =    10.1297    Adj R-squared =  0.0674

         Source |  Partial SS       df          MS         F      Prob>F

          Model |    1579.22         1      1579.22     15.39    0.0001

          treat |    1579.22         1      1579.22     15.39    0.0001

       Residual |   20316.78       198       102.61

          Total |      21896       199    110.03015
```

The results are presented in the form of an ANOVA table that decomposes the total variation in **opt** scores into two parts—that which can be explained by the treatment group assignment (labeled **treat**) and the unexplained variance (labeled **Residual**). Sometimes, this decomposition is described as between groups (that is, treatment) and within groups (that is, residual). The table also includes a row labeled **Model**, which we will ignore for now. The key part of the output shows that the *F*-value for treatment group assignment is 15.39 with a *p*-value of 0.0001. The effect of treatment group assignment is significant.

> **Video tutorial: Factor variables**
> See a video showing the basics of using factor variables in Stata at
> http://www.stata.com/sbs/factor-variables.

Let's take a moment to consider the output from the `anova` command starting with
the body of the output, the ANOVA source table. The source table lists four sources of
variance, `Model`, `treat`, `Residual`, and `Total`. For each source of variance, the output
shows the sum of squares (labeled `Partial SS`), and the degrees of freedom (labeled
`df`), and the mean squares (labeled `MS`). Let's consider each of these (but I will present
them in a different order from the one listed in the table).

- `Total`. This row (at the bottom) represents the total variation in the DV (opt).
  It shows the sum of squares, degrees of freedom, and mean squares. The degrees
  of freedom is the number of observations minus 1 ($\text{df}_{\text{Total}} = 199$). The MS total is
  110.03, which is what we found for the variance of optimism via the `summarize`
  command. This shows that the MS total is just another name for the variance.
- `treat`. This row (the second row) shows the sum of squares for `treat` (that is,
  SS_{treat}) is 1579.22. Because there are 2 treatment groups, this effect has 1 degree
  of freedom (that is, $\text{df}_{\text{treat}} = 1$). The mean square for treat (MS_{treat}) is computed
  by dividing SS_{treat} by df_{treat}, yielding 1579.22 divided by 1, or 1579.22.
- `Residual`. The residual sum of squares ($\text{SS}_{\text{Residual}}$) is 20316.78, which is equal
  to the SS_{Total} minus SS_{treat} (that is, $21896 - 1579.22$), which yields 20316.78.
  The $\text{df}_{\text{Residual}}$ can likewise be computed by taking the df_{Total} minus df_{treat} (that
  is, $199 - 1$) to yield 198. The $\text{MS}_{\text{Residual}}$ is computed by taking the $\text{SS}_{\text{Residual}}$
  (20316.78) divided by the $\text{df}_{\text{Residual}}$ (198), yielding 102.61.
- `Model`. This represents the total amount of variation explained by all the IVs in
  the model. Because we have only one IV, results shown for `treat` and `Model` are
  the same.

The F statistic for `treat` is computed by dividing the MS_{treat} (1579.22) by the
$\text{MS}_{\text{Residual}}$ (102.61), yielding 15.39. Stata calculates the likelihood of this F-value arising
by chance when the means for the two groups are actually equal in the population, and
finds that the probability of such an extreme F-value arising by chance is 0.0001.

Let's now consider the upper portion of the output. I deferred discussing this portion
of the output because it draws information from the ANOVA source table. The output
shows that the number of observations is 200. The output also shows the R-squared
for the model, which is the proportion of variance in the DV that is explained by the
entire model (all predictors). We can manually compute the R-squared by dividing the
SS_{Model} (1579.22) by the SS_{Total} (21896), yielding 0.0721. Because `treat` is the only
variable in the model, we can say that this is the percentage of the total variance that

is explained by treatment group assignment.[1] The adjusted R-squared is shown below the R-squared in the output. This value, 0.0674, accounts for the number of predictors in the model.

The output also shows that the `Root MSE` is 10.13. This value can be computed by taking the square root of $MS_{Residual}$ (that is, the square root of 102.61 is 10.13). This value is used when computing CIs for the group means.

Speaking of means, let's display the mean of `opt` for the treatment and control groups. After we run the `anova` command, the `margins` command, shown below, computes the means of the DV as a function of `treat`.

```
. margins treat, nopvalues

Adjusted predictions                             Number of obs    =        200

Expression    : Linear prediction, predict()
```

	Margin	Delta-method Std. Err.	[95% Conf. Interval]	
treat				
Con	44.29	1.012966	42.29241	46.28759
OT	49.91	1.012966	47.91241	51.90759

Note: The nopvalues option

Usually, we like to see p-values. They tell us whether effects of interest are significant. The `margins` command normally displays p-values that test the null hypothesis that the displayed mean equals zero. The p-values of such tests are not very interesting and can mislead us into believing that we suddenly have interesting results because the output displays significant p-values. Thus I frequently use the `nopvalues` option with the `margins` command to suppress the display of p-values when the underlying null hypothesis test is not interesting.

This output shows the mean of `opt` for the control group is 44.29, compared with 49.91 for the treatment group. Note these values match what we found from the `ttest` command. The right portion of the output shows the confidence interval for each of these means. For example, the 95% confidence interval for the mean optimism for the control group is 42.29 to 46.29.

---

1. If there are additional IVs in the model, the R-squared represents the proportion of variance explained by all the IVs in the model.

Underneath the hood! Confidence intervals

How did the `margins` command compute the confidence intervals for the means? It computes the standard error of the mean using the delta method, which in this case is equivalent to dividing the root mean squared error by the square root of N for each group. For the control group, this is 10.13/sqrt(100), which yields 1.013. The sample size (N) for the optimism therapy group is also 100, so the standard error is the same for this group. The standard error is then multiplied by the t-value associated with 198 degrees of freedom and a 95% confidence level (which is 1.972), which yields 1.997. Adding this value to the mean for the control group yields the upper confidence limit, 46.287. Subtracting that value yields the lower confidence limit, 42.293.

We can graph the means and confidence intervals computed via the `margins` command by using the `marginsplot` command, shown below.

```
. marginsplot
  Variables that uniquely identify margins: treat
```

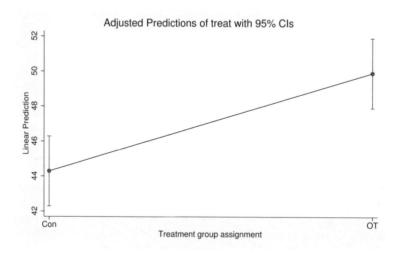

Figure 4.1: Mean optimism by treatment group assignment

Note! Margins and marginsplot are a team

The `margins` and `marginsplot` commands work together as a team. Following the `margins` command, the `marginsplot` command graphs the means and confidence intervals computed by the `margins` command. But, if you execute any other command after `margins`, but before `marginsplot`, you will get the following error:

```
. marginsplot
previous command was not margins
r(301);
```

This error is saying that the last command you issued was not the `margins` command. The solution is to run the `margins` command followed by the `marginsplot` command (with no other commands in between).

4.3.1 Computing effect sizes

If we were writing this up for publication, we would want to include an estimate of the strength of the treatment effect. Such measures are called an "effect size". A common effect size measure is Cohen's d. In terms of this example, Cohen's d would represent the degree to which the treatment is superior to the control, represented using standardized units (that is, sd = 1). The formula for Cohen's d is shown below in (4.1).

$$d = \frac{\overline{x}_1 - \overline{x}_2}{s} \tag{4.1}$$

We can compute the difference in means from the last `margins` command (that is, by taking 44.29 minus 49.91). We can obtain the standard deviation via the root mean squared error from the output of the `anova` command (10.1297). Using the `display` command below, we see that Cohen's d for the effect of optimism therapy (versus control) is -0.55.

```
. display (44.29 - 49.91) / 10.1297
-.55480419
```

Rather than computing Cohen's d manually, we can use the `esize twosample` command to perform this computation. This shows the value of Cohen's d (which is the same as our hand computation) and also shows a 95% confidence interval. It also shows an alternative measure of effect size called Hedges's g.

```
. esize twosample opt, by(treat)
Effect size based on mean comparison
                                   Obs per group:
                                           Con =          100
                                            OT =          100

     Effect Size   |    Estimate       [95% Conf. Interval]
    ---------------+------------------------------------------
       Cohen´s d   |   -.5548064      -.8366402    -.2716222
       Hedges´s g  |   -.5527018      -.8334665    -.2705918
```

By including the **all** option, we see that the **esize** command shows five different measures of effect size. My experience is that the most commonly reported effect size is Cohen's *d*. For more details and references about these different effect sizes, you can type **help esize**.

```
. esize twosample opt, by(treat) all
Effect size based on mean comparison
                                   Obs per group:
                                           Con =          100
                                            OT =          100

      Effect Size  |    Estimate       [95% Conf. Interval]
    ---------------+------------------------------------------
        Cohen´s d  |   -.5548064      -.8366402    -.2716222
        Hedges´s g |   -.5527018      -.8334665    -.2705918
    Glass´s Delta 1|   -.5588313       -.845396    -.2696437
    Glass´s Delta 2|   -.5508673      -.8371509    -.2619931
   Point-Biserial r|   -.2685585      -.3875675    -.1352412
```

We can also describe the strength of the effect using the proportion of variance that the IV explains in the DV. The results from the **anova** command show that the R-squared value is 0.0721. This means that the treatment group assignment explains 7.21% of the variance in **opt**. The adjusted R-squared is 0.0674, which adjusts for the number of the predictors in the model.

Using the **estat esize** command, we can obtain the percentage of explained variance as well as a confidence interval for that estimate, as shown below.

```
. estat esize
Effect sizes for linear models

        Source  |   Eta-Squared     df    [95% Conf. Interval]
    ------------+-------------------------------------------------
         Model  |    .0721237        1    .0181106     .1489302
    ------------+-------------------------------------------------
         treat  |    .0721237        1    .0181106     .1489302
```

The `estat esize` command uses the term "eta-squared" to describe the percentage of variance explained.[2] The output shows that the variable `treat` explains 7.21% of the variance in optimism, with a 95% confidence interval ranging from 1.8% to 14.9%.

A more common (and appropriate) value to report is called omega-squared,[3] which is adjusted for the number of predictors. We can request computation of omega-squared by using the `estat esize` command, adding the `omega` option as shown below.

```
. estat esize, omega
Effect sizes for linear models
```

Source	Omega-Squared	df	[95% Conf. Interval]	
Model	.0674374	1	.0131516	.1446318
treat	.0674374	1	.0131516	.1446318

This output shows us that after we adjust for the number of predictors, `treat` explains 6.7% of the variance in `opt`. Further, we are 95% confident that the population value lies between 1.3% and 14.5%.

Video tutorial: Effect sizes

You can take a quick video tour of effect sizes at http://www.stata.com/sbs/effect-size. In addition, *The Stata Blog* describes and illustrates effect sizes; see http://blog.stata.com/2013/09/05/measures-of-effect-size-in-stata-13/.

4.4   Comparing three groups using ANOVA

In the previous section, we saw an example in which Professor Cheer compared an optimism therapy group with a control group to assess the effectiveness of the therapy. A second study is conducted in which a third group is added, a group that receives traditional therapy. After finding that the optimism was greater in the optimism therapy group as compared with the control group, she is now interested in comparing optimism therapy with traditional therapy (to see whether it is also superior to traditional therapy in increasing optimism). She is also interested in assessing the effect of traditional therapy versus control (to assess how effective traditional therapy is in increasing optimism).

---

2. Note how the value of eta-squared matches the value of R-squared from the output of the `anova` command. In the regression tradition, the percentage of explained variance is called R-squared; in the ANOVA tradition, it is called eta-squared. These are two different names for the exact same thing.

3. Note how the value of omega-squared matches the value of adjusted R-squared from the output of the `anova` command. In the ANOVA tradition, the adjusted R-squared is called omega-squared. Like with R-squared and eta-squared, these are just two different names for the exact same thing.

To recap, we see this study has three groups: 1) control (Con); 2) traditional therapy (TT); and 3) optimism therapy (OT). An ANOVA will be used to assess the effect of treatment group assignment (the IV) on optimism (the DV). The F test produced by ANOVA will test the following null hypothesis,

$$H_0: \mu_{\text{Con}} = \mu_{\text{TT}} = \mu_{\text{OT}}$$

where μ_{Con} is the population mean of the control group (group 1), μ_{TT} is the population mean of the traditional therapy group (group 2), and μ_{OT} is the population mean of the optimism therapy group (group 3). This test of the equality of the three population means does not directly test the questions that Professor Cheer has in mind. In particular, she wants to test the following two specific hypotheses.[4]

Hypothesis 1. Is optimism therapy superior to traditional therapy (that is, is the mean optimism greater for those in optimism therapy than that for those in traditional therapy)? We can test this via the following null hypothesis:

$$H_0\#1: \mu_{\text{OT}} = \mu_{\text{TT}}$$

Hypothesis 2. Is the traditional therapy group superior to the control group? That is, is the mean optimism for the traditional therapy group greater than the mean optimism for the control group? We can test this using the following null hypothesis:

$$H_0\#2: \mu_{\text{TT}} = \mu_{\text{Con}}$$

The dataset for this example is used below, and the first five observations are displayed. The variable `treat` indicates the treatment assignment, coded as follows: 1 = control (Con), 2 = traditional therapy (TT), 3 = optimism therapy (OT). The variable `opt` is the optimism score at the end of the study.

```
. use opt-3
. list in 1/5
```

	id	opt	treat
1.	1	41.0	Con
2.	2	45.0	TT
3.	3	45.0	OT
4.	4	45.0	Con
5.	5	48.0	Con

---

4. Although the first study already established that the optimism therapy group was superior to the control group, she is still interested in the comparison of the optimism therapy group with the control group but more as a replication of an established result than one of her primary hypotheses. As we analyze these data, we will also examine this question as well.

We now run an ANOVA comparing the mean of the DV (optimism) among the three treatment groups using the command below.

```
. anova opt treat

                              Number of obs =       300    R-squared     =  0.0523
                              Root MSE      =  10.1165     Adj R-squared =  0.0459

          Source |  Partial SS        df        MS          F     Prob>F

           Model |  1676.3467          2     838.17333     8.19   0.0003

           treat |  1676.3467          2     838.17333     8.19   0.0003

        Residual |  30396.09         297     102.34374

           Total |  32072.437        299     107.26567
```

The `anova` command displays an F test that tests the overall null hypothesis regarding the equality of the means of the three groups, repeated below.

$$H_0 : \mu_{\text{Con}} = \mu_{\text{TT}} = \mu_{\text{OT}}$$

The F-value is 8.19 with a p-value of 0.0003. Based on this p-value, we can reject the overall null hypothesis of the equality of the means of the three treatment groups. We will soon consider the two hypotheses of interest for this study. But first let's compute (and graph) the mean of optimism by treatment group assignment.

We can use the `margins` command below to show the mean of the DV (`opt`) by treatment group assignment (`treat`).

```
. margins treat, nopvalues
Adjusted predictions                                 Number of obs      =       300
Expression    : Linear prediction, predict()

                        Delta-method
              Margin     Std. Err.      [95% Conf. Interval]

       treat
        Con    44.71     1.011651       42.71909    46.70091
         TT    49.49     1.011651       47.49909    51.48091
         OT    49.93     1.011651       47.93909    51.92091
```

We can then use the `marginsplot` command to make a graph showing the mean of the DV (`optimism`) by treatment group assignment (see figure 4.2). This graph shows the mean of optimism by the three levels of `treat`, that is, for the control group, traditional therapy group, and optimism therapy group.

```
. marginsplot
Variables that uniquely identify margins: treat
```

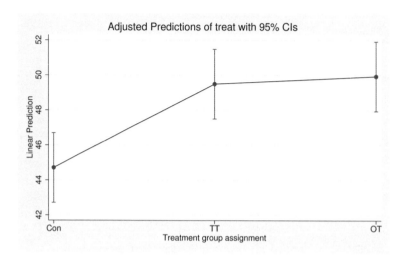

Figure 4.2: Mean optimism by treatment group assignment

Let's now address the two null hypotheses of interest by using the `contrast` command.

4.4.1 Testing planned comparisons using contrast

Let's test the two null hypotheses we specified before the start of the study. Let's start by testing the first null hypothesis, which compares optimism therapy (group 3) with traditional therapy (group 2). We can perform this test using the `contrast` command shown below. (I will discuss the syntax of `contrast` extensively in chapter 5.)

```
. contrast r3b2.treat
Contrasts of marginal linear predictions
Margins         : asbalanced
```

	df	F	P>F
treat	1	0.09	0.7586
Denominator	297		

	Contrast	Std. Err.	[95% Conf. Interval]	
treat (OT vs TT)	.44	1.43069	-2.375575	3.255575

Because I specified `r3b2.treat`, this `contrast` command compares group 3 with group 2. The lower portion of the output shows that the difference in the means for those in optimism therapy (OT) versus those in traditional therapy (TT) is 0.44 with a 95% confidence interval of -2.38 to 3.26. The upper portion of the output shows that the difference in the means is not statistically significant $[F(1, 297) = 0.09, p = 0.7586]$. In regard to our first null hypothesis, $H_0 \#1$, the results are not consistent with our prediction that optimism would be greater in the optimism therapy group than in the traditional therapy group.

Now consider our second hypothesis, which compares the traditional therapy group (group 2) with the control group (group 1). This comparison is tested with the `contrast` command below. By specifying `r2b1.treat` with `contrast`, we obtain a comparison of group 2 with group 1.

```
. contrast r2b1.treat
Contrasts of marginal linear predictions
Margins         : asbalanced
```

	df	F	P>F
treat	1	11.16	0.0009
Denominator	297		

	Contrast	Std. Err.	[95% Conf. Interval]	
treat (TT vs Con)	4.78	1.43069	1.964425	7.595575

The lower portion of the output shows that the difference in the mean optimism for the traditional therapy (TT) versus control (Con) groups is 4.78 with a 95% confidence interval of 1.96 to 7.60. The upper portion of the output shows that this difference in means is statistically significant [$F(1,297) = 11.16$, $p = 0.0009$]. The mean optimism for the traditional therapy group is significantly greater than the mean optimism for the control group. Thus we can reject $H_0\#2$ and say the results are consistent with our prediction that traditional therapy yields greater optimism than the control.

Note: One-tailed tests

Consider the second hypothesis, which predicts that average optimism will be greater in the traditional therapy group than in the control group. This hypothesis is framed in a one-sided manner: that we predict that traditional therapy will yield greater optimism than the control. Some might argue that it would be appropriate to test such a one-sided prediction using a one-sided test. If we made a one-sided prediction and feel that we can persuade reviewers that a one-sided test is justified, the p-value can be divided by two if the results are in the predicted direction. For example, say that we wanted to report the results of the test of the second hypothesis as a one-sided test. In that case, we could report that the mean optimism for the traditional therapy group is significantly greater than the mean optimism for the control group [$F(1,297) = 11.16$, $p = 0.00045$, one-tailed]. Note how the p-value is divided by two and reported as a one-tailed value. If you report one-sided tests, it is important to have a very strong basis for selecting a one-tailed test. As Murnane and Willett (2011, 106) write, "Only when you can mount a compelling defense of the argument that a particular policy or intervention can have only a directed impact (positive or negative) on the outcomes of interest, in the population, is the use of one-tailed tests justified". Without a compelling defense of a one-tailed test, you may risk receiving critical comments regarding the suitability of your analytic methods.

Professor Cheer was also interested in replicating the results of the first study, which found that optimism was greater in the optimism therapy group than in the control group. Specifying `r3b1.treat` with `contrast` provides a comparison of the third group (optimism therapy) with the first group (control).

```
. contrast r3b1.treat
Contrasts of marginal linear predictions
Margins         : asbalanced
```

	df	F	P>F
treat	1	13.31	0.0003
Denominator	297		

	Contrast	Std. Err.	[95% Conf. Interval]
treat (OT vs Con)	5.22	1.43069	2.404425 8.035575

These results are consistent with the first study: optimism therapy is superior to the control [$F(1, 297) = 13.31$, $p = 0.0003$]. The difference in the means between the optimism therapy group versus the control group is 5.22 (95% CI = [2.40, 8.04]).

As you might imagine, Professor Cheer was glad to replicate the finding that optimism therapy yielded greater optimism than the control but was disappointed to find that optimism therapy was no better than traditional therapy. This finding will be further explored in chapter 7, in which the study will be expanded to consider the role of preexisting depression in explaining the efficacy of optimism therapy. But before we embark on that topic, let's consider the computation of effect sizes for this study.

4.4.2   Computing effect sizes for planned comparisons

When we were considering the comparison of two groups (optimism therapy versus control), I illustrated two types of effect sizes. The first type was Cohen's d (or one of its variants), and the second type was eta-squared (and its cousin omega-squared). When we computed Cohen's d, we used the `esize twosample` command. If we apply this command in this case, where `opt` has three levels, we obtain an error because, as implied by the name of the command, `treat` is expected to have only two groups.

```
. esize twosample opt, by(treat) cohensd
more than 2 groups found, only 2 allowed
r(420);
```

Because Cohen's d represents an effect size comparing two groups, `estat twosample` expects a comparison of two groups. However, we can use the `esize twosample` command with an `if` specification to restrict the analysis to just those two groups, as illustrated below:[5]

---

5. Note how the `if` specification precedes the comma. Also, the `inlist()` function is true if the value of `treat` is 1 or 3.

```
. esize twosample opt if inlist(treat,1,3), by(treat)
Effect size based on mean comparison
                                  Obs per group:
                                           Con =          100
                                            TT =            0
                                            OT =          100

       Effect Size │   Estimate      [95% Conf. Interval]
    ───────────────┼──────────────────────────────────────
        Cohen's d  │  -.5176895     -.7988822    -.2352309
       Hedges's g  │  -.5157256     -.7958516    -.2343386
```

Cohen's d for the difference between the optimism therapy group and control group is -0.518, with a confidence interval ranging from -0.80 to -0.24. The values are negative because it computed the mean for group 1 (control) minus group 3 (optimism therapy). To report the treatment effect of optimism therapy (versus control), we would report the difference as 0.52 (95% CI $= [0.24, 0.80]$). It is important to note that this effect size is computed only with respect to the optimism therapy and control groups, with the traditional therapy group omitted entirely from the computation.

We can compute an effect size expressed as the proportion of variance in optimism that is explained by treatment group assignment. This measure of effect size is much more general than Cohen's d because it can be used in a wider variety of analyses (including ANOVA with more than two groups, factorial ANOVA, analysis of covariance, and regression models). Using the `estat esize` command with the `omega` option, we see that treatment group assignment explains 4.6% of the variance in optimism (95% CI $= [0.5\%, 9.9\%]$).

```
. estat esize, omega
Effect sizes for linear models

             Source │ Omega-Squared      df      [95% Conf. Interval]
       ─────────────┼────────────────────────────────────────────────
              Model │   .0458855          2     .0049216      .098858

              treat │   .0458855          2     .0049216      .098858
```

This reflects the effect size with respect to the comparison of all three treatment groups. However, I would like an estimate of the proportion of variance that corresponds to the `contrast` command, which compared the optimism therapy group with the control group. In other words, I want a blend of the `estat esize` command with the `contrast` command. Referring back to the `contrast r3b1.treat` command comparing the optimism therapy group with the control group (group 3 versus 1), we see that the F-value is 13.31, the numerator degrees of freedom is 1, and the denominator degrees of freedom is 297. We can supply these three pieces of information to the `esizei` command to compute the eta-squared and omega-squared associated with that contrast. The `esizei` command is issued below by supplying the numerator degrees of freedom (1) followed by the denominator degrees of freedom (297) followed by the *F*-value (13.31).

```
. esizei 1 297 13.31
Effect sizes for linear models
```

Effect Size	Estimate	[95% Conf. Interval]	
Eta-Squared	.0428926	.0091701	.0957584
Omega-Squared	.03967	.0058339	.0927138

Using omega-squared, we can say that the comparison of optimism therapy with the control explains 3.97% of the variance in optimism (95% CI = [0.6%, 9.3%]).

4.5 Estimation commands and postestimation commands

Let's take a moment to talk more about the commands we have been using in this chapter, in particular the `anova`, `margins`, `marginsplot`, `contrast`, and `estat esize` commands. I want to illustrate that the order in which you issue these commands is important by showing what happens if we issue these commands in different orders.

But first, we need to wipe the slate clean, as if we are starting a brand-new Stata session. Instead of quitting Stata and starting it again, we can issue the `clear all` command. This command *wipes the slate clean*, as though we freshly started Stata.

```
. clear all
```

Let's now use `opt-3.dta`, the one we used in the most recent analysis.

```
. use opt-3
```

Now suppose that we type the `estat esize` command.

```
. estat esize
last estimates not found
r(301);
```

Stata issues an error message, even though we found that this exact same command had worked previously. We are asking Stata to tell us the percentage of variance explained, but we have not run a model (remember that we used the `clear all` command to start fresh). The error message `last estimates not found` is Stata's way of saying, "I cannot compute the percentage of variance explained until you first fit an appropriate model" (using a command like `anova`).

Let's try using the `margins treat` command.

```
. margins treat
last estimates not found
r(301);
```

Again we encounter the same issue. The `margins` command works only after we have fit a model (such as using the `anova` command).

To use Stata jargon, we call the `anova` command an "estimation command". We call it this because it fits a statistical model. Once you fit a model, you can then use follow-up commands (for example, `margins` or `estat esize`) to perform further computations based on the most recently fit model. Let's illustrate this below by first using the `anova` command to fit a model.

```
. anova opt treat
```

	Number of obs =	300	R-squared	=	0.0523
	Root MSE =	10.1165	Adj R-squared	=	0.0459

Source	Partial SS	df	MS	F	Prob>F
Model	1676.3467	2	838.17333	8.19	0.0003
treat	1676.3467	2	838.17333	8.19	0.0003
Residual	30396.09	297	102.34374		
Total	32072.437	299	107.26567		

Stata has now saved information about the estimation results from this `anova` command that can be used by follow-up commands (such as `margins`, `marginsplot`, or `estat esize`). So if we now issue a command like `estat esize`, Stata reports the percentage of variance explained with respect to the `anova` model we ran above.

```
. estat esize, omega
Effect sizes for linear models
```

Source	Omega-Squared	df	[95% Conf. Interval]	
Model	.0458855	2	.0049216	.098858
treat	.0458855	2	.0049216	.098858

These follow-up commands (such as `estat esize` and `margins`) are called, in Stata jargon, "postestimation commands". These commands report information based on the most recently issued estimation command (in this case, `anova opt treat`). They are called postestimation commands because they are run after an estimation command.

Suppose we now run a command that is not a postestimation command, like the `tabulate` command. Will that ruin our ability to subsequently use a postestimation command like `estat esize`? Let's try it out. First, we run a `tabulate` command below.

```
. tabulate treat
Treatment
    group
assignment        Freq.      Percent        Cum.

        Con        100        33.33        33.33
         TT        100        33.33        66.67
         OT        100        33.33       100.00

      Total        300       100.00
```

Now, we run the `estat esize` command. It still worked! Even though we ran the `tabulate` command, the `estat esize` command still works, reporting the percentage of explained variance with respect to the `anova` command.

```
. estat esize
Effect sizes for linear models
```

Source	Eta-Squared	df	[95% Conf. Interval]	
Model	.0522675	2	.0115777	.1048857
treat	.0522675	2	.0115777	.1048857

Running commands like `tabulate`, `summarize`, `describe`, and `esize twosample` does not interfere with our ability to continue to use postestimation commands like `estat esize`. The `tabulate` command, for example, is not an estimation command (meaning it does not fit a statistical model).

Note! Is the ttest command an estimation command?

Let's consider the `ttest` command. You might think that this is an estimation command because it performs a statistical test comparing the equality of means between two groups. But actually, Stata does not consider the `ttest` command to be an estimation command. So commands like `margins` or `estat esize` cannot be used after the `ttest` command to further elaborate on the results of the `ttest` command. You can issue such commands, but they would refer back to the most recent estimation command.

Let's now consider a special postestimation command, the `marginsplot` command. In the previous section, we ran the `marginsplot` command to display a graph of `opt` by `treat`. Let's try running the `marginsplot` command below.

```
. marginsplot
previous command was not margins
r(301);
```

This error message is telling us that the `marginsplot` command has to follow the `margins` command. So let's try this again by issuing the `margins` command to obtain the means broken down by `treat` immediately followed by the `marginsplot` command.

```
. margins treat, nopvalues
Adjusted predictions                              Number of obs    =        300
Expression    : Linear prediction, predict()
```

	Margin	Delta-method Std. Err.	[95% Conf. Interval]	
treat				
Con	44.71	1.011651	42.71909	46.70091
TT	49.49	1.011651	47.49909	51.48091
OT	49.93	1.011651	47.93909	51.92091

```
. marginsplot
    Variables that uniquely identify margins: treat
```

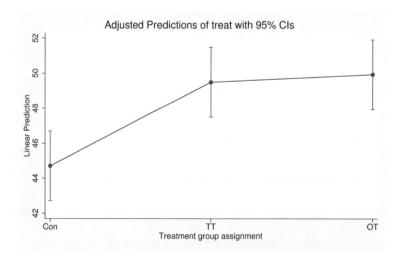

Figure 4.3: Mean optimism by treatment group assignment

This worked, creating the graph shown in figure 4.3. But note that the `marginsplot` command has to immediately follow the `margins` command. If we issue any command between the `margins` and `marginsplot` command, Stata gives us an error, as illustrated below.

```
. margins married
. summarize married
. marginsplot
previous command was not margins
r(301);
```

To recap, we see that the `anova` command is a special kind of command called an estimation command. So far we have seen one estimation command, the `anova` command for running an ANOVA. In this book, we will also see the `regress` command (for performing linear regressions) as well as a few other specialized regression commands.

After you issue an estimation command, you can issue postestimation commands such as `margins`, `marginsplot`, `contrast`, or `estat esize`. You will get an error message if you run a postestimation command like `margins` without having first issued an `estimation` command.

After you run an estimation command like `anova`, subsequent postestimation commands like `margins` or `estat esize` will report results pertaining to that `anova` command. You can run other nonestimation commands (like `describe` or `summarize`), and Stata will still be able to run additional postestimation commands. Such postestimation commands will continue to be based on the most recent estimation command.

If we run a different estimation command (`anova` or some other estimation command like `regress`), subsequent postestimation commands will report on the most recent estimation command.

Note! What postestimation commands can I use?

What if you wanted to see a list of the postestimation commands you can use after the `anova` command? You can type `help anova postestimation`, and a help file will show the postestimation commands that you can use after the `anova` command. Or if you wanted to see all the postestimation commands available after the `regress` command, you can type `help regress postestimation`. After you run an estimation command, such as the `regress` command, it would be useful if Stata had a command that would show you the appropriate postestimation tools that are available following the estimation command you used (for example, after the `regress` command). Such a command now exists in Stata 14 (and later)—it is called the `postest` command. Say that you run a regression using the `regress` command; you can then type `postest`, and the *Postestimation Selector* is displayed. This shows just the postestimation tools (and all the postestimation tools) that are appropriate following the `regress` command. You can then use point-and-click methods to select and execute the postestimation tools of your choosing. For more details about the *Postestimation Selector*, type `help postest`. Also, for more details about commonly used postestimation commands, see chapter 21.

4.6   Interpreting confidence intervals

The `marginsplot` command displays means and confidence intervals that were computed from the most recent `margins` command. Sometimes, these confidence intervals might tempt you into falsely believing that they tell us about differences among groups. To illustrate this point, let's use the analysis from section 4.4 predicting optimism from treatment group assignment. The dataset `opt-3.dta` is used below, and the `anova` command is shown, but the output is omitted.

```
. use opt-3, clear
. anova opt treat
  (output omitted )
```

The `margins` command is used to estimate the adjusted means of optimism by `treat`. The output also includes the 95% confidence interval for each adjusted mean. For example, in the control group (group 1), we are 95% confident that the mean of optimism is between 42.7 and 46.7.

```
. margins treat, nopvalues
Adjusted predictions                          Number of obs    =        300
Expression   : Linear prediction, predict()
```

	Margin	Delta-method Std. Err.	[95% Conf. Interval]	
treat				
Con	44.71	1.011651	42.71909	46.70091
TT	49.49	1.011651	47.49909	51.48091
OT	49.93	1.011651	47.93909	51.92091

We can graph the adjusted means and confidence intervals computed by the `margins` command using the `marginsplot` command shown below.

```
. marginsplot
  Variables that uniquely identify margins: treat
```

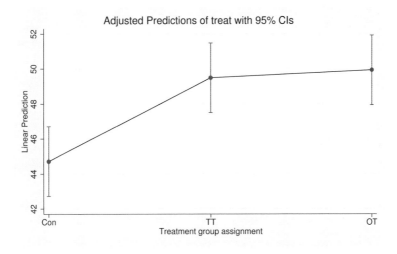

Figure 4.4: Mean optimism by treatment group assignment

Figure 4.4 is a graphical representation of the adjusted means and confidence intervals computed by the `margins` command. The confidence intervals are an important part of the graph because they reflect the precision of the estimate of each group mean. However, their purpose is not for informing comparisons between means. Our eye might be tempted to use the overlap (or lack of overlap) of confidence intervals between groups to draw conclusions about the significance of the differences between groups, but the most accurate way to perform such comparisons is via the `contrast` command. For example, the `contrast` command below shows that the comparison of the traditional therapy group with the control group is statistically significant ($t = 3.34$, $p = 0.001$).

```
. contrast r2b1.treat, nowald pveffects
Contrasts of marginal linear predictions
Margins        : asbalanced
```

	Contrast	Std. Err.	t	P>|t|
treat				
(TT vs Con)	4.78	1.43069	3.34	0.001

Using the **contrast** command, like the one shown above, is the most precise and appropriate way to assess the significance of group differences. In fact, this is a useful segue to the following chapter, which focuses on using the **contrast** command for performing specific comparisons between groups.

4.7   Closing thoughts

In this section, we have seen how you can use the **anova** command to perform a one-way ANOVA testing the effect of an IV on a DV. We also saw how the **margins** and **marginsplot** commands make it easy to compute and graph the adjusted means of the outcome as a function of the IV. We briefly considered the **contrast** command for making specific comparisons among means. The **contrast** command allows you to test a wide array of specific comparisons among group means. The following chapter (chapter 5) is devoted to exploring the use of the **contrast** command to help you form meaningful contrasts among the levels of the IV.

For more details about options you can use with the **margins** command, see section 21.3. Section 21.4 contains more details about customizing the appearance of the graphs created by the **marginsplot** command. For further help with these commands, see **help margins** and **help marginsplot**.

5 Contrasts for a one-way ANOVA

5.1 Chapter overview

This chapter is devoted to showing you how to perform specific comparisons among groups (contrasts) after performing a one-way analysis of variance (ANOVA). Such contrasts allow you to test for the exact pattern of results that you hypothesized before the start of your study. This chapter illustrates many different kinds of contrasts that you can perform using Stata's built-in contrast operators. Further, it shows how you can perform custom contrasts (if you want to perform a contrast that is not among the built-in contrasts).

5.2   Introducing contrasts

It turns out that Professor Cheer is interested in studying happiness as well as studying optimism. She is interested in exploring the effect of marital status on happiness. She decides to explore this question using the General Social Survey (GSS) dataset from the year 2012. The `use` command reads this dataset into memory. (Section 1.1 shows how to download the example datasets.)

```
. use gss2012_sbs, clear
```

Let's start by looking at a frequency distribution of the variable `marital`, which contains the marital status of the respondent. The distribution of this variable is shown with the user-written `fre` command (see section 1.1 to see how to download this command).

```
. fre marital
```
marital — marital status

		Freq.	Percent	Valid	Cum.
Valid	1 married	900	45.59	45.59	45.59
	2 widowed	163	8.26	8.26	53.85
	3 divorced	317	16.06	16.06	69.91
	4 separated	68	3.44	3.44	73.35
	5 never married	526	26.65	26.65	100.00
	Total	1974	100.00	100.00	

This shows that 1,974 people responded, and their responses were coded as married, widowed, divorced, separated, or never married. The `fre` command shows the numeric codes associated with each of these responses.

Let's now use the `fre` command to show the frequency distribution for the variable `happy7`, which is the happiness rating of the respondent on a 1 to 7 scale. We can see that a response of 1 represents `Completely unhappy` and that a response of 7 represents `Completely happy`. For this question, there are 1,284 valid responses and 690 missing values. This large number of missing values reflects the fact that this question was only asked of some of the respondents.

```
. fre happy7
```

happy7 — how happy R is (recoded)

			Freq.	Percent	Valid	Cum.
Valid	1	Completely unhappy	5	0.25	0.39	0.39
	2	Very unhappy	16	0.81	1.25	1.64
	3	Fairly unhappy	35	1.77	2.73	4.36
	4	Neither happy nor unhappy	77	3.90	6.00	10.36
	5	Fairly happy	440	22.29	34.27	44.63
	6	Very happy	563	28.52	43.85	88.47
	7	Completely happy	148	7.50	11.53	100.00
		Total	1284	65.05	100.00	
Missing	.c	Cannot choose	11	0.56		
	.i	Inapplicable	672	34.04		
	.n	No Answer	7	0.35		
		Total	690	34.95		
Total			1974	100.00		

Let's now perform a one-way ANOVA in which **happy7** is the DV[1] and marital status (**marital**) is the IV.

```
. anova happy7 marital
```

	Number of obs =	1,284	R-squared	=	0.0408
	Root MSE =	.977102	Adj R-squared =		0.0378

Source	Partial SS	df	MS	F	Prob>F
Model	51.89968	4	12.97492	13.59	0.0000
marital	51.89968	4	12.97492	13.59	0.0000
Residual	1221.0972	1,279	.95472807		
Total	1272.9969	1,283	.99220334		

The F statistic above tests the null hypothesis that the mean of happiness is equal for all five levels of marital status. We can express this null hypothesis as shown below.

$$H_0 : \mu_1 = \mu_2 = \mu_3 = \mu_4 = \mu_5$$

The overall test of this null hypothesis is significant: $F(4, 1279) = 13.59$, $p < 0.001$. We can reject the null hypothesis that the average happiness is equal among the five marital status groups.

---

1. Some might be bothered by analyzing a Likert scale like **happy7** as though it were measured on an interval scale. For the sake of these examples, let's assume that **happy7** is measured on an interval scale.

5.2.1   Computing and graphing means

Let's use the `margins` command to compute the mean of happiness by marital status. This output shows, for example, that the mean of happiness for those who are married (group 1) is 5.71 with a 95% confidence interval (CI) of 5.63 to 5.79.

```
. margins marital, nopvalues
Adjusted predictions                            Number of obs   =      1,284
Expression   : Linear prediction, predict()

                      Delta-method
             Margin    Std. Err.      [95% Conf. Interval]

     marital
     married   5.714038   .0406773      5.634237      5.79384
     widowed   5.429907     .09446      5.244593      5.61522
    divorced   5.252475   .0687486      5.117603     5.387348
   separated   5.119048   .1507701      4.823264     5.414831
never married  5.365169   .0517863      5.263573     5.466764
```

The `marginsplot` command can be used to make a graph of the means and CIs reported by the `margins` command. This produces the graph shown in figure 5.1.

```
. marginsplot
  Variables that uniquely identify margins: marital
```

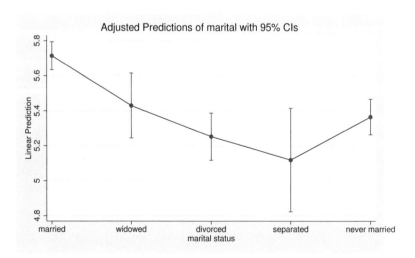

Figure 5.1: Mean happiness by marital status

5.2.2   Making contrasts among means

Let's probe this finding in more detail. Suppose that Professor Cheer predicted (before even seeing the data) that those who are married will be happier than each of the four other marital status groups. We can frame this as four separate null hypotheses, shown below.

$$H_0\#1\colon \mu_2 = \mu_1$$
$$H_0\#2\colon \mu_3 = \mu_1$$
$$H_0\#3\colon \mu_4 = \mu_1$$
$$H_0\#4\colon \mu_5 = \mu_1$$

The first null hypothesis states that the mean happiness is the same for group 2 and group 1 (widowed versus married). The second states that the mean happiness is the same for groups 3 and 1 (divorced versus married). The third states that the mean happiness is the same for groups 4 and 1 (separated versus married). Finally, the fourth states that the mean happiness is the same for groups 5 and 1 (never married versus married).

We can test each of these four null hypotheses using the contrast command shown below. (I will discuss the syntax of the contrast command shortly.)

```
. contrast r.marital
Contrasts of marginal linear predictions
Margins      : asbalanced
```

	df	F	P>F
marital			
(widowed vs married)	1	7.63	0.0058
(divorced vs married)	1	33.39	0.0000
(separated vs married)	1	14.52	0.0001
(never married vs married)	1	28.07	0.0000
Joint	4	13.59	0.0000
Denominator	1279		

	Contrast	Std. Err.	[95% Conf. Interval]	
marital				
(widowed vs married)	-.2841316	.1028462	-.4858973	-.0823659
(divorced vs married)	-.4615629	.0798813	-.6182756	-.3048502
(separated vs married)	-.5949905	.156161	-.9013504	-.2886306
(never married vs married)	-.3488696	.0658518	-.478059	-.2196801

This contrast command compares each marital status group to group 1. The first test compares those who are widowed with those who are married. The upper portion of the output shows the F statistic for the test of this hypothesis. This test is statistically significant: $F(1, 1279) = 7.63$, $p = 0.0058$. The lower portion of the output shows that the average difference in the happiness for those who are widowed versus married is -0.28, (95% CI $= [-0.49, -0.08]$). Those who are widowed are significantly less happy than those who are married. Put another way, those who are married are significantly happier than those who are widowed. We can reject $H_0\#1$ and say the results are consistent with our prediction that those who are married are significantly happier.

Let's now consider the output for the second, third, and fourth contrasts. These test the second, third, and fourth null hypotheses contrasting those who are divorced versus married, separated versus married, and never married versus married. The upper portion of the `contrast` output shows that each of these contrasts is statistically significant ($ps \leq 0.001$). Further, the differences in the means (as shown in the lower portion of the output) are always negative, indicating that those who are married are happier than the group they are compared with. Thus we can reject the second, third, and fourth null hypotheses.

To summarize, we can reject all four null hypotheses. Further, each difference was in the predicted direction. Those who are married are significantly happier than those who are widowed, significantly happier than those who are divorced, significantly happier than those who are separated, and significantly happier than those who have never been married.

5.2.3   Graphing contrasts

Let's create a graph that visually depicts these contrasts. We first use the `margins` command to replicate the results we found above via the `contrast` command. (This will allow us to then use the `marginsplot` command to graph the results from the `margins` command.)

```
. margins r.marital
Contrasts of adjusted predictions
Expression    : Linear prediction, predict()
```

	df	F	P>F
marital			
(widowed vs married)	1	7.63	0.0058
(divorced vs married)	1	33.39	0.0000
(separated vs married)	1	14.52	0.0001
(never married vs married)	1	28.07	0.0000
Joint	4	13.59	0.0000
Denominator	1279		

	Contrast	Delta-method Std. Err.	[95% Conf. Interval]	
marital				
(widowed vs married)	-.2841316	.1028462	-.4858973	-.0823659
(divorced vs married)	-.4615629	.0798813	-.6182756	-.3048502
(separated vs married)	-.5949905	.156161	-.9013504	-.2886306
(never married vs married)	-.3488696	.0658518	-.478059	-.2196801

Now we can use the `marginsplot` command (below) to create a graph illustrating the differences in the means (comparing each group with those who are unmarried) along with the 95% confidence interval for each difference. This graph (see figure 5.2) shows, for example, that the estimate of the difference in the means for widowed versus married is -0.28 and that the 95% confidence interval is -0.49 to -0.08. This confidence interval excludes 0, indicating that this difference is significant at the 5% level.

Note: Keeping up to date

Stata is frequently updated with minor additions and fixes that you can obtain, with respect to your current version, free of charge. Best of all, obtaining such updates is almost effortless. When you start Stata, it will ask itself if seven or more days have passed since it last checked for updates. If so, it will check for updates and offer to download and install them. That is all there is to it! You can customize this behavior if you like, such as by changing the update interval. Type `help update` for more information.

```
. marginsplot, xlabel(, labsize(small)) xscale(range(1.7 5.3))
  Variables that uniquely identify margins: marital
```

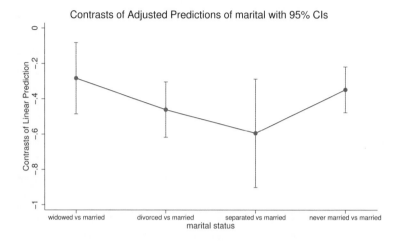

Figure 5.2: Reference group contrasts comparing each group with those who are married

---

Tip! Customizing graphs created by marginsplot

In the `marginsplot` command above, I added the `xlabel()` and `xscale()` options to make the labels on the x axis display more compactly. (Try running the `marginsplot` command without these options, and you will see how the labels for the x axis bleed off the edge of the graph.) Such options to customize the display of graphs using the `marginsplot` command are discussed in section 21.4.

5.2.4   Options with the margins and contrast commands

Let's briefly consider some options that we can use with the `margins` and `contrast` commands for the purpose of saving space on the printed page.

Below we use the `margins` command to compute means of the dependent variable (DV) by levels of the independent variable (IV). The `margins` command computes the mean of the DV (`happiness`) by marital status.

```
. margins marital
Adjusted predictions                          Number of obs    =       1,284
Expression    : Linear prediction, predict()
```

	Margin	Delta-method Std. Err.	t	P>|t|	[95% Conf. Interval]	
marital						
married	5.714038	.0406773	140.47	0.000	5.634237	5.79384
widowed	5.429907	.09446	57.48	0.000	5.244593	5.61522
divorced	5.252475	.0687486	76.40	0.000	5.117603	5.387348
separated	5.119048	.1507701	33.95	0.000	4.823264	5.414831
never mar..	5.365169	.0517863	103.60	0.000	5.263573	5.466764

The `margins` command shows the mean, standard error, *t*-value, *p*-value, and 95% confidence interval. For those who are married, the average is 5.71 and the standard error is 0.04. The 95% confidence interval is 5.63 to 5.79. The *t*-value and *p*-value test the null hypothesis that the mean equals zero, which is seldom an interesting hypothesis test.

Also note that the output shows the value labels for the levels of `marital`; however, in this example, the label for the last group, never married, was truncated. Using the `nopvalues` option, we can omit the *t*-value and *p*-value from the output (which are not interesting anyways) and leave more room for these labels.

```
. margins marital, nopvalues
Adjusted predictions                          Number of obs    =       1,284
Expression    : Linear prediction, predict()
```

	Margin	Delta-method Std. Err.	[95% Conf. Interval]	
marital				
married	5.714038	.0406773	5.634237	5.79384
widowed	5.429907	.09446	5.244593	5.61522
divorced	5.252475	.0687486	5.117603	5.387348
separated	5.119048	.1507701	4.823264	5.414831
never married	5.365169	.0517863	5.263573	5.466764

Now consider the `contrast` command below. Note how there are two tables in the output. The first table provides significance tests of each of the contrasts. The second table provides estimates of the contrasts, with a standard error and confidence interval.

```
. contrast r.marital
Contrasts of marginal linear predictions
Margins       : asbalanced
```

	df	F	P>F
marital			
(widowed vs married)	1	7.63	0.0058
(divorced vs married)	1	33.39	0.0000
(separated vs married)	1	14.52	0.0001
(never married vs married)	1	28.07	0.0000
Joint	4	13.59	0.0000
Denominator	1279		

	Contrast	Std. Err.	[95% Conf.	Interval]
marital				
(widowed vs married)	-.2841316	.1028462	-.4858973	-.0823659
(divorced vs married)	-.4615629	.0798813	-.6182756	-.3048502
(separated vs married)	-.5949905	.156161	-.9013504	-.2886306
(never married vs married)	-.3488696	.0658518	-.478059	-.2196801

We can specify the `nowald` option to omit the upper table and save space. While we do have a confidence interval for each contrast, we no longer have significance tests (that is, p-values) associated with each test.

```
. contrast r.marital, nowald
Contrasts of marginal linear predictions
Margins       : asbalanced
```

	Contrast	Std. Err.	[95% Conf.	Interval]
marital				
(widowed vs married)	-.2841316	.1028462	-.4858973	-.0823659
(divorced vs married)	-.4615629	.0798813	-.6182756	-.3048502
(separated vs married)	-.5949905	.156161	-.9013504	-.2886306
(never married vs married)	-.3488696	.0658518	-.478059	-.2196801

We can add the `effects` option, which displays both the t-value and the p-value associated with each contrast as well as the confidence interval for the contrast. This output is comprehensive, but it is so wide that the description of each contrast takes more than one printed line in this book. When you run this command on your computer, the output will likely display with one contrast per line, creating a very concise form of output.

```
. contrast r.marital, nowald effects
Contrasts of marginal linear predictions
Margins        : asbalanced
```

	Contrast	Std. Err.	t	P>|t|	[95% Conf. Interval]	
marital (widowed vs married)	-.2841316	.1028462	-2.76	0.006	-.4858973	-.0823659
(divorced vs married)	-.4615629	.0798813	-5.78	0.000	-.6182756	-.3048502
(separated vs married)	-.5949905	.156161	-3.81	0.000	-.9013504	-.2886306
(never ma.. vs married)	-.3488696	.0658518	-5.30	0.000	-.478059	-.2196801

Instead, we can use the **pveffects** option, which displays a *t*-value and *p*-value associated with each contrast but omits the display of the confidence interval for the contrast, as illustrated below. I will sometimes use these options to show a compact form of output while also including a test statistic and significance value for each contrast.

```
. contrast r.marital, nowald pveffects
Contrasts of marginal linear predictions
Margins        : asbalanced
```

	Contrast	Std. Err.	t	P>|t|
marital				
(widowed vs married)	-.2841316	.1028462	-2.76	0.006
(divorced vs married)	-.4615629	.0798813	-5.78	0.000
(separated vs married)	-.5949905	.156161	-3.81	0.000
(never married vs married)	-.3488696	.0658518	-5.30	0.000

You can exert the exact same control over the output from the **margins** command by including these options as suboptions within the **contrast()** option. For example, the **nowald** and **pveffects** suboptions are specified within the **contrast** option below.

```
. margins r.marital, contrast(nowald pveffects)

Contrasts of adjusted predictions

Expression    : Linear prediction, predict()
```

	Contrast	Delta-method Std. Err.	t	P>|t|
marital				
(widowed vs married)	-.2841316	.1028462	-2.76	0.006
(divorced vs married)	-.4615629	.0798813	-5.78	0.000
(separated vs married)	-.5949905	.156161	-3.81	0.000
(never married vs married)	-.3488696	.0658518	-5.30	0.000

5.2.5 Computing effect sizes for contrasts

When you report significant findings, it is a good practice to include a measure of effect size. A common effect-size measure to report for ANOVA models is omega-squared, which is analogous to adjusted R-squared in a regression model. Let's illustrate how we can compute omega-squared for the example looking at happiness and marital status. First, let's repeat the ANOVA command.

```
. anova happy7 marital
```

		Number of obs =	1,284	R-squared	=	0.0408
		Root MSE =	.977102	Adj R-squared =		0.0378

Source	Partial SS	df	MS	F	Prob>F
Model	51.89968	4	12.97492	13.59	0.0000
marital	51.89968	4	12.97492	13.59	0.0000
Residual	1221.0972	1,279	.95472807		
Total	1272.9969	1,283	.99220334		

And let's repeat the **contrast** command in which we compare each level of marital status with those who are married.

```
. contrast r.marital
Contrasts of marginal linear predictions
Margins      : asbalanced
```

	df	F	P>F
marital			
(widowed vs married)	1	7.63	0.0058
(divorced vs married)	1	33.39	0.0000
(separated vs married)	1	14.52	0.0001
(never married vs married)	1	28.07	0.0000
Joint	4	13.59	0.0000
Denominator	1279		

	Contrast	Std. Err.	[95% Conf. Interval]	
marital				
(widowed vs married)	-.2841316	.1028462	-.4858973	-.0823659
(divorced vs married)	-.4615629	.0798813	-.6182756	-.3048502
(separated vs married)	-.5949905	.156161	-.9013504	-.2886306
(never married vs married)	-.3488696	.0658518	-.478059	-.2196801

Say that we wanted to report the contrast of those who are married versus divorced. The average difference in happiness for divorced versus married people is -0.46 (95% CI $= [-0.62, -0.30]$) and is statistically significant, $F(1, 1279) = 33.39$, $p < 0.001$. It is recommend that you supplement the reporting of this finding with an estimate of the effect size. To that end, let's use `esizei`, which I introduced in section 4.4.2. This command uses numerator degrees of freedom, denominator degrees of freedom, and the F-value to compute eta-squared and omega-squared (with 95% confidence intervals). As shown below, the `esizei` command is issued by supplying the numerator degrees of freedom, the denominator degrees of freedom, and the F-value. The `esizei` command then computes eta-squared and omega-squared for this contrast and confidence intervals for both effect-size estimates. Using omega-squared, we can report that the contrast of those who are married versus widowed explains 2.47% of the variance in happiness (95% CI $= [1.04\%, 4.40\%]$).

```
. esizei 1 1279 33.39
Effect sizes for linear models
```

Effect Size	Estimate	[95% Conf. Interval]	
Eta-Squared	.0254421	.0111734	.0447873
Omega-Squared	.0246802	.0104003	.0440405

5.2.6   Summary

This section has introduced the concept of contrasts for making comparisons among groups in the context of a one-way ANOVA. In all of these examples, we specified `r.marital` after the `contrast` and `margins` commands. You might be rightly asking, What is this `r.` prefix and what does it mean? Stata calls this a "contrast operator", and this is just one of many contrast operators that you can choose from. The `r.` contrast operator compares each group with a reference group (which, by default, is group 1).

The `r.` contrast operator tested contrasts that corresponded to the null hypotheses based on the research questions for this study. It compared the happiness of each group with the reference group (married). There are many other kinds of questions that we could have asked, questions implying different kinds of contrasts among the groups. Fortunately, Stata offers a wide variety of contrast operators that answers many interesting research questions. These are described in the following section.

5.3   Overview of contrast operators

The examples I have shown so far have illustrated one of the contrast operators you can use with the `contrast` (or `margins`) command, the `r.` contrast operator. Table 5.1 provides an overview of the contrast operators that can be used with the `contrast` and `margins` commands. This table provides a brief description of the contrast operator and shows the section of this chapter in which the contrast operator is discussed.[2]

---

2. The custom contrast operator is left empty. It takes the form {*varname numlist*}, described in more detail in section 5.9.

Table 5.1: Summary of contrast operators

Operator	Section	Description
r.	5.4	differences from the reference (base) level; the default
g.	5.5	differences from the balanced grand mean
a.	5.6	differences from the next level (adjacent contrasts)
ar.	5.6.1	differences from the previous level (reverse adjacent contrasts)
h.	5.7	differences from the balanced mean of subsequent levels (Helmert contrasts)
j.	5.7.1	differences from the balanced mean of previous levels (reverse Helmert contrasts)
p.	5.8	orthogonal polynomial in the level values
q.	5.8	orthogonal polynomial in the level sequence
	5.9	custom contrasts
gw.	5.10	differences from the observation-weighted grand mean
hw.	5.10	differences from the observation-weighted mean of subsequent levels
jw.	5.10	differences from the observation-weighted mean of previous levels
pw.	5.10	observation-weighted orthogonal polynomial in the level values
qw.	5.10	observation-weighted orthogonal polynomial in the level sequence

The following sections discuss each of these contrast operators in turn. Examples are provided illustrating their use and how to interpret the results. This discussion begins with a more detailed discussion of the r. contrast operator.

5.4 Compare each group against a reference group

This section provides further examples illustrating the r. contrast operator for making reference group contrasts. Let's continue with the example from section 5.2 predicting happiness from marital status. The anova command for that analysis is repeated below, but the output is omitted.

```
. use gss2012_sbs
. anova happy7 marital
  (output omitted)
```

5.4.1   Selecting a specific contrast

Suppose we wanted to focus just on the contrast of those who have never been married with those who are married (group 5 versus 1). We can specify the r5. contrast operator, and this yields a contrast of group 5 (the group we specified) compared with the reference group (group 1).

```
. contrast r5.marital, nowald pveffects
Contrasts of marginal linear predictions
Margins       : asbalanced
```

	Contrast	Std. Err.	t	P>|t|
marital				
(never married vs married)	-.3488696	.0658518	-5.30	0.000

If we wanted to compare only those who are divorced with those who are married, we could have specified r3.marital. This would have shown just the contrast of group 3 with group 1.

Suppose you wanted to focus on the contrast of group 3 with group 1 (divorced versus married) and group 5 with group 1 (never married versus married). You can perform those two contrasts by specifying the r(3 5). contrast operator. This compares each of the groups within the parentheses with the reference group (group 1).

```
. contrast r(3 5).marital, pveffects
Contrasts of marginal linear predictions
Margins       : asbalanced
```

	df	F	P>F
marital			
(divorced vs married)	1	33.39	0.0000
(never married vs married)	1	28.07	0.0000
Joint	2	23.41	0.0000
Denominator	1279		

	Contrast	Std. Err.	t	P>|t|
marital				
(divorced vs married)	-.4615629	.0798813	-5.78	0.000
(never married vs married)	-.3488696	.0658518	-5.30	0.000

The output shows the test of the contrast of those who are divorced versus married (group 3 versus 1) and the contrast of those who have never been married versus those who are married (group 5 versus 1). The upper portion of the output includes the joint test of these two contrasts, $F(2, 1279) = 23.41$, $p = 0.0000$. This jointly tests the contrast of groups 3 versus 1 and 5 versus 1 in a single test. In other words, this tests the equality of the means for groups 1, 3, and 5. This is a useful way to test the overall equality of the means for a subset of groups.

5.4.2 Selecting a different reference group

Suppose that instead we wanted to compare each group with a different reference group. For example, let's compare each group with those who have never been married, group 5. We can specify the `rb5.` contrast operator, which requests reference group contrasts using group 5 as the baseline (reference) group.

```
. contrast rb5.marital, nowald pveffects
Contrasts of marginal linear predictions
Margins      : asbalanced
```

	Contrast	Std. Err.	t	P>|t|
marital				
(married vs never married)	.3488696	.0658518	5.30	0.000
(widowed vs never married)	.064738	.1077242	0.60	0.548
(divorced vs never married)	-.1126933	.0860709	-1.31	0.191
(separated vs never married)	-.2461209	.1594159	-1.54	0.123

Only the first of these contrasts is statistically significant (comparing those who are married versus never married), $p = 0.000$. Each of the other comparisons is not significant. For example, the contrast of those who are separated versus those who have never been married (group 4 versus 5) is not significant ($p = 0.123$).

5.4.3 Selecting a contrast and reference group

You can specify both the reference group and the contrasts to be made at one time. For example, let's compare the happiness of those who are divorced (group 3) with that of those who are separated (group 4). This shows no significant difference in happiness between those who are divorced versus separated ($p = 0.421$).

```
. contrast r3b4.marital, nowald pveffects
Contrasts of marginal linear predictions
Margins      : asbalanced
```

	Contrast	Std. Err.	t	P>|t|
marital				
(divorced vs separated)	.1334276	.1657045	0.81	0.421

More than one specific contrast can be specified at once by using parentheses. The `contrast` command below compares group 1 with group 5 and group 3 with group 5. People who are married are significantly happier than those who have never been married ($p = 0.000$), but those who are divorced are not significantly different from those who have never been married ($p = 0.191$).

```
. contrast r(1 3)b5.marital, nowald pveffects
Contrasts of marginal linear predictions
Margins        : asbalanced
```

	Contrast	Std. Err.	t	P>|t|
marital				
(married vs never married)	.3488696	.0658518	5.30	0.000
(divorced vs never married)	-.1126933	.0860709	-1.31	0.191

5.5   Compare each group against the grand mean

This section illustrates the **g.** contrast operator, which compares each group with the grand mean of all groups. Continuing the example from the previous section regarding marital status and happiness, Professor Cheer is interested in comparing the mean happiness of each marital status group with the grand mean of happiness.

Let's begin this analysis by using the GSS dataset and running the **anova** command below. The output of the **anova** command is the same as that from the previous section, so it is omitted.

```
. use gss2012_sbs
. anova happy7 marital
  (output omitted)
```

The following **margins** command uses the **g.** contrast operator to compare the mean happiness of each marital status group with the grand mean.[3]

---

3. I chose the **margins** command (instead of the **contrast** command) so that we can graph the results using the **marginsplot** command.

```
. margins g.marital, contrast(nowald pveffects)

Contrasts of adjusted predictions

Expression   : Linear prediction, predict()
```

	Contrast	Delta-method Std. Err.	t	P>|t|
marital				
(married vs mean)	.3379109	.0512003	6.60	0.000
(widowed vs mean)	.0537793	.0835602	0.64	0.520
(divorced vs mean)	-.123652	.066817	-1.85	0.064
(separated vs mean)	-.2570796	.1235624	-2.08	0.038
(never married vs mean)	-.0109587	.0569015	-0.19	0.847

The mean of those who are married (group 1) is 0.34 units greater than the grand mean, and this difference is significant. The contrast comparing those who are separated with the grand mean is also significant ($p = 0.038$). Those who are separated are significantly less happy than the grand mean.

Let's now graph these differences, along with the confidence intervals, using the **marginsplot** command shown below.[4] This produces the graph shown in figure 5.3. When the confidence interval for a contrast excludes 0, the difference is significant at the 5% level.

---

4. I added the `xdim(, nolabels)` option to display the group number on the x axis. Without the option, the labels are very long and overlap. For example, the label for the first group would be married versus mean. You can see more options that you can use with the **marginsplot** command in section 21.4.

```
. marginsplot, yline(0) xdim(, nolabels)
  Variables that uniquely identify margins: marital
```

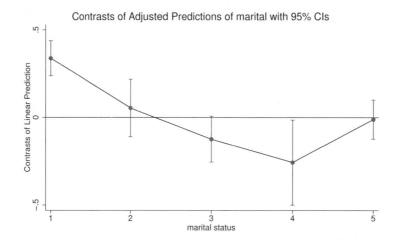

Figure 5.3: Contrasts comparing each group with the grand mean

5.5.1   Selecting a specific contrast

Suppose we wanted to focus only on the contrast of those who are married (group 1) with the grand mean. Using the **g1.** contrast operator, we perform just the contrast of group 1 with the grand mean.

```
. contrast g1.marital, nowald pveffects
Contrasts of marginal linear predictions
Margins        : asbalanced
```

	Contrast	Std. Err.	t	P>|t|
marital				
(married vs mean)	.3379109	.0512003	6.60	0.000

As we have seen previously, we can use parentheses to specify more than one contrast. In the example below, we compare the first group with the grand mean, and we compare the third group with the grand mean.

```
. contrast g(1 3).marital, nowald pveffects
Contrasts of marginal linear predictions
Margins       : asbalanced
```

	Contrast	Std. Err.	t	P>|t|
marital				
(married vs mean)	.3379109	.0512003	6.60	0.000
(divorced vs mean)	-.123652	.066817	-1.85	0.064

5.6   Compare adjacent means

This section illustrates contrasts that compare the means of adjacent groups, for example, group 1 versus 2, group 2 versus 3, group 3 versus 4, and so forth. These kinds of contrasts are especially useful for studies where you expect a nonlinear relationship between an ordinal or interval IV and DV. For example, consider a hypothetical study about the dosage of a new pain medication where the researchers expect that at a certain dosage level, the effects of the medication will kick-in and lead to a statistically significant reduction in pain. To study this, researchers give people in pain different dosages of the medication, and their pain level is measured on a 100-point scale, where 100 is the worst pain and 0 is no pain. The medication dosages range from 0mg to 250mg, incrementing by 50mg, and yield 6 dosage groups: 0 mg, 50 mg, 100 mg, 150 mg, 200 mg, and 250 mg. The most general null hypothesis that could be tested is that the average pain is equal across all six dosage groups:

$$H_0: \mu_1 = \mu_2 = \mu_3 = \mu_4 = \mu_5 = \mu_6$$

Let's start by testing this hypothesis. The dataset for this example, `pain.dta`, is used below. The DV is stored in the variable `pain`, and the dosage is stored in the variable `dosegrp`, which is coded as follows: 1 = 0 mg, 2 = 50 mg, 3 = 100 mg, 4 = 150 mg, and 6 = 250 mg. The `anova` command is used to predict `pain` from the `dosegrp` categories.

```
. use pain
. anova pain i.dosegrp
                    Number of obs =      180    R-squared     =  0.4602
                    Root MSE      =  10.4724    Adj R-squared =  0.4447

        Source |  Partial SS      df       MS         F     Prob>F

         Model |  16271.694        5   3254.3389    29.67  0.0000

       dosegrp |  16271.694        5   3254.3389    29.67  0.0000

      Residual |  19082.633      174   109.67031

         Total |  35354.328      179   197.51021
```

The test associated with `dosegrp` tests the null hypothesis above and is significant $[F(5, 174) = 29.67, p < 0.001]$. We can reject the overall null hypothesis. Let's use the `margins` command and the `marginsplot` command to display and graph the predicted mean of `pain` by `dosegrp`. The graph of the means is shown in figure 5.4.

```
. margins dosegrp, nopvalues
Adjusted predictions                          Number of obs     =        180
Expression    : Linear prediction, predict()
```

	Margin	Delta-method Std. Err.	[95% Conf. Interval]	
dosegrp				
Zero (control)	71.83333	1.911982	68.05967	75.607
50 mg	70.6	1.911982	66.82634	74.37366
100 mg	72.13333	1.911982	68.35967	75.907
150 mg	70.4	1.911982	66.62634	74.17366
200 mg	54.7	1.911982	50.92634	58.47366
250 mg	48.3	1.911982	44.52634	52.07366

```
. marginsplot
    Variables that uniquely identify margins: dosegrp
```

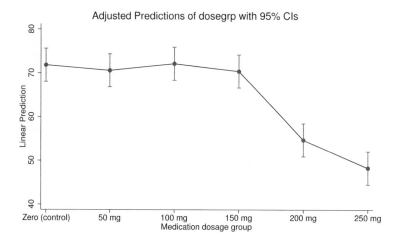

Figure 5.4: Mean pain rating by dosage group

For this study, the research question of interest focuses on the test of each dosage against the previous dosage to determine the dosage that leads to a statistically significant decrease in pain. This leads us to five specific null hypotheses.

$$H_0\#1\colon \mu_1 = \mu_2$$
$$H_0\#2\colon \mu_2 = \mu_3$$
$$H_0\#3\colon \mu_3 = \mu_4$$
$$H_0\#4\colon \mu_4 = \mu_5$$
$$H_0\#5\colon \mu_5 = \mu_6$$

Let's now test each of these hypotheses using the `contrast` command with the `a.` contrast operator to compare each dosage with the adjacent (subsequent) dosage.

```
. margins a.dosegrp, contrast(nowald pveffects)
Contrasts of adjusted predictions
Expression   : Linear prediction, predict()
```

	Contrast	Delta-method Std. Err.	t	P>|t|
dosegrp				
(Zero (control) vs 50 mg)	1.233333	2.703952	0.46	0.649
(50 mg vs 100 mg)	-1.533333	2.703952	-0.57	0.571
(100 mg vs 150 mg)	1.733333	2.703952	0.64	0.522
(150 mg vs 200 mg)	15.7	2.703952	5.81	0.000
(200 mg vs 250 mg)	6.4	2.703952	2.37	0.019

The contrasts of groups 1 versus 2, 2 versus 3, and 3 versus 4 are not significant. The contrast of group 4 versus 5 (150 mg versus 200 mg) is significant ($p = 0.0000$). Likewise, the contrast of group 6 versus 5 (250 mg versus 200 mg) is significant ($p = 0.0190$).

Because we used the `margins` command to estimate these differences, we can graph the differences using the `marginsplot` command. The resulting graph is shown in figure 5.5.

```
. marginsplot, yline(0)
   Variables that uniquely identify margins: dosegrp
```

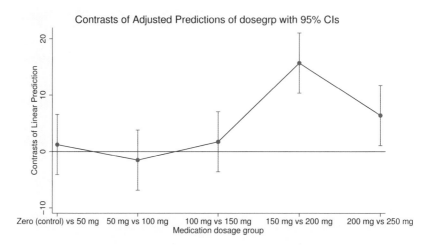

Figure 5.5: Contrasts of each dosage with the previous dosage

Figure 5.5 shows that the adjacent group contrasts for groups 1 versus 2, 2 versus 3, and 3 versus 4 are not significant. It also shows that the contrast of 4 versus 5 and 5 versus 6 is significant. This pain medication appears to begin its effectiveness with the dosage given to group 5, 200 mg.

5.6.1 Reverse adjacent contrasts

In looking at the estimated differences in the means from the previous section, I notice that the contrasts focus on the contrast of a lower dose to a higher dose. For example, the contrast of group 2 with group 1 compares 50 mg with 0 mg. The results might be clearer if we reverse the order of the contrasts. The `ar.` contrast operator, shown below, provides adjacent group contrasts in reverse order.

```
. contrast ar.dosegrp, nowald pveffects
Contrasts of marginal linear predictions
Margins        : asbalanced
```

	Contrast	Std. Err.	t	P>|t|
dosegrp				
(50 mg vs Zero (control))	-1.233333	2.703952	-0.46	0.649
(100 mg vs 50 mg)	1.533333	2.703952	0.57	0.571
(150 mg vs 100 mg)	-1.733333	2.703952	-0.64	0.522
(200 mg vs 150 mg)	-15.7	2.703952	-5.81	0.000
(250 mg vs 200 mg)	-6.4	2.703952	-2.37	0.019

By supplying the `ar.` contrast operator, we can form the contrasts in reverse order, comparing the higher group (dosage) with the adjacent lower group (dosage). By comparing each group with the previous group, we can interpret the difference in the means can be interpreted as the reduction in pain comparing the higher dosage with the lower dosage.

5.6.2   Selecting a specific contrast

When making adjacent group contrasts, you can select a specific contrast. For example, using the `a.` contrast operator, we can specify particular contrasts of interest. We can select the contrast of group 1 versus its adjacent category (that is, group 2) as shown below. Note that the `a1` operator contrasts the first group with its adjacent group. If we had specified `a3`, this would contrast group 3 with the subsequent category (group 4).

```
. contrast a1.dosegrp, nowald pveffects
Contrasts of marginal linear predictions
Margins        : asbalanced
```

	Contrast	Std. Err.	t	P>|t|
dosegrp				
(Zero (control) vs 50 mg)	1.233333	2.703952	0.46	0.649

We can also select contrasts using the `ar.` contrast operator. The `ar.` contrast operator forms contrasts with the previous group, so specifying `ar3.` (as shown below) contrasts group 3 with the previous group (group 2).

```
. contrast ar3.dosegrp, nowald pveffects
```
Contrasts of marginal linear predictions
Margins   : asbalanced

	Contrast	Std. Err.	t	P>|t|
dosegrp (100 mg vs 50 mg)	1.533333	2.703952	0.57	0.571

We can combine selected contrasts as well. Suppose we wanted to test the equality of the mean pain ratings for the first four groups. The individual contrasts of the adjacent means suggest that the overall test of the equality of these means would be nonsignificant, but we have not formally performed such a test. Let's perform this test using the contrast command, shown below.

```
. contrast a(1 2 3).dosegrp
```
Contrasts of marginal linear predictions
Margins   : asbalanced

	df	F	P>F
dosegrp			
(Zero (control) vs 50 mg)	1	0.21	0.6489
(50 mg vs 100 mg)	1	0.32	0.5714
(100 mg vs 150 mg)	1	0.41	0.5223
Joint	3	0.21	0.8918
Denominator	174		

	Contrast	Std. Err.	[95% Conf. Interval]	
dosegrp				
(Zero (control) vs 50 mg)	1.233333	2.703952	-4.103433	6.570099
(50 mg vs 100 mg)	-1.533333	2.703952	-6.870099	3.803433
(100 mg vs 150 mg)	1.733333	2.703952	-3.603433	7.070099

This contrast compares groups 1 versus 2, groups 2 versus 3, and groups 3 versus 4. Each of these contrasts is nonsignificant, but let's focus on the joint test. That test is also not significant ($p = 0.8918$). The joint test simultaneously tests all the specified contrasts, providing a test of the null hypothesis of the equality of the means for groups 1, 2, 3, and 4. We could cite this statistical test to indicate that the test of the equality of the pain ratings for the first four dosage groups was not significant.

5.7   Comparing with the mean of subsequent and previous levels

This section describes contrasts that compare each group mean with the mean of the subsequent groups (also known as Helmert contrasts). It also illustrates contrasts that compare each group mean with the mean of the previous groups (also known as reverse Helmert contrasts). For example, consider a follow-up to the pain study described in section 5.6. The previous study focused on the lowest dosages needed to achieve significant pain reduction. This hypothetical study focuses on determining the dosage at which no significant pain reductions are achieved. The participants in this study all suffer from pain and are given 1 of 6 different medication dosages: 300 mg, 400 mg, 500 mg, 600 mg, 800 mg, or 1,000 mg. In this example, the variable `dosage` contains the actual size of the dosage (in mg). That is, the variable `dosage` contains the values 300, 400, 500, 600, 800, or 1,000. The dataset for this example, `pain2.dta`, is used below, and the `tabulate` command shows the tabulation of the variable `dosage`.

```
. use pain2
. tabulate dosage
```

Medication dosage in mg	Freq.	Percent	Cum.
300	30	16.67	16.67
400	30	16.67	33.33
500	30	16.67	50.00
600	30	16.67	66.67
800	30	16.67	83.33
1000	30	16.67	100.00
Total	180	100.00	

The overall null hypothesis regarding dosage is that the average pain is equal across all six dosage groups:

$$H_0 : \mu_{300} = \mu_{400} = \mu_{500} = \mu_{600} = \mu_{800} = \mu_{1000}$$

Let's start by testing this overall null hypothesis using the `anova` command below.

```
. use pain2
. anova pain i.dosage

                        Number of obs =      180    R-squared     =  0.2052
                        Root MSE      =  10.5056    Adj R-squared =  0.1824

         Source |  Partial SS        df        MS          F     Prob>F

          Model |  4958.8667          5   991.77333       8.99   0.0000

         dosage |  4958.8667          5   991.77333       8.99   0.0000

       Residual |  19204.133        174   110.36858

          Total |     24163        179   134.98883
```

We can reject the overall null hypothesis of the equality of the six means, $F(5, 174) = 8.99$, $p < 0.001$.

Let's use the `margins` and `marginsplot` commands to show and graph the means by the six levels of `dosage`. The graph of the means is shown in figure 5.6. Note how the spacing of the dosages on the x axis reflects the actual dosage.

```
. margins dosage, nopvalues
Adjusted predictions                         Number of obs     =        180
Expression    : Linear prediction, predict()

                           Delta-method
                 Margin    Std. Err.     [95% Conf. Interval]

       dosage
          300   43.83333    1.91806     40.04768    47.61899
          400       37.6    1.91806     33.81434    41.38566
          500   31.86667    1.91806     28.08101    35.65232
          600   29.63333    1.91806     25.84768    33.41899
          800   30.63333    1.91806     26.84768    34.41899
         1000   29.43333    1.91806     25.64768    33.21899
```

```
. marginsplot
  Variables that uniquely identify margins: dosage
```

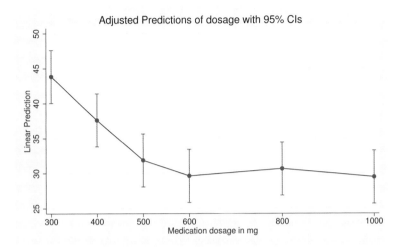

Figure 5.6: Mean pain rating by dosage

For this study, we want to establish the dosage threshold where no statistically significant pain reduction is achieved when compared with higher dosages. Thus the mean pain rating at each dosage is compared with the mean pain rating for all the higher dosages. At some dosage, the pain rating will not be significantly different from those receiving higher dosages. We can express this using the following null hypotheses.

$$H_0 \#1 \colon \mu_{300} = \mu_{>300}$$
$$H_0 \#2 \colon \mu_{400} = \mu_{>400}$$
$$H_0 \#3 \colon \mu_{500} = \mu_{>500}$$
$$H_0 \#4 \colon \mu_{600} = \mu_{>600}$$
$$H_0 \#5 \colon \mu_{800} = \mu_{1000}$$

Let's now test each of these null hypotheses below using the `margins` command combined with the `h.` contrast operator.

```
. margins h.dosage, contrast(nowald pveffects)
Contrasts of adjusted predictions
Expression    : Linear prediction, predict()
```

	Contrast	Delta-method Std. Err.	t	P>|t|
dosage				
(300 vs > 300)	12	2.101129	5.71	0.000
(400 vs > 400)	7.208333	2.144456	3.36	0.001
(500 vs > 500)	1.966667	2.214784	0.89	0.376
(600 vs > 600)	-.4	2.349134	-0.17	0.865
(800 vs 1000)	1.2	2.712546	0.44	0.659

The **h.** contrast operator compares the mean of each group with the mean of the subsequent groups. The comparisons are specified in terms of the actual dosage, for example, (300 **versus** > 300). The contrast of 300 mg versus more than 300 mg is significant ($p = 0.000$), as is the contrast of 400 mg versus more than 400 mg ($p = 0.001$). The contrast of those receiving 500 mg with those receiving more than 500 mg is not significant ($p = 0.376$). The subsequent contrasts are also all not significant. Referring to the null hypotheses, we would reject the first and second null hypotheses, and we would not reject the third through fifth null hypotheses. In other words, those receiving 400 mg experience significantly more pain than those receiving 500 mg or more. Comparing those receiving 500 mg with those receiving 600 mg or more, we find no statistically significant difference in pain.

We can visualize these contrasts using the **marginsplot** command shown below. This creates the graph in figure 5.7, which shows each of the contrasts with a confidence interval for the contrast. When the confidence interval for the contrast excludes 0, the difference is significant at the 5% level. (The **xlabel()** option is used to make the labels of the x axis more readable.)

```
. marginsplot, yline(0) xlabel(, angle(45))
  Variables that uniquely identify margins: dosage
```

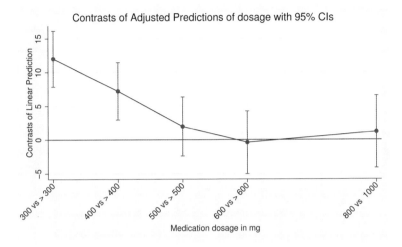

Figure 5.7: Contrasts of the mean pain of each group with subsequent groups

5.7.1 Comparing with the mean of previous levels

We have seen how the h. operator compares the mean of each group with the mean of the subsequent groups. You might be interested in making the same kind of contrast but in the opposite direction. That is, you might be interested in comparing the mean of each group with the mean of the previous groups. The j. contrast operator is used for such contrasts.

Although it does not make much sense in the context of this example, the contrast command is shown below to illustrate the use of the j. contrast operator.

```
. contrast j.dosage, nowald pveffects
Contrasts of marginal linear predictions
Margins        : asbalanced
```

	Contrast	Std. Err.	t	P>|t|
dosage				
(400 vs 300)	-6.233333	2.712546	-2.30	0.023
(500 vs < 500)	-8.85	2.349134	-3.77	0.000
(600 vs < 600)	-8.133333	2.214784	-3.67	0.000
(800 vs < 800)	-5.1	2.144456	-2.38	0.018
(1000 vs <1000)	-5.28	2.101129	-2.51	0.013

When you use the j. contrast operator, the mean of each group is compared with the mean of the previous groups. In this case, all the contrasts are significant.

5.7.2   Selecting a specific contrast

Returning to the h. contrast operator, I will show how you can select specific contrasts. Say we wanted to focus just on the contrast of those whose value of dosage was 400 with those who have higher values of dosage. We can specify h400.dosage as shown in the contrast command below.[5]

```
. contrast h400.dosage, nowald pveffects
Contrasts of marginal linear predictions
Margins       : asbalanced
```

	Contrast	Std. Err.	t	P>|t|
dosage				
(400 vs >400)	7.208333	2.144456	3.36	0.001

Alternatively, we might want to focus just on the contrasts of 400 mg versus the subsequent groups and 500 mg versus the subsequent groups. We can perform these contrasts as shown below.

```
. contrast h(400 500).dosage, nowald pveffects
Contrasts of marginal linear predictions
Margins       : asbalanced
```

	Contrast	Std. Err.	t	P>|t|
dosage				
(400 vs >400)	7.208333	2.144456	3.36	0.001
(500 vs >500)	1.966667	2.214784	0.89	0.376

This selection process can be used with the j. operator as well. If we wanted to select the contrast of 500 mg versus the mean of the previous groups, we could specify j500.dosage as shown below.

---

5. We specify h400. because that is the actual value stored in the variable dosage. Specifying the contrast in this way is related to the coding of the variable dosage and not to the fact that we are using the h. contrast operator.

```
. contrast j500.dosage, nowald pveffects
Contrasts of marginal linear predictions
Margins        : asbalanced
```

	Contrast	Std. Err.	t	P>|t|
dosage				
(500 vs <500)	-8.85	2.349134	-3.77	0.000

5.8   Polynomial contrasts

Let's consider the use of polynomial contrasts for assessing nonlinear trends (for example, quadratic, cubic, quartic, etc.). Let's refer back to the example from section 5.6 that looked at the relationship between the dosage of pain medication and pain ratings using the pain.dta dataset. In that dataset, the medication dosages ranged from 0 mg to 250 mg, and were recorded in the variable dosegrp. This variable was coded $1 = 0$ mg, $2 = 50$ mg, $3 = 100$ mg, $4 = 150$ mg, $5 = 200$ mg, and $6 = 250$ mg. Let's use the pain.dta dataset followed by the anova command to predict pain from i.dosage. The anova output is omitted to save space.

```
. use pain
. anova pain i.dosegrp
  (output omitted)
```

We can specify q.dosegrp with the contrast command to compute tests of polynomial trend with respect to the levels of dosegrp. (We use the noeffects option to save space and focus on the results of the Wald tests.)

```
. contrast q.dosegrp, noeffects
Contrasts of marginal linear predictions
Margins        : asbalanced
```

	df	F	P>F
dosegrp			
(linear)	1	109.12	0.0000
(quadratic)	1	29.25	0.0000
(cubic)	1	0.00	0.9824
(quartic)	1	8.39	0.0043
(quintic)	1	1.62	0.2048
Joint	5	29.67	0.0000
Denominator	174		

The tests of linear trend are significant, $F(1, 174) = 109.12$, $p < 0.001$; the test of quadratic trend is also significant, $F(1, 174) = 29.25$, $p <= 0.001$. The test of cubic trend is not significant ($p = 0.982$), but the test of quartic trend is significant, $F(1, 174) = 8.39$, $p = 0.0043$.

Suppose you wanted to fit the relationship between the IV and DV using a linear term and wanted to assess whether there are significant nonlinear trends in the relationship between the IV and DV. You can use the `contrast` command shown below to test just the nonlinear terms.

```
. contrast q(2/6).dosegrp, noeffects
Contrasts of marginal linear predictions
Margins        : asbalanced
```

	df	F	P>F
dosegrp			
(quadratic)	1	29.25	0.0000
(cubic)	1	0.00	0.9824
(quartic)	1	8.39	0.0043
(quintic)	1	1.62	0.2048
Joint	4	9.81	0.0000
Denominator	174		

The output above shows that the joint test of all the nonlinear terms (powers 2 through 6) is significant, $F(4, 174) = 9.81$, $p < 0.0001$. In such a case, it would be inadvisable to fit the relationship between the IV and DV using only a linear fit.

The `q.` contrast operator assumes that the levels of `dosegrp` are equidistant from each other. Indeed, the levels of `dosegrp` range from 0 mg to 250 mg in 50 mg increments. Let's consider an example using the `pain2.dta` dataset (from section 5.7), where the levels of `dosage` were not equidistant. In the `pain2.dta` dataset, the dosages were recorded as the actual dosage in mg, that is, 300, 400, 500, 600, 800, or 1,000. The dataset `pain2.dta` is used below, and the `anova` command is used to predict `pain` from the `dosage` groups. The output of the `anova` command is omitted to save space.

```
. use pain2
. anova pain i.dosage
  (output omitted )
```

We use the `contrast` command and apply the `p.` contrast operator to `dosage`. This tests the polynomial trends based on the actual dosage, accounting for the differing gaps among the levels of `dosage`.

```
. contrast p.dosage, noeffects
Contrasts of marginal linear predictions
Margins        : asbalanced
```

	df	F	P>F
dosage			
(linear)	1	28.37	0.0000
(quadratic)	1	13.44	0.0003
(cubic)	1	2.60	0.1084
(quartic)	1	0.47	0.4956
(quintic)	1	0.05	0.8202
Joint	5	8.99	0.0000
Denominator	174		

The tests of linear and quadratic trends are significant, while the cubic, quartic, and quintic trends are not significant. Let's form a test just of the cubic, quartic, and quintic trends using the `contrast` command below.

```
. contrast p(3/6).dosage, noeffects
Contrasts of marginal linear predictions
Margins        : asbalanced
```

	df	F	P>F
dosage			
(cubic)	1	2.60	0.1084
(quartic)	1	0.47	0.4956
(quintic)	1	0.05	0.8202
Joint	3	1.04	0.3759
Denominator	174		

The joint test of the cubic, quartic, and quintic trends is not significant, $F(3, 174) = 1.04$, $p = 0.3759$. These tests show that there are significant linear and quadratic trends in the relationship between `pain` and `dosage`. In addition, the joint test of the cubic, quartic, and quintic trends is not significant. Taken together, these tests suggest that we could be justified in modeling `pain` as a linear and quadratic function of `dosage`.

Tip! Minding your Ps and Qs

In this example, the levels of dosage are not equidistant, and we would have obtained different results by specifying `q.dosage` compared with specifying `p.dosage`. In the previous example, where the levels of `dosegrp` were equidistant, specifying `p.dosegrp` yields the same results as `q.dosegrp`.

5.9   Custom contrasts

In the previous sections (namely, sections 5.4 to 5.8), we have seen a number of contrast operators that can be used to form contrasts among the levels of an IV. Specifically, we have seen the `r.`, `a.`, `ar.`, `g.`, `h.`, `j.`, `p.`, and `q.` contrast operators. These contrast operators automate the process of forming contrasts for most of the kinds of contrasts you might make. For those times when you want to make another kind of contrast, you can specify a custom contrast. Let's illustrate the use of custom contrasts using the analysis we saw in section 5.2. Using the GSS dataset, let's predict happiness from marital status. The output from the `anova` command is the same as section 5.2, so it is omitted to save space.

```
. use gss2012_sbs, clear
. anova happy7 marital
  (output omitted)
```

Let's start by illustrating how to perform custom contrasts using simple examples that compare one group with another group. For the first example, let's compare the mean of group 1 (married) with group 5 (not married). The custom contrast is enclosed within curly braces by specifying the variable name followed by the contrast coefficients. The contrast coefficients map to the levels (groups) of the variable. In this example, the contrast coefficient of 1 is applied to group 1, and -1 is applied to group 5. (A contrast coefficient of 0 is applied to groups 2, 3, and 4.) The result is a contrast of group 1 minus group 5. The `contrast` command computes difference in the means for group 1 versus group 5 as 0.35, and that difference is significant ($t = 5.30$, $p = 0.000$).

```
. contrast {marital 1 0 0 0 -1}, nowald pveffects
Contrasts of marginal linear predictions
Margins        : asbalanced
```

	Contrast	Std. Err.	t	P>|t|
marital				
(1)	.3488696	.0658518	5.30	0.000

Let's switch the above contrast. Let's compare group 5 (not married) with group 1 (married), as shown below. Note how the results are exactly the same as the previous results, except that the sign changes from positive to negative.

```
. contrast {marital -1 0 0 0 1}, nowald pveffects
Contrasts of marginal linear predictions
Margins      : asbalanced
```

	Contrast	Std. Err.	t	P>|t|
marital				
(1)	-.3488696	.0658518	-5.30	0.000

Now, let's compare those who are divorced (group 4) with those who are separated (group 3). We perform this contrast by using a contrast coefficient of 1 for group 4 and -1 for group 3. The difference in the means for group 4 versus group 3 is -0.13; this test is not significant ($p = 0.421$).

```
. contrast {marital 0 0 -1 1 0}, nowald pveffects
Contrasts of marginal linear predictions
Margins      : asbalanced
```

	Contrast	Std. Err.	t	P>|t|
marital				
(1)	-.1334276	.1657045	-0.81	0.421

All the above contrasts could have been performed using the r. contrast operator but were useful for getting us familiar with how to specify custom contrasts. Let's now consider a contrast that cannot be performed with one of the standard contrast operators. Say that we want to compare those who are married (group 1) with the average of those who are separated and divorced (groups 3 and 4). We can form that contrast as shown below. The contrast is statistically significant. The mean for group 1 is significantly different from the average of the means for groups 3 and 4. Put another way, the mean for group 1 minus the average of the means of groups 3 and 4 equals 0.53, and that contrast is significantly different from 0 ($t = 5.72$, $p = 0.000$).

```
. contrast {marital 1 0 -.5 -.5 0}, nowald pveffects
Contrasts of marginal linear predictions
Margins      : asbalanced
```

	Contrast	Std. Err.	t	P>|t|
marital				
(1)	.5282767	.0922992	5.72	0.000

Suppose we want to compare those who are married (group 1) with the average of those who are widowed, divorced, or separated (groups 2, 3, and 4). We can perform that contrast as shown below. The mean of those who are married is 0.45 units higher than the average of the means for those are widowed, divorced, or separated. This contrast is significant ($t = 5.92$, $p = 0.000$).

```
. contrast {marital 1 -.3333333 -.3333333 -.3333333 0}, nowald pveffects
Contrasts of marginal linear predictions
Margins          : asbalanced
```

	Contrast	Std. Err.	t	P>|t|
marital				
(1)	.4468955	.0754781	5.92	0.000

Note! Contrasts must sum to zero

The contrast coefficients that we specify in a custom contrast must sum to zero. In the previous example, the contrast coefficients for groups 2, 3, and 4 are expressed as -.3333333, using 7 digits after the decimal point. Although the sum of the coefficients for that custom contrast is not exactly zero, it is close enough to zero to satisfy the contrast command. Had we used six or fewer digits, the sum of the coefficients would be sufficiently different from zero for the contrast command to complain with the following error:

```
invalid contrast vector
r(198);
```

This error arises when the sum of the coefficients is not sufficiently close to zero.

Let's form a contrast of the average of those who are married (group 1) or separated (group 4) with the average of those who are widowed (group 2) or divorced (group 3). Note how the coefficients for groups 1 and 4 are specified as 0.5 and the coefficients for groups 2 and 3 are specified as -0.5.

```
. contrast {marital .5 -.5 -.5 .5 0}, nowald pveffects
Contrasts of marginal linear predictions
Margins          : asbalanced
```

	Contrast	Std. Err.	t	P>|t|
marital				
(1)	.075352	.0975132	0.77	0.440

To formulate the contrast coefficients, you may want to write out the null hypothesis that you want to test. For example, let's write out the null hypothesis corresponding to the contrast above:

$$H_0\colon (\mu_1 + \mu_4)/2 = (\mu_2 + \mu_3)/2$$

Let's now rewrite this showing the coefficient multiplied by each mean. That yields the following:

$$H_0\colon (1/2) \times \mu_1 + (1/2) \times \mu_4 = (1/2) \times \mu_2 + (1/2) \times \mu_3$$

Then, let's solve this for zero by moving the right side of the equation to the left side of the equal sign and making those coefficients negative.

$$H_0\colon (1/2) \times \mu_1 + (1/2) \times \mu_4 + -(1/2) \times \mu_2 + -(1/2) \times \mu_3(1/2) = 0$$

We can then sort the groups to yield the following:

$$H_0\colon (1/2) \times \mu_1 + -(1/2) \times \mu_2 + -(1/2) \times \mu_3 + (1/2) \times \mu_4 = 0$$

We then use these contrast coefficients in the `contrast` command, repeated below. Note that a value of 0 is inserted with respect to group 5.

```
. contrast {marital .5 -.5 -.5 .5 0}, pveffects
(output omitted)
```

5.10   Weighted contrasts

The examples I have presented so far have sidestepped the issue of how to account for unequal sample sizes. For example, say that we are comparing group 1 with groups 2 and 3 combined (that is, married versus the combination of those previously married and never married). So far, the examples I have shown estimate the mean for groups 2 and 3 combined by obtaining the mean for group 2 and the mean for group 3 and then by averaging those means. Stata calls this the "unweighted" approach because it gives equal weight to the groups even if their sample sizes are different. Instead, we could weight the means for groups 2 and 3 proportionate to their sample size. Stata calls this the "weighted" approach because the means are weighted in proportion to their observed sample size.

Let's illustrate this using the GSS dataset, predicting happiness from the three level marital status coded as follows: 1 = married, 2 = previously married, and 3 = never married. The `anova` command is used below to predict happiness from marital status (but the output is omitted to save space).

```
. use gss2012_sbs
. anova happy7 marital3
(output omitted)
```

We can obtain the average happiness by marital status using the `margins` command, shown below.

```
. margins marital3, nopvalues
Adjusted predictions                          Number of obs    =     1,284
Expression    : Linear prediction, predict()
```

	Margin	Delta-method Std. Err.	[95% Conf. Interval]	
marital3				
Married	5.714038	.0407055	5.634181	5.793895
Prevmaried	5.290598	.05219	5.188211	5.392986
Never married	5.365169	.0518222	5.263503	5.466834

Let's use the `h.` contrast operator to compare each group with the mean of the subsequent groups, using the unweighted approach. Let's focus our attention on the contrast of group 1 with groups 2 and 3 (that is, `Married vs >Married`).

```
. margins h.marital3, contrast(nowald pveffects)
Contrasts of adjusted predictions
Expression    : Linear prediction, predict()
```

	Contrast	Delta-method Std. Err.	t	P>|t|
marital3				
(Married vs >Married)	.3861547	.0548568	7.04	0.000
(Prevmaried vs Never married)	-.0745702	.0735482	-1.01	0.311

The `margins` command with the `h.` contrast operator estimates the difference in the means for group 1 versus groups 2 and 3 as 0.3861547. We can manually compute this estimate by taking the mean of group 1 minus the average of the means from groups 2 and 3, displayed below.

```
. display 5.714038 - (5.290598 + 5.365169)/2
.3861545
```

With a tiny bit of algebra, we can express this a little bit differently, as shown below. The fractions (1/2) indicate that groups 2 and 3 are weighted equally.

```
. display 5.714038 + -(1/2)*5.290598 + -(1/2)*5.365169
.3861545
```

Let's compare this with the weighted approach, which weights the mean of groups 2 and 3 by their sample size. The `hw.` contrast operator is used to obtain the weighted estimate.

```
. margins hw.marital3, contrast(nowald pveffects)

Contrasts of adjusted predictions

Expression    : Linear prediction, predict()
```

	Contrast	Delta-method Std. Err.	t	P>|t|
marital3				
(Married vs >Married)	.385891	.0548562	7.03	0.000
(Prevmaried vs Never married)	-.0745702	.0735482	-1.01	0.311

Instead of weighting groups 2 and 3 equally—by (1/2)—`margins` now weights them by their individual sample size divided by the combined sample size. The N for group 2 is 351, and the N for group 3 is 356, and the combined N for the two groups is 707. The estimate of the difference using the weighted approach is computed as shown below. We weight the mean for group 2 by (351/707) and the mean for group 3 by (356/707).

```
. display 5.714038 + -(351/707)*5.290598 + -(356/707)*5.365169
.38589081
```

Although this example has focused on the `h.` contrast operator, this issue arises for all the contrast operators that involve contrasts among more than two groups, namely, the `g.`, `h.`, `j.`, `p.`, and `q.` operators. Used without the `w`, these will provide unweighted estimates. If you include the `w` by specifying `gw.`, `hw.`, `jw.`, `pw.`, or `qw.`, then the weighted estimates are computed.

5.11   Pairwise comparisons

Sometimes, you want to test all the pairwise comparisons that can be formed for a factor variable. The `pwcompare` command can be used to form such comparisons. Let's illustrate the `pwcompare` command to form pairwise comparisons of the happiness ratings among the five marital status groups.

```
. use gss2012_sbs

. anova happy7 marital
  (output omitted)

. pwcompare marital
```

Pairwise comparisons of marginal linear predictions

Margins   : asbalanced

	Contrast	Std. Err.	Unadjusted [95% Conf. Interval]	
marital				
widowed vs married	-.2841316	.1028462	-.4858973	-.0823659
divorced vs married	-.4615629	.0798813	-.6182756	-.3048502
separated vs married	-.5949905	.156161	-.9013504	-.2886306
never married vs married	-.3488696	.0658518	-.478059	-.2196801
divorced vs widowed	-.1774313	.1168292	-.4066293	.0517667
separated vs widowed	-.3108589	.1779166	-.6598993	.0381815
never married vs widowed	-.064738	.1077242	-.2760736	.1465976
separated vs divorced	-.1334276	.1657045	-.4585102	.191655
never married vs divorced	.1126933	.0860709	-.0561623	.2815489
never married vs separated	.2461209	.1594159	-.0666245	.5588664

It might not be immediately obvious how the pairwise comparisons are ordered. Recall that marital is coded as follows: 1 = married, 2 = widowed, 3 = divorced, 4 = separated, and 5 = never married. You can see that groups 2 through 5 (widowed through never married) are compared with group 1 (the married group). Next, groups 3 through 5 (divorced through never married) are compared with group 2 (widowed). Then, groups 4 and 5 (separated and never married) are compared with group 3 (divorced). Finally, group 5 (never married) is compared with group 4 (separated).

Let's interpret the results of the first pairwise comparison. We can tell that this comparison is significant because the confidence interval excludes zero. To make it easier to identify significant results, we can specify the pveffects option, which will display t-values and p-values in lieu of the confidence interval, as shown below.

```
. pwcompare marital, pveffects
Pairwise comparisons of marginal linear predictions
Margins      : asbalanced
```

	Contrast	Std. Err.	Unadjusted t	P>|t|
marital				
widowed vs married	-.2841316	.1028462	-2.76	0.006
divorced vs married	-.4615629	.0798813	-5.78	0.000
separated vs married	-.5949905	.156161	-3.81	0.000
never married vs married	-.3488696	.0658518	-5.30	0.000
divorced vs widowed	-.1774313	.1168292	-1.52	0.129
separated vs widowed	-.3108589	.1779166	-1.75	0.081
never married vs widowed	-.064738	.1077242	-0.60	0.548
separated vs divorced	-.1334276	.1657045	-0.81	0.421
never married vs divorced	.1126933	.0860709	1.31	0.191
never married vs separated	.2461209	.1594159	1.54	0.123

The output now makes it clear that the first four comparisons are significant and that the remaining comparisons are not significant. The significant comparisons each involve those who are married. In each case, the group being compared with those who are married has lower happiness. Or, put more clearly, those who are married are (significantly) happier than each of the other groups.

Specifying the `groups` option displays group codes that signify groups not significantly different from each other. Groups that share a letter in common are not significantly different from each other, and groups that do not share a letter in common are significantly different. Those who are widowed, divorced, separated, and never married all share the letter **A** in common; thus all the pairwise comparisons involving these four groups are not significant. Further, the married group has no letter, indicating it is significantly different from the remaining groups. This is consistent with the results we saw in the previous `pwcompare` command. As you can see, the `groups` option creates output that provides a concise summary of what pairwise differences are significant and nonsignificant.

```
. pwcompare marital, groups
Pairwise comparisons of marginal linear predictions
Margins      : asbalanced
```

	Margin	Std. Err.	Unadjusted Groups
marital			
married	5.714038	.0406773	
widowed	5.429907	.09446	A
divorced	5.252475	.0687486	A
separated	5.119048	.1507701	A
never married	5.365169	.0517863	A

```
Note: Margins sharing a letter in the group label
      are not significantly different at the 5%
      level.
```

So far, these pairwise comparisons have not adjusted for multiple comparisons. The `mcompare()` option permits you to select a method for adjusting for multiple comparisons. The default method is to perform no adjustment for multiple comparisons. Four methods are provided that can be used with balanced or unbalanced data: Tukey's method, Bonferroni's method, Šidák's method, and Scheffé's method. These can be selected by specifying `tukey`, `bonferroni`, `sidak`, or `scheffe` within the `mcompare()` option. Three additional methods are provided but require balanced data. These are the Student–Newman–Keuls, Duncan, and Dunnett methods. These can be specified using the `snk`, `duncan`, or `dunnett` option.

Let's adjust for multiple comparisons by using the `mcompare(tukey)` option. Let's also include the `pveffects` option to show the t-values and p-values for each comparison. Using the Tukey adjustment, we find the same pattern of significant results (that is, the first four contrasts are significant).

```
. pwcompare marital, mcompare(tukey) pveffects
Pairwise comparisons of marginal linear predictions
Margins       : asbalanced
```

	Number of Comparisons
marital	10

	Contrast	Std. Err.	Tukey t	P>|t|
marital				
widowed vs married	-.2841316	.1028462	-2.76	0.046
divorced vs married	-.4615629	.0798813	-5.78	0.000
separated vs married	-.5949905	.156161	-3.81	0.001
never married vs married	-.3488696	.0658518	-5.30	0.000
divorced vs widowed	-.1774313	.1168292	-1.52	0.550
separated vs widowed	-.3108589	.1779166	-1.75	0.405
never married vs widowed	-.064738	.1077242	-0.60	0.975
separated vs divorced	-.1334276	.1657045	-0.81	0.929
never married vs divorced	.1126933	.0860709	1.31	0.685
never married vs separated	.2461209	.1594159	1.54	0.534

Let's adjust for multiple comparisons by using the `mcompare(sidak)` option. Using this criterion, we see that the first contrast (widowed versus married) is no longer significant ($p = 0.057$). The other three contrasts that were previously significant still remain significant.

Note: Job outlook for statisticians

Perhaps you are reading this book because you are thinking about a career in the field of statistics. In that case, you might be wondering about the job prospects in that field. You can find more information from the U.S. Bureau of Labor Statistics at http://www.bls.gov/ooh/math/statisticians.htm.

```
. pwcompare marital, mcompare(sidak) pveffects
Pairwise comparisons of marginal linear predictions
Margins        : asbalanced
```

	Number of Comparisons
marital	10

				Sidak
	Contrast	Std. Err.	t	P>|t|
marital				
widowed vs married	-.2841316	.1028462	-2.76	0.057
divorced vs married	-.4615629	.0798813	-5.78	0.000
separated vs married	-.5949905	.156161	-3.81	0.001
never married vs married	-.3488696	.0658518	-5.30	0.000
divorced vs widowed	-.1774313	.1168292	-1.52	0.749
separated vs widowed	-.3108589	.1779166	-1.75	0.570
never married vs widowed	-.064738	.1077242	-0.60	1.000
separated vs divorced	-.1334276	.1657045	-0.81	0.996
never married vs divorced	.1126933	.0860709	1.31	0.879
never married vs separated	.2461209	.1594159	1.54	0.730

Suppose we preferred to use the Bonferroni adjustment. We can do this by using the `mcompare(bonferroni)` option, as shown below. Although the output is not shown (to save space), the Bonferroni adjustment also indicates that the comparison of widowed versus married is no longer significant ($p = 0.058$). The three contrasts that were previously significant remain significant.

```
. pwcompare marital, mcompare(bonferroni) pveffects
  (output omitted)
```

Finally, let's consider the Scheffé method, using the `mcompare(scheffe)` option. To save space, I have omitted the output. The results using the Scheffé method show the same pattern of significance as the Bonferroni and Šidák methods. The comparison of widowed versus married is not significant ($p = 0.107$), and the three other significant contrasts remain significant.

```
. pwcompare marital, mcompare(scheffe) pveffects
  (output omitted)
```

When you have balanced data, you can choose among any of the multiple-comparison methods available via the `mcompare()` option.[6] To illustrate, we will consider an example using balanced data, that is, in which all groups have equal cell sizes. Let's use the example from section 5.6, which looked at pain ratings as a function of medica-

---

6. That is, you can specify `bonferroni`, `sidak`, `scheffe`, `tukey`, `snk`, `duncan`, or `dunnett`.

tion dosage group (`dosegrp`). The `dosegrp` variable was coded as follows: $1 = 0$ mg, $2 = 50$ mg, $3 = 100$ mg, $4 = 150$ mg, $5 = 200$ mg, and $6 = 250$ mg. The `anova` command predicting `pain` from `dosegrp` is shown below.

```
. use pain
. anova pain dosegrp

                       Number of obs =        180   R-squared     =  0.4602
                       Root MSE      =    10.4724   Adj R-squared =  0.4447

       Source │ Partial SS          df          MS          F    Prob>F
    ───────────┼──────────────────────────────────────────────────────
        Model │ 16271.694            5   3254.3389      29.67    0.0000

      dosegrp │ 16271.694            5   3254.3389      29.67    0.0000

     Residual │ 19082.633          174   109.67031
    ───────────┼──────────────────────────────────────────────────────
        Total │ 35354.328          179   197.51021
```

Let's use the `pwcompare` command to form pairwise comparisons of all dosage groups, using Tukey's method of adjustment for multiple comparisons. I include the `pveffects` option to display t-values and p-values in lieu of the confidence intervals for the effects.

```
. pwcompare dosegrp, mcompare(tukey) pveffects
Pairwise comparisons of marginal linear predictions

Margins        : asbalanced

                │  Number of
                │ Comparisons
    ─────────────┼─────────────
        dosegrp │          15
```

			Tukey	
	Contrast	Std. Err.	t	P>|t|
dosegrp				
50 mg vs Zero (control)	-1.233333	2.703952	-0.46	0.997
100 mg vs Zero (control)	.3	2.703952	0.11	1.000
150 mg vs Zero (control)	-1.433333	2.703952	-0.53	0.995
200 mg vs Zero (control)	-17.13333	2.703952	-6.34	0.000
250 mg vs Zero (control)	-23.53333	2.703952	-8.70	0.000
100 mg vs 50 mg	1.533333	2.703952	0.57	0.993
150 mg vs 50 mg	-.2	2.703952	-0.07	1.000
200 mg vs 50 mg	-15.9	2.703952	-5.88	0.000
250 mg vs 50 mg	-22.3	2.703952	-8.25	0.000
150 mg vs 100 mg	-1.733333	2.703952	-0.64	0.988
200 mg vs 100 mg	-17.43333	2.703952	-6.45	0.000
250 mg vs 100 mg	-23.83333	2.703952	-8.81	0.000
200 mg vs 150 mg	-15.7	2.703952	-5.81	0.000
250 mg vs 150 mg	-22.1	2.703952	-8.17	0.000
250 mg vs 200 mg	-6.4	2.703952	-2.37	0.174

As shown in the output, a total of 15 comparisons are made. For each comparison a *p*-value (reflecting the Tukey adjustment for multiple comparisons) is shown.

To create a more concise summary of these results, we can use the `groups` option.

```
. pwcompare dosegrp, mcompare(tukey) groups
Pairwise comparisons of marginal linear predictions

Margins        : asbalanced
```

	Number of Comparisons
dosegrp	15

	Margin	Std. Err.	Tukey Groups
dosegrp			
Zero (control)	71.83333	1.911982	A
50 mg	70.6	1.911982	A
100 mg	72.13333	1.911982	A
150 mg	70.4	1.911982	A
200 mg	54.7	1.911982	B
250 mg	48.3	1.911982	B

```
Note: Margins sharing a letter in the group label
      are not significantly different at the 5%
      level.
```

The output above shows that groups 5 and 6 (200 mg and 250 mg) both belong to group B and are not significantly different. Also, groups 1, 2, 3, and 4 (0 mg, 50 mg, 100 mg, and 150 mg) all belong to group A and are not significantly different. This is reflected in the more detailed output from the previous `pwcompare` command that shows these groups not to be significantly different.

You can find more information about the `pwcompare` command by seeing `help pwcompare`.

Tip! The pwmean command

Instead of using the `anova` command followed by the `pwcompare` command, you can use the `pwmean` command. For example, the `pwmean` command below reports the same results as the previous `pwcompare` command without needing to previously run the `anova` command.

```
. pwmean pain, over(dosegrp) mcompare(tukey) groups
```

5.12   Closing thoughts

In this section, we have seen how you can understand and dissect the effect of an IV on the DV. The `margins` and `marginsplot` commands make it easy to compute and graph the means of the DV as a function of the IV. Further, the `contrast` and `margins` commands can be used with contrast operators to form meaningful contrasts among the levels of the IV. There are many built-in contrast operators that you can use, and you can also specify custom contrast operators. Further, if you use the `margins` command to form such contrasts, the `marginsplot` command can be used to graph the contrasts.

For more details about options that you can use with the `contrast` command, see section 21.2. For more details about options you can use with the `margins` command, see section 21.3. Section 21.4 contains more details about customizing the appearance of the graphs created by the `marginsplot` command. For more information about pwcompare, see section 21.5. For further help with these commands, you can also type `help contrast`, `help margins`, `help marginsplot`, and `help pwcompare`.

6   Analysis of covariance

6.1   Chapter overview

Analysis of covariance (ANCOVA) is a blend of analysis of variance (ANOVA) and regression. The focus is on the group differences due to the independent variable (IV) (like ANOVA) but after accounting for the contribution of the continuous predictors. In the context of ANCOVA, the continuous predictors are called covariates. In a sense, ANCOVA includes any model that includes a combination of categorical and continuous predictors where you are testing for differences on the categorical predictors. Traditionally, ANCOVA has been used in two different ways.

1) **Designed experiment with randomization.** In a designed experiment, ANCOVA can reduce residual error variance compared with ANOVA. One can draw causal conclusions about the effect of group assignment on the dependent variable (DV), and the covariate reduces the residual variance (increasing the power and precision of the estimates), thus reducing the required sample size needed to detect a given effect.

2) In the analysis of observational data. With observational data, it is not possible to draw causal conclusions about the relationship between the IV and DV because of the presence of potential confounding variables. If important confounding variables are measured and included in the dataset, ANCOVA is a statistical attempt to adjust for the naturally occurring confounding.

Although the mechanics of such analyses are the same, the underlying motivations are different. Sections 6.2 and 6.3 provide examples illustrating the use of ANCOVA in experiments, while section 6.4 describes its use with observational data.

Note: More about regression

Because ANCOVA is a blend of ANOVA and regression, this chapter sometimes performs analyses using the `anova` command and sometimes using the `regress` command. You might benefit by flipping forward to some of the later chapters for related information. For more information about running ANOVA models via the `regress` command, see chapter 10. Also see part IV for more information on regression models, especially chapter 10, which gives more information about multiple regression and the `regress` command.

When you run an ANCOVA model, it is important to check the assumptions of the model. For the sake of space, I will not be performing any checks. You can see chapter 18 for methods for checking assumptions regarding the relationship between the covariate and the DV. Additionally, ANCOVA assumes that the slope terms associated with the covariate are the same for all levels of the IV. This is called the homogeneity of slopes (or homogeneity of regression) assumption. Chapter 8 shows how to assess the homogeneity of slopes assumption and how to fit models in which this assumption does not hold.

Video tutorial: ANCOVA

See a video demonstration of running an ANCOVA at
http://www.stata.com/sbs/ancova.

6.2   Example 1: ANCOVA with an experiment using a pretest

Let's return to the example from section 4.3 in which an optimism therapy group was compared with a control group using optimism as the DV. Suppose that Professor Cheer wants to perform a follow-up study, but the DV will be happiness. Before beginning the study, Professor Cheer performs a power analysis to determine the appropriate sample size and to see whether she can afford to run the study given budgetary considerations. Given Professor Cheer's desire to find a standard effect size of 0.5, she finds that the required sample size would be $N = 128$ (64 per group). See sections 11.2.2 and 11.2.4 for more information about how such a power analysis is performed using Stata.

Professor Cheer is dismayed because this sample size is beyond the scope of her budget. She could use a smaller sample size, but then she would risk spending lots of effort in performing the study only to find no significant result because the power of the study was not sufficient. She mentions this dilemma to her friend (Professor Marge Innovera), who suggests that she modify the design of her study to use a pretest–posttest design in which happiness is measured both at the pretest and at posttest. Professor Innovera explains that the pretest measure of happiness is likely to be correlated with the posttest and can be used as a covariate that can reduce the required sample size. As the correlation between the covariate and the DV increases, the required sample size decreases.[1] In short, using a pretest that is well correlated with the posttest as a covariate, Professor Cheer can use a smaller sample size while still maintaining the desired level of power.

Professor Cheer believes that the correlation between happiness measured at pretest and posttest would be 0.7, but for the sake of the power analysis, she assumes a more conservative value of 0.6. Using this pretest–posttest design with a covariate that has a correlation of 0.6 with the DV, Professor Innovera determines that a sample size of $N = 82$ (41 per group) can be used while still achieving the same level of power to detect a standard effect size of 0.5. The Stata commands for performing this power analysis are illustrated in section 11.4.1.

Now imagine that Professor Cheer has performed the study and collected the data. The data from the study are stored in the dataset named `hap-ancova1.dta`. Note that there are two treatment groups, each containing an N of 41 as suggested by the power analysis above. The group labeled `OT` received optimism therapy, while the `Con` group was a control group.

```
. use hap-ancova1
. tabulate treat
```

Treatment group assignment	Freq.	Percent	Cum.
Con	41	50.00	50.00
OT	41	50.00	100.00
Total	82	100.00	

Using the pretest–posttest design, Professor Cheer measured happiness before the start of the study (`hap_pre`) and again at the end of the study (`hap_post`). Let's assess the correlation between these two measures.

---

1. This is achieved because the covariate explains more of the variance in the DV, thus reducing the residual error variance and increasing statistical power.

```
. corr hap_pre hap_post
(obs=82)

             |  hap_pre hap_post
   ----------+------------------
     hap_pre |   1.0000
    hap_post |   0.5780   1.0000
```

As expected, the two measures of happiness are substantially correlated to about the same degree predicted in the power analysis (0.60).

So now let's run the ANCOVA, in which the pretest is used as a covariate. We can fit such models focusing on the ANOVA aspect using the `anova` command, or we can focus on the regression aspect using the `regress` command. Below we fit the model using the `anova` command. Note how the covariate is specified as `c.hap_pre`; we use the `c.` prefix to indicate that this is a continuous variable.

```
. anova hap_post treat c.hap_pre

                     Number of obs =        82    R-squared     =  0.3749
                     Root MSE      =   7.85825    Adj R-squared =  0.3591

         Source |   Partial SS        df        MS          F    Prob>F
    ------------+------------------------------------------------------
          Model |   2925.9707         2    1462.9854      23.69  0.0000
                |
          treat |   318.81289         1    318.81289       5.16  0.0258
        hap_pre |   2700.4097         1    2700.4097      43.73  0.0000
                |
       Residual |   4878.4195        79    61.752146
    ------------+------------------------------------------------------
          Total |   7804.3902        81    96.350497
```

The `anova` output shows us that after we account for the happiness measured at the start of the study, the differences in the happiness between the two groups (at the end of the study) is significant $[F(1, 79) = 5.16, p = 0.0258]$. We can quantify this difference by examining the adjusted means, computed below using the `margins` command. After we account for happiness at the pretest, the mean posttest happiness for the control condition is 51.5, compared with 55.4 for the optimism therapy group.

```
. margins treat, nopvalues

Predictive margins                              Number of obs     =        82
Expression     : Linear prediction, predict()

             |            Delta-method
             |    Margin    Std. Err.    [95% Conf. Interval]
    ---------+----------------------------------------------------
       treat |
         Con |   51.48863   1.228184     49.04399    53.93327
          OT |   55.4382    1.228184     52.99356    57.88284
```

If we want to quantify the effect of optimism therapy (versus control), we can apply the r. contrast operator to treat, as shown in the margins command below.

```
. margins r.treat, contrast(nowald effects)
Contrasts of predictive margins
Expression    : Linear prediction, predict()
```

	Contrast	Delta-method Std. Err.	t	P>|t|	[95% Conf. Interval]	
treat (OT vs Con)	3.949567	1.738231	2.27	0.026	.489704	7.40943

This shows us that the difference in the posttest optimism, after we adjust for pretest optimism, between optimism therapy and the control group is 3.95 (95% CI = [0.49, 7.41]).[2]

In writing this up for publication, we would want to include a measure of the effect size. Below we use the estat esize command to compute omega-squared. For variety, we specify the level(90) option to obtain 90% confidence intervals (CIs).

```
. estat esize, omega level(90)
Effect sizes for linear models
```

Source	Omega-Squared	df	[90% Conf. Interval]	
Model	.3590885	2	.206948	.4654293
treat	.0494611	1	0	.1499655
hap_pre	.3481616	1	.2069899	.461934

The output shows omega-squared for the overall model (0.3591). The effect of treat explains 4.9% (90% CI = [0%, 15.0%]) of the variance in happiness at the posttest, after we adjust for happiness at the pretest.

Note: Effect sizes for contrasts

In this example, the estat esize command tells us about the effect size associated with the two-level variable treat. If treat had three or more levels, we might have used the contrast command to perform specific contrasts. In such a case, you could use the esizei command to obtain the omega-squared associated with a particular contrast (see section 4.4.2 for more details).

---

2. Note how the p-value from the margins command mirrors the p-value from the anova command ($p = 0.026$ from the margins command and $p = 0.0258$ from the anova command). These are the same (aside from rounding) because they are both testing the same null hypothesis, testing the equality of the posttreatment means after adjusting for the covariates.

As I noted before, we could have performed this analysis with the `regress` command instead of the `anova` command. The structure of `regress` is much the same as `anova`. The key differences are that we include `i.treat` to specify that this is a categorical (factor) variable. We can omit the `c.` prefix from `hap_pre` because all variables are assumed to be continuous unless prefixed by the `i.` prefix.

```
. regress hap_post i.treat hap_pre

      Source |       SS           df       MS            Number of obs   =        82
-------------+----------------------------------         F(2, 79)        =     23.69
       Model |  2925.97071          2   1462.98536       Prob > F        =    0.0000
    Residual |  4878.41953         79   61.752146        R-squared       =    0.3749
-------------+----------------------------------         Adj R-squared   =    0.3591
       Total |  7804.39024         81   96.3504968       Root MSE        =    7.8583

    hap_post |      Coef.   Std. Err.      t    P>|t|     [95% Conf. Interval]
-------------+----------------------------------------------------------------
       treat |
         OT  |   3.949567   1.738231     2.27   0.026     .489704    7.40943
     hap_pre |   .6030754   .0911975     6.61   0.000    .4215514   .7845994
       _cons |   20.54792   4.883424     4.21   0.000    10.82771   30.26813
```

The `regress` command provides much the same kind of output as the `anova` command, except that it provides regression coefficients as well as associated standard errors, significance tests, and CIs for each of the effects in the model. Unlike `anova`, `regress` reports an estimate of the regression coefficient relating happiness at pretest to happiness at posttest, $B = 0.60$ (95% CI = $[0.42, 0.78]$), and a test showing this coefficient is significant $[t(79) = 6.61, p < 0.001]$. When you report the output of an ANCOVA for publication, it would be useful to report this information about the regression coefficients for each of the covariates.

With respect to the factor variables (like `treat`), they are broken up into a series of dummy variables, and the coefficients associated with each of the dummy variables are presented. It can be cumbersome to interpret the results of such dummy variables, which is why I recommend using the `margins` and `contrast` commands to extract any information you need with respect to the factor variables. The `margins` and `contrast` commands work exactly the same way after the `regress` command as they do following the `anova` command and will yield identical results (given the same DV, IV, and covariates).

For example, the `contrast` command is used below to estimate the effect of `treat`, which is significant $[F(1, 79) = 5.16, p = 0.0258]$. These results match those found for the effect of `treat` from the `anova` command.

```
. contrast treat
```
Contrasts of marginal linear predictions
Margins   : asbalanced

	df	F	P>F
treat	1	5.16	0.0258
Denominator	79		

Note: Alternative analysis

Another way to analyze a pretest–posttest design would be to use a repeated measures ANOVA. However, the ANCOVA strategy actually has greater statistical power than the repeated measures ANOVA (see Rausch, Maxwell, and Kelley [2003]). This is illustrated at the conclusion of section 11.4.1.

Just for fun, let's see what would have happened if we analyzed our data ignoring the covariate (that is, ignoring happiness measured before the start of the study). This shows us what might have happened if we had not included the covariate as part of our design.

```
. anova hap_post treat
```

| | | Number of obs = | 82 | R-squared | = | 0.0289 |
| | | Root MSE | = 9.73321 | Adj R-squared | = | 0.0168 |
Source	Partial SS	df	MS	F		Prob>F
Model	225.56098	1	225.56098	2.38		0.1268
treat	225.56098	1	225.56098	2.38		0.1268
Residual	7578.8293	80	94.735366			
Total	7804.3902	81	96.350497			

Using this traditional ANOVA, we do not find significant differences between the two groups [$F(1, 80) = 2.38$, $p = 0.1268$], whereas we found significant differences using the ANCOVA. The covariate led to a substantial reduction in the residual error variance, which increased the F-value in the ANCOVA. In the ANOVA, the $MS_{Residual}$ is 94.74, compared with the $MS_{Residual}$ in the ANCOVA which was 61.75. This underscores the benefit of including the covariate for increasing the power of the test of the treatment effect.

6.3   Example 2: Experiment using covariates

As we saw in the previous example, using the pretest as a covariate was an excellent way to increase the power of a study. However, this strategy can make us question the generalizability of our results. As discussed in section 11.4.2, we might be concerned that measuring happiness at the pretest might make people focus on their level of happiness more than they normally do, which might artificially increase the treatment effect. If so, this would undermine the external validity of our results (Shadish, Cook, and Campbell 2002).

We can still obtain the benefits of using a covariate while preserving the generalizability of our results by selecting other covariates that are related to happiness. In section 11.4.2, we will see that we could include three covariates (satisfaction with family life, health, and how often respondents socialize with friends); inclusion of these covariates would allow us to use $N = 96$ (48 per group) while having sufficient power to detect a Cohen's d effect size of 0.5 using 80% power.

On the basis of this power analysis, Professor Cheer conducted a second study with a total of 96 participants. Before randomization (for example, after the participants received their informed consent information but before randomization), participants were asked to provide information about themselves. This included the measures we intend to use as covariates, namely, measures of their satisfaction with their family life (`satfam`), health (`health`), and how often they socialize with friends (`socfrend`). Then, half the participants (48) were randomly assigned to receive optimism therapy and the other half assigned to the control group. At the conclusion of the study, happiness was measured for all participants.

The dataset with the results of this study is stored in `hap-ancova2.dta`. The `use` command reads this file into memory, and the `describe` command shows all the variables in this dataset.

```
. use hap-ancova2

. describe

Contains data from hap-ancova2.dta
  obs:            96
  vars:            6                              25 Aug 2014 01:02
  size:        2,304
```

variable name	storage type	display format	value label	variable label
id	float	%9.0g		ID variable
hap	float	%9.0g		Happiness score
treat	float	%9.0g	treat	Treatment group assignment
satfam	float	%21.0g	sat	Satisfaction with family life
health	float	%21.0g	sat	Self reported satisfaction with health
socfrend	float	%35.0g	freq	Socialize with friends, how often

```
Sorted by:
```

The variable `treat` contains the treatment group assignment, 1 for control and 2 for optimism therapy. As we can see, there are 48 participants in each group.

```
. fre treat

treat — Treatment group assignment
```

		Freq.	Percent	Valid	Cum.
Valid	1 Con	48	50.00	50.00	50.00
	2 OT	48	50.00	50.00	100.00
	Total	96	100.00	100.00	

Below we can see summary statistics for the DV (`hap`) as well as the three covariates, satisfaction with family life (`satfam`), health (`health`), and how often respondents socialize with friends (`socfrend`).

```
. summarize hap satfam health socfrend
```

Variable	Obs	Mean	Std. Dev.	Min	Max
hap	96	50.11458	10.36377	28	81
satfam	96	3.854167	1.414058	1	7
health	96	3.708333	1.03534	2	7
socfrend	96	4.59375	1.165993	2	7

Before performing the ANCOVA, we will first perform an analysis just with the covariates so that we can assess the degree to which the covariates are related to the DV.

```
. anova hap c.satfam c.health c.socfrend
                         Number of obs =        96    R-squared     =  0.2878
                         Root MSE      =   8.88757    Adj R-squared =  0.2646

            Source |   Partial SS       df        MS          F      Prob>F

             Model |   2936.7613         3     978.92043     12.39   0.0000

            satfam |   1165.1555         1     1165.1555     14.75   0.0002
            health |   934.06437         1     934.06437     11.83   0.0009
          socfrend |   290.43862         1     290.43862      3.68   0.0583

          Residual |   7266.9783        92     78.988894

             Total |    10203.74        95     107.40779
```

The combination of these three covariates explains about 29% of the variance in the DV.[3] The covariates `satfam` and `health` are statistically significant. The p-value for `socfrend` is 0.0583. Although it is not significant, we will include `socfrend` in the ANCOVA analysis because our selection of that variable was based on prior research (our analysis involving the General Social Survey [GSS dataset]) and not strictly based on whether it was significant in the current analysis.

We should do a more thorough checking of assumptions with respect to these predictors (as described in chapter 18), but for the sake of brevity, let's assume that these checks have been performed and are satisfactory. So let's now run the ANCOVA in which the DV is `hap`, the covariates are `satfam`, `health`, and `socfrend`, and the IV is `treat`. This analysis is performed using the `anova` command below. Note the covariates are prefixed with `c.` to indicate they are continuous variables.

```
. anova hap c.satfam c.health c.socfrend treat
                         Number of obs =        96    R-squared     =  0.3299
                         Root MSE      =   8.66808    Adj R-squared =  0.3005

            Source |   Partial SS       df        MS          F      Prob>F

             Model |   3366.4018         4     841.60045     11.20   0.0000

            satfam |   1043.4505         1     1043.4505     13.89   0.0003
            health |   1133.2569         1     1133.2569     15.08   0.0002
          socfrend |   285.62132         1     285.62132      3.80   0.0543
             treat |    429.6405         1      429.6405      5.72   0.0188

          Residual |   6837.3378        91     75.13558

             Total |    10203.74        95     107.40779
```

The ANCOVA results show that after we adjust for the covariates, the effect of treatment on happiness is significant [$F(1, 91) = 5.72$, $p = 0.0188$].

---

3. This exceeds the percentage of variance assumed in the power analysis performed in section 11.4.2, where we assumed the covariates would explain 25% of the variance in the DV.

Note: Does the order matter?

In the `anova` command, the variable `treat` was entered last. Does the order matter? Could we have entered `treat` as the first variable? The order does not matter. The variable `treat` could have been entered first, last, or any where in between. This is because the `anova` command assesses the contribution of each variable after adjusting for all other variables in the model.

To better understand the effect of treatment group assignment on happiness, we will use the `margins` command (below) to obtain the adjusted means of happiness by treatment group assignment.

```
. margins treat
Predictive margins                              Number of obs     =       96
Expression    : Linear prediction, predict()
```

	Margin	Delta-method Std. Err.	t	P>|t|	[95% Conf. Interval]	
treat						
Con	47.96332	1.261744	38.01	0.000	45.45702	50.46962
OT	52.26585	1.261744	41.42	0.000	49.75955	54.77215

The average happiness of those receiving optimism therapy was 52.27 compared with 47.96 for those in the control condition.

We can express the effect of the treatment in terms of the percentage of the variance in happiness that it explains (via omega-squared).[4] The `estat esize` command with the `omega` option is used below. It shows that after we adjust for the covariates, the treatment explains 4.88% of the variance in happiness (95% CI $= [0.0\%$ to $16.16\%]$).

```
. estat esize, omega
Effect sizes for linear models
```

Source	Omega-Squared	df	[95% Conf. Interval]	
Model	.3004643	4	.1179849	.4147738
satfam	.1228703	1	.0189234	.2541799
health	.1327531	1	.0243156	.2652529
socfrend	.0295503	1	.	.1326677
treat	.048783	1	0	.1615929

We could ask Stata to report the results as though we had used the `regress` command instead of the `anova` command by simply issuing the command `regress`. After

---

4. Recall that omega-squared is just another name for adjusted R-squared.

you run an `anova` command, you can type just `regress`, and Stata will show you the
results using a regression-style output, as shown below.

```
. regress

      Source |       SS           df       MS              Number of obs   =        96
-------------+----------------------------------           F(4, 91)        =     11.20
       Model |  3366.4018          4   841.600449           Prob > F        =    0.0000
    Residual |  6837.33779        91   75.1355801           R-squared       =    0.3299
-------------+----------------------------------           Adj R-squared   =    0.3005
       Total |  10203.7396        95   107.407785           Root MSE        =    8.6681

------------------------------------------------------------------------------
         hap |      Coef.   Std. Err.      t    P>|t|     [95% Conf. Interval]
-------------+----------------------------------------------------------------
      satfam |    2.37615   .6376182     3.73   0.000      1.1096     3.6427
      health |   3.439477   .8856273     3.88   0.000     1.680287   5.198666
    socfrend |   1.499186   .7689234     1.95   0.054    -.0281857   3.026558
             |
       treat |
          OT |   4.302534   1.799262     2.39   0.019      .72852    7.876547
       _cons |   19.16363   5.086194     3.77   0.000     9.060528   29.26673
------------------------------------------------------------------------------
```

Rather than displaying sums of squares for the covariates, we now see regression
coefficients. For example, we see that happiness increases by 2.38 units for every one-
unit increase in family satisfaction. For a publication, we may want to include the
results with respect to the covariates as displayed by the `regress` command above.

6.4 Example 3: Observational data

Suppose we were interested in knowing whether being married causes happiness. It is
unlikely that we could perform an experiment in which we could randomly assign some
people to get married while randomly assigning other people not to get married.

Suppose we use survey data to collect information about marital status and happiness and run a regression that finds that being married is related to happiness. Is it just
being married, or is it some other factor that accompanies marriage, that is the cause of
greater happiness? For example, those who are married may be older or more educated
or have more children, and these may be the key factors leading to greater happiness.
We could try to account for such factors by including age, education, and number of
children as additional predictors (covariates). Then the effects associated with being
married would be adjusted for age, education, and number of children.

Let's consider this question using the GSS dataset. Below we use the GSS dataset and
keep just the cases in which the variable `happy7` is not missing. The `count` command
shows that 1,284 observations remain after we eliminate observations where happiness
was missing.

```
. use gss2012_sbs, clear
. keep if !missing(happy7)
(690 observations deleted)
. count
  1,284
```

Using the `fre` command below, we see that among these 1,284 observations, 577 were married and 707 were unmarried.

```
. fre married
married — marital: married=1, unmarried=0 (recoded)
```

		Freq.	Percent	Valid	Cum.
Valid	0 Unmarried	707	55.06	55.06	55.06
	1 Married	577	44.94	44.94	100.00
	Total	1284	100.00	100.00	

6.4.1   Model 1: No covariates

For these models, let's use the `regress` command. Let's start by examining the association between marital status and happiness.

```
. * Model 1
. regress  happy7 i.married
```

Source	SS	df	MS	Number of obs	=	1,284
				F(1, 1282)	=	49.48
Model	47.3107209	1	47.3107209	Prob > F	=	0.0000
Residual	1225.68616	1,282	.956073451	R-squared	=	0.0372
				Adj R-squared	=	0.0364
Total	1272.99688	1,283	.99220334	Root MSE	=	.97779

happy7	Coef.	Std. Err.	t	P>|t|	[95% Conf.	Interval]
married						
Married	.385891	.0548568	7.03	0.000	.2782721	.49351
_cons	5.328147	.0367736	144.89	0.000	5.256004	5.40029

We see that being married is significantly related to happiness ($p < 0.001$). The coefficient for `married` shows that the happiness of those who are married is 0.386 units higher than that of those who are not married. The variable `married` explains 3.72% of the variance in happiness (using R-squared), or 3.64% using the adjusted R-squared.

6.4.2   Model 2: Demographics as covariates

Perhaps the difference is not really due to being married but to differences in demographic characteristics between those who are and are not married. Let's include such demographic variables regarding gender, age, education, and number of children, as shown in the model below.

```
. * Model 2
. regress  happy7 i.married i.female age educ children
```

Source	SS	df	MS			
				Number of obs	=	1,282
				F(5, 1276)	=	12.22
Model	58.1605884	5	11.6321177	Prob > F	=	0.0000
Residual	1214.33629	1,276	.95167421	R-squared	=	0.0457
				Adj R-squared	=	0.0420
Total	1272.49688	1,281	.993362123	Root MSE	=	.97554

happy7	Coef.	Std. Err.	t	P>|t|	[95% Conf. Interval]	
married						
Married	.3610625	.0563424	6.41	0.000	.2505287	.4715964
female						
Female	.0346811	.0549526	0.63	0.528	-.0731263	.1424885
age	-.0027304	.001667	-1.64	0.102	-.0060008	.0005401
educ	.0247194	.0093185	2.65	0.008	.0064381	.0430006
children	.0411291	.0185344	2.22	0.027	.0047679	.0774903
_cons	5.03851	.1520095	33.15	0.000	4.740294	5.336726

The regression coefficient for `married` remains significant ($p < 0.001$) and indicates that after we adjust for these demographic characteristics, those who are married are 0.361 units happier than those who are not married. This is similar to the unadjusted regression coefficient from model 1.

6.4.3   Model 3: Demographics, socializing as covariates

Studies indicate that social connections are an important part of happiness. Perhaps those who married have more social connections, leading them to be happier. Let's include the variables `socrel`, `soccommun`, `socfrend`, and `socbar`, which reflect socializing with relatives, neighbors, and friends and at bars.

```
. * Model 3
. regress  happy7 i.married i.female age educ children
> socrel socommun socfrend socbar
```

Source	SS	df	MS		
				Number of obs	= 1,277
				F(9, 1267)	= 11.42
Model	95.2064245	9	10.5784916	Prob > F	= 0.0000
Residual	1174.03398	1,267	.926625085	R-squared	= 0.0750
				Adj R-squared	= 0.0684
Total	1269.24041	1,276	.994702513	Root MSE	= .96261

happy7	Coef.	Std. Err.	t	P>|t|	[95% Conf. Interval]	
married						
Married	.4064634	.0563013	7.22	0.000	.2960093	.5169175
female						
Female	.0326347	.0552228	0.59	0.555	-.0757034	.1409729
age	-.0002636	.0017269	-0.15	0.879	-.0036515	.0031243
educ	.0162513	.0094535	1.72	0.086	-.0022949	.0347976
children	.0453883	.0186229	2.44	0.015	.0088532	.0819234
socrel	.0364189	.0169974	2.14	0.032	.0030729	.069765
socommun	.0312075	.0139082	2.24	0.025	.0039218	.0584932
socfrend	.071874	.0186172	3.86	0.000	.0353501	.108398
socbar	.0093567	.0179825	0.52	0.603	-.0259221	.0446356
_cons	4.430552	.1835766	24.13	0.000	4.070404	4.7907

The variables `socrel`, `socommun`, and `socfrend` are all significantly (and positively) related to happiness, but `socbar` is not. After we account for these variables, those who are married are 0.406 units happier than those who are not married, and this difference is significant ($p < 0.001$). In fact, this difference is even greater than the difference we found using model 2 (in which the coefficient was 0.361). Using the `tabstat` command, we find that those who are unmarried actually have higher values on all of these socializing variables.

```
. tabstat socrel socommun  socfrend, by(married)
Summary statistics: mean
  by categories of: married (marital: married=1, unmarried=0 (recoded))
```

married	socrel	socommun	socfrend
Unmarried	4.617021	3.454031	4.217822
Married	4.488735	3.05913	3.795494
Total	4.559282	3.276911	4.028037

Because each of these socializing variables is positively related to happiness, the adjusted mean of happiness is raised for those who are unmarried and is lowered for those who are married, accentuating the difference in the adjusted means. This is an interesting result because we often think that the inclusion of covariates always reduces group differences. But, as we see here, it is possible to have covariates that can actually accentuate group differences.

6.4.4   Model 4: Demographics, socializing, health as covariates

Let's add another covariate, health status. The model below includes all the covariates from the previous model and also includes self-rated health as an additional covariate. After we adjust for demographics, socializing, and health, the happiness of those who are married is 0.309 units greater than that of those who are not married. The inclusion of health status above and beyond demographics and socializing reduced the disparity in happiness between those who are married and unmarried, but the difference remains as 0.309 units and is still significant.

```
. * Model 4
. regress  happy7 i.married i.female age educ children
> socrel socommun socfrend socbar health
```

Source	SS	df	MS		Number of obs	=	1,272
					F(10, 1261)	=	23.10
Model	194.796405	10	19.4796405		Prob > F	=	0.0000
Residual	1063.16507	1,261	.843112667		R-squared	=	0.1549
					Adj R-squared	=	0.1481
Total	1257.96148	1,271	.989741525		Root MSE	=	.91821

happy7	Coef.	Std. Err.	t	P>|t|	[95% Conf. Interval]	
married						
Married	.3094437	.0544948	5.68	0.000	.2025333	.4163541
female						
Female	.0466346	.0528062	0.88	0.377	-.0569631	.1502323
age	.0022315	.0016667	1.34	0.181	-.0010383	.0055013
educ	.0006243	.0092063	0.07	0.946	-.0174371	.0186857
children	.0482414	.0177857	2.71	0.007	.0133486	.0831342
socrel	.0314293	.0162649	1.93	0.054	-.0004799	.0633385
socommun	.0348071	.0132891	2.62	0.009	.0087359	.0608783
socfrend	.0583137	.0178489	3.27	0.001	.0232969	.0933306
socbar	.000468	.0172004	0.03	0.978	-.0332766	.0342127
health	.2633656	.0244093	10.79	0.000	.2154783	.311253
_cons	3.748184	.1859945	20.15	0.000	3.383291	4.113076

We could continue to add additional variables that could account for the difference in happiness between those who are and are not married, but let's consider this model (model 4) as our final model. For this model, we can estimate the adjusted mean of happiness by marital status using the `margins` command below.

```
. margins married, nopvalues

Predictive margins                              Number of obs    =      1,272
Model VCE    : OLS

Expression   : Linear prediction, predict()
```

	Margin	Delta-method Std. Err.	[95% Conf. Interval]	
married				
Unmarried	5.366594	.035514	5.296921	5.436267
Married	5.676038	.0395569	5.598433	5.753642

We can also estimate the unadjusted means of happiness by marital status by specifying the `over(married)` option, shown below.

```
. margins, over(married) nopvalues

Predictive margins                              Number of obs    =      1,272
Model VCE    : OLS

Expression   : Linear prediction, predict()
over         : married
```

	Margin	Delta-method Std. Err.	[95% Conf. Interval]	
married				
Unmarried	5.332382	.0346804	5.264345	5.40042
Married	5.718039	.038426	5.642653	5.793424

We see that the average (unadjusted) happiness for those who are unmarried is 5.33 versus 5.72 for those who are married. After we adjust for demographics, socializing, and health, the adjusted mean of happiness is 5.37 for those who are unmarried and 5.68 for those who are married.

Using the `estat esize, omega` command, we get an estimate of the percentage of variance explained by each of the predictors. After we adjust for all the covariates, the variable `married` explains 2.4% of the variance in happiness (95% CI = $[0.997\%, 4.35\%]$).

```
. estat esize, omega
Effect sizes for linear models
```

Source	Omega-Squared	df	[95% Conf. Interval]	
Model	.1481486	10	.1074886	.1780739
married	.0241597	1	.0099664	.0435235
female	0	1	.	.0055644
age	.0006276	1	.	.0077584
educ	0	1	.	.0004294
children	.005012	1	0	.0162282
socrel	.0021617	1	.	.0110759
soccommun	.0046222	1	0	.0155663
socfrend	.0076071	1	.000549	.020426
socbar	0	1	.	.
health	.0837906	1	.0568123	.114001

Let's recap what we found. When using model 1 (which had no covariates), we found the difference in happiness between those who are and are not married is 0.39 and explains 3.64% of the variance in happiness (using adjusted R-squared, which is the equivalent of omega-squared). Using model 4, which we adjusted for demographics, socializing, and health, we found the difference in happiness between those who are and are not married is 0.31. In this covariate-adjusted model, being married (versus not married) explains 2.4% of the variance in happiness (using omega-squared). The covariates explain a portion of the differences in happiness due to marriage but not all of it.

6.5   Some technical details about adjusted means

This section provides some technical details about how adjusted means are computed using the `margins` command. It is very possible that you are not interested in the level of detail contained in this section, so I can summarize the contents by saying that the `margins` command supports two methods for adjusting for the effects of co-variates. When you are using linear models (like `anova` and `regress`), the adjusted means computed using these two different methods yield the exact same results. If you are computing adjusted means using such linear models, then you are probably not interested in the details of this section.

On the other hand, suppose you are using models such as logistic regression, Poisson regression, or any model in which the DV is modeled using a nonlinear link function of the predictors. When you use such nonlinear models, the concept of adjusted means from an ANOVA perspective becomes more complicated. The complication arises from the fact that there are two general methods that can be used to compute adjusted means. This section peeks under the hood of the `margins` command to learn more about how it computes adjusted means. I start with linear models, showing the different methods and their equivalence in the case of linear models. Then, we can see how the methods differ in the case of a nonlinear model, using logistic regression as an example.

Let's consider how we would compute adjusted means using the kind of formula we would see in an ANOVA textbook. We will see how this method is algebraically equivalent to a second method that Stata calls the "marginal value at the mean". We will then compare that with a third method called the "average marginal value".[5] These two sentences sound very similar, but the key to the difference is in terms of the order in which the covariates and fitted values were averaged. In short, the marginal value at the mean obtains the mean of the covariates and then uses the regression equation to compute fitted values (when the covariates are held constant at their mean). Compare this with the average marginal value, which computes the fitted value for each observation and then averages all the fitted values. This is explained in more detail below.

Let's return to the example from the previous section comparing the happiness of those who are married with that of those who are not married. For simplicity, let's begin with a model that includes a single covariate, health status. Below we see the ANCOVA, using marital status as the IV and health as the covariate.

```
. use gss2012_sbs, clear

. keep if !missing(happy7,married,health)
(695 observations deleted)

. regress happy7 i.married health
```

Source	SS	df	MS	Number of obs	=	1,279
				F(2, 1276)	=	90.20
Model	156.288821	2	78.1444103	Prob > F	=	0.0000
Residual	1105.43753	1,276	.866330351	R-squared	=	0.1239
				Adj R-squared	=	0.1225
Total	1261.72635	1,278	.987266314	Root MSE	=	.93077

happy7	Coef.	Std. Err.	t	P>|t|	[95% Conf. Interval]	
married						
Married	.3027384	.0527839	5.74	0.000	.1991857	.4062912
health	.2645904	.0234416	11.29	0.000	.2186022	.3105786
_cons	4.467503	.084339	52.97	0.000	4.302045	4.632962

---

5. The term "marginal value" might be foreign. Think of it as the fitted value from the regression model.

The `margins` command yields the following adjusted means.

```
. margins married

Predictive margins                              Number of obs    =      1,279
Model VCE      : OLS

Expression    : Linear prediction, predict()

                          Delta-method
                Margin   Std. Err.      t    P>|t|    [95% Conf. Interval]

      married
    Unmarried   5.368435  .0351924  152.55  0.000     5.299393    5.437476
      Married   5.671173  .0390368  145.28  0.000      5.59459    5.747757
```

The `tabstat` command below shows the overall mean of `happy7` and `health` and the means by marital status. These values will be used for manually computing the adjusted means below.

```
. tabstat happy7 health, by(married)

Summary statistics: mean
  by categories of: married (marital: married=1, unmarried=0 (recoded))

    married │    happy7    health

  Unmarried │  5.333333   3.27234
    Married │  5.714286  3.567944

      Total │    5.5043  3.405004
```

Let's manually compute the adjusted means by marital status. There are a number of different ways that we can compute the adjusted means (and obtain the exact same answer). Three methods are illustrated below.

6.5.1 Computing adjusted means: Method 1

Let's consider how we would manually compute the adjusted means of happiness using the kind of formula we would find in a textbook discussing ANCOVA. Let's start by defining some symbols and terms and associating them with the values computed previously.

- $\overline{Y}_{\text{Unmarried}}$—The (unadjusted) mean of the DV among those who are unmarried (that is, 5.333333).
- $\overline{Y}_{\text{Married}}$—The (unadjusted) mean of the DV among those who are married (that is, 5.714286).
- $\overline{X}_{\text{Unmarried}}$—The mean of the covariate among those who are unmarried (that is, 3.27234).

- $\overline{X}_{\texttt{Married}}$—The mean of the covariate among those who are married (that is, 3.567944).
- $\overline{X}_{\text{Overall}}$—The (overall) mean of the covariate (that is, 3.405004).
- B—The slope term for the covariate from `regress` (that is, 0.2645904).

We can then compute the adjusted mean of the DV, $\overline{Y}'_{\texttt{Unmarried}}$, as shown in (6.1) below.

$$\overline{Y}'_{\texttt{Unmarried}} = \overline{Y}_{\texttt{Unmarried}} - B(\overline{X}_{\texttt{Unmarried}} - \overline{X}_{\text{Overall}}) \qquad (6.1)$$

The `display` command uses (6.1) to compute the adjusted mean of the DV for those who are unmarried.

```
. display 5.333333 - .2645904*(3.27234 - 3.405004)
5.3684346
```

We can then compute the adjusted mean of the DV, $\overline{Y}'_{\texttt{Married}}$, as shown in (6.2) below.

$$\overline{Y}'_{\texttt{Married}} = \overline{Y}_{\texttt{Married}} - B(\overline{X}_{\text{Married}} - \overline{X}_{\text{Overall}}) \qquad (6.2)$$

The `display` command uses (6.2) to compute the adjusted mean of the DV for those who are married.

```
. display 5.714286 - .2645904*(3.567944 - 3.405004)
5.6711736
```

6.5.2   Computing adjusted means: Method 2

The second method we will illustrate is what Stata calls the "marginal value at the mean". This method, which is algebraically equivalent to method 1, uses the regression model for computing the adjusted means. Let's first define some symbols and terms and associate them with their values computed above.

- B_0 - The intercept term from the `regress` command (that is, 4.467503).
- $B_{\texttt{Married}}$ - The coefficient associated with the IV `married` from the `regress` command (that is, 0.3027384).
- $B_{\texttt{Health}}$ - The coefficient associated with the covariate `health` from the `regress` command (that is, 0.2645904).
- $\overline{\texttt{Health}}$ - The overall mean of the covariate `health` (that is, 3.405004).

The adjusted mean for unmarried people, $\overline{Y}'_{\texttt{Unmarried}}$, is obtained by computing the fitted value of the regression model for unmarried people while holding each covariate (that is, health) at the mean, as shown in (6.3) below.

$$\overline{Y}'_{\texttt{Unmarried}} = B_0 + B_{\texttt{Married}} \times 0 + B_{\texttt{Health}} \times \overline{\texttt{Health}} \qquad (6.3)$$

We can then use the `display` command to compute the adjusted mean of the DV, $\overline{Y}'_{\text{Unmarried}}$, as shown below. (Note that the coefficient for `married` is multiplied by 0, because this is the adjusted mean for the unmarried group.)

```
. display 4.467503 + .3027384*0 + .2645904*3.405004
5.3684344
```

The adjusted mean for married people, $\overline{Y}'_{\text{Married}}$, is obtained by computing the fitted value of the regression model for married people while holding each covariate (that is, health) at the mean, as shown in (6.4) below.

$$\overline{Y}'_{\text{Married}} = B_0 + B_{\text{Married}} \times 1 + B_{\text{Health}} \times \overline{\text{Health}} \tag{6.4}$$

We can then use the `display` command to compute the adjusted mean of the DV, $\overline{Y}'_{\text{Married}}$, as shown below.

```
. display 4.467503 + .3027384*1 + .2645904*3.405004
5.6711728
```

If we use the `margins` command shown below, Stata computes the adjusted means using this method (method 2). Note that `at((mean) health)` is included to indicate that the variable `health` should be held constant at its mean. The output shows us that `health` has been held constant at its mean, 3.405004.

```
. margins married, at((mean) health) nopvalues
Adjusted predictions                      Number of obs    =     1,279
Model VCE    : OLS

Expression   : Linear prediction, predict()
at           : health           =    3.405004 (mean)

-------------------------------------------------------------------
             |            Delta-method
             |    Margin   Std. Err.    [95% Conf. Interval]
-------------+-----------------------------------------------------
     married |
   Unmarried |  5.368435   .0351924     5.299393    5.437476
     Married |  5.671173   .0390368      5.59459    5.747757
-------------------------------------------------------------------
```

Stata produces the same adjusted means if we specified the overall mean of health ourselves using the following syntax:

```
. margins married, at(health=3.405004)
  (output omitted)
```

By default, the `margins` command uses a different approach, illustrated below.

6.5.3 Computing adjusted means: Method 3

The default method that Stata uses for obtaining adjusted means is by computing what they call the "average marginal value". Let's start by using the margins command to compute the adjusted means via the average marginal value method (the default method).

```
. margins married, nopvalues
Predictive margins                              Number of obs     =      1,279
Model VCE    : OLS

Expression   : Linear prediction, predict()

------------------------------------------------------------------------------
                     |            Delta-method
                     |    Margin   Std. Err.     [95% Conf. Interval]
---------------------+--------------------------------------------------------
             married |
           Unmarried |   5.368435   .0351924      5.299393    5.437476
             Married |   5.671173   .0390368       5.59459    5.747757
------------------------------------------------------------------------------
```

The average marginal value method uses the regression equation to compute the fitted value for each observation. The regression equation for this model (based on the results from the regress command) is repeated in (6.5).

$$\widehat{Y} = 4.467503 + 0.3027384 \times \texttt{married} + 0.2645904 \times \texttt{health} \tag{6.5}$$

To compute the adjusted mean for those who are unmarried, we set the value of married to 0, and the fitted values from the regression equation are computed for all observations. The replace command is used to assign a value of 0 to married for all observations. Then, the generate command creates a variable yhat_unmarried, which contains the fitted value for each observation.

```
. replace married = 0
(574 real changes made)
. generate yhat_unmarried = 4.467503 + .3027384*married + .2645904*health
```

The fitted values (stored in yhat_unmarried) are then averaged to yield the adjusted mean for those who are unmarried. The summarize command computes the mean of yhat_unmarried.

```
. summarize yhat_unmarried
    Variable |       Obs        Mean    Std. Dev.       Min        Max
-------------+--------------------------------------------------------
yhat_unmar~d |     1,279    5.368434    .2964415    4.732093   5.790455
```

The same procedure is used to compute the adjusted mean for those who are married by specifying that married equals one and then computing the predicted value for all observations. The `summarize` command averages across all observations to yield the adjusted mean for those who are married.

```
. replace married = 1
(1,279 real changes made)
. generate yhat_married = 4.467503 + .3027384*married + .2645904*health
. summarize yhat_married
     Variable |        Obs        Mean    Std. Dev.        Min         Max
--------------+------------------------------------------------------------
 yhat_married |      1,279    5.671173    .2964415    5.034832    6.093194
```

This is called the "average marginal value" because the marginal value is computed for each observation and then these marginal values are averaged.

I hope that you have noticed that the adjusted means that we obtained using method 2 (marginal value at the mean) and method 3 (average marginal value) are identical (within rounding error). This might lead you to wonder why Stata offers two different methods when they both lead to the same results. Actually, the two methods do not always lead to the same results. For instance, the two methods differ when using logistic regression. Although this book does not delve into logistic regression and other nonlinear models, I wanted to briefly show how the two methods of computing adjusted means can yield different results when using nonlinear models.

6.5.4   Differences between method 2 and method 3

We have seen that the adjusted mean that we get when using the marginal value at the mean (method 2) is identical to the result computed using the average marginal value (method 3). This is true for models that model the DV as a linear function of the predictors (for example, ANOVA, regression). However, these two methods produce different results when the predicted value is a nonlinear function of the predictors. For example, predicted probabilities from a logistic regression model are a nonlinear function of the predictors. Let's illustrate this with a simple example using a logistic regression model in which the DV is a 0 or 1, indicating if the person is very happy (1 = very happy), the IV is `married`, and the covariate is `health`. As we see below, the odds of a married person being very happy is 2.31 times greater than the odds of an unmarried person being very happy.

```
. use gss2012_sbs, clear

. logit vhappy i.married health, or
Iteration 0:   log likelihood = -785.24809
Iteration 1:   log likelihood = -716.88983
Iteration 2:   log likelihood = -715.19614
Iteration 3:   log likelihood = -715.19151
Iteration 4:   log likelihood = -715.19151
Logistic regression                          Number of obs     =       1,287
                                             LR chi2(2)        =      140.11
                                             Prob > chi2       =      0.0000
Log likelihood = -715.19151                  Pseudo R2         =      0.0892
```

vhappy	Odds Ratio	Std. Err.	z	P>|z|	[95% Conf. Interval]	
married						
Married	2.314472	.2989977	6.50	0.000	1.796753	2.981366
health	1.712151	.1063081	8.66	0.000	1.51597	1.93372
_cons	.0420623	.0102619	-12.99	0.000	.0260752	.0678514

Let's calculate the probability of a person being happy as a function of marital status using the marginal value at the mean (method 2) and the average marginal value (method 3).

First, we compute the probability of a person being happy as a function of marital status using method 2. The results below show that the predicted probability of being happy for an unmarried person with average health is 0.207, compared with 0.377 for a married person with average health.

```
. * marginal value at the mean: (method 2)
. margins married, at((mean) health) nopvalues
Adjusted predictions                         Number of obs     =       1,287
Model VCE      : OIM

Expression     : Pr(vhappy), predict()
at             : health          =    3.399378 (mean)
```

	Margin	Delta-method Std. Err.	[95% Conf. Interval]	
married				
Unmarried	.2074138	.0156633	.1767143	.2381132
Married	.3772107	.0211886	.3356819	.4187396

Now let's compute the probability of being happy as a function of marital status using method 3. The results below show the predicted probability of being happy for those who are married is 0.224. This is computed by holding marital status constant at 0 (unmarried), computing the predicted probability of being happy for each person in the sample, and then averaging these predicted probabilities. The average of these predicted probabilities is 0.224. Repeating this process, holding marital status constant at 1 (married), yields an average predicted probability of being happy of 0.387.

```
. * average marginal value: (method 3)
. margins married, nopvalues

Predictive margins                           Number of obs    =     1,287
Model VCE    : OIM

Expression   : Pr(vhappy), predict()

                          Delta-method
                Margin    Std. Err.      [95% Conf. Interval]

      married
   Unmarried    .2238518   .0155295      .1934144    .2542891
     Married    .3871423   .0194666      .3489883    .4252962
```

As you can see, the `margins` command yields different predicted probabilities when using method 2 versus using method 3.

6.5.5   Adjusted means: Summary

As we have seen in this section, Stata offers two methods for computing adjusted means. When using linear models like `anova` or `regress`, we do not need to be very concerned about these different methods: they yield the exact same value for the adjusted means. But when we use logistic regression, the marginal value at the mean differs from the average marginal value because the probability that the DV equals one is not a linear function of the predictors. In other words, we cannot simply generalize the concepts involved in computing adjusted means from linear models to nonlinear models.

6.6   Closing thoughts

This section has illustrated the use of ANCOVA with an example drawn from the tradition of experimental design and another drawn from the tradition of analyzing survey data. In both cases, the mechanics of running the analyses are the same. However, there are a couple of key differences between these two types of analyses. First, although we can draw causal conclusions about the relationship between the IV and DV when using ANCOVA after analyzing a randomized study, we cannot make such conclusions when using ANCOVA when analyzing survey data. Second, if you are using ANCOVA to increase power for a randomized study, it is important to plan for the inclusion of the covariates as you consider the data to be collected as well as the sample size requirements for the study. See chapter 11 for more details about the use of covariates in the computation of sample size and power analysis.

7 Two-way factorial between-subjects ANOVA

7.1 Chapter overview

This chapter shows you how to perform two-way factorial between-subjects analysis of variance (ANOVA) using Stata. The chapter focuses on how to test for interactions, with a special emphasis on how to understand and dissect those interactions. A key feature of such interactions is the number of levels of each independent variable (IV) (that is, each between-subjects factor). In fact, this is such a key feature that interactions are often named after the number of levels for each IV. If both IVs have two levels, then the interaction is called a "two-by-two interaction", and the design is often called a "two-by-two design".

The ways that you can dissect an interaction depends on the number of levels of each IV. This chapter focuses on three types of interactions: two-by-two interactions, two-by-three interactions, and three-by-three interactions. These three types cover the most common interactions that researchers will encounter. Further, these interactions form the building blocks that can be used for analyzing larger interactions. The tools used to dissect two-by-three interactions generalize to two-by-X interactions. Likewise, the section on three-by-three interactions can be generalized to any kind of two-way interaction.

Four examples are presented in this chapter: one illustrates a two-by-two design (see section 7.2); two illustrate a two-by-three design (see sections 7.3.1 and 7.3.2); and one illustrates a three-by-three design (see section 7.4). For each design, I present each of the techniques that you could use for dissecting and understanding the interaction. You do not need to apply the techniques in the order in which they are presented, nor do you need to apply all the techniques that are illustrated. You can pick and choose the techniques that are most applicable to your study to dissect your interaction. Further, I recommend that you create an analysis plan (before examining your data) in which you describe the predicted pattern of results and how you plan to dissect the interaction to test for the exact pattern of results that you anticipate.

The examples in this chapter continue the hypothetical research studies we saw in chapter 4 in which a research psychologist, Professor Cheer is interested in increasing optimism. Her research focuses on the effectiveness of different kinds of therapy for increasing optimism. Her first study found that individuals receiving optimism therapy had greater optimism than those in a control group (see sections 4.2 and 4.3). Her second study added a third group, traditional therapy. This second study found that both optimism therapy and traditional therapy yielded greater optimism than the control. However, she was disappointed to find that optimism therapy was not significantly different from traditional therapy (see section 4.4).

After pondering the result of the last study, Professor Cheer believes that optimism might be more malleable for those with low levels of depression than for those with high levels of depression. If so, then optimism therapy might be more effective with those with low levels of depression. This chapter presents four studies aimed at understanding how the efficacy of optimism therapy depends on the level of depression at the start of the study.

The first example uses a two-by-two design with two levels of treatment (optimism therapy versus no therapy [control]) and two levels of depression status (*minimal/mild* versus *moderate/severe*). The second example extends this example by using a two-by-three design; it uses the same two levels of treatment (optimism therapy versus no therapy [control]) but including three levels of depression status (*minimal, mild/moderate,* and *severe*). The third example also uses a two-by-three design but instead focuses on three types of treatment (optimism therapy, traditional therapy, and no therapy [control]) and two levels of depression (*minimal/mild* versus *moderate/severe*). The fourth example illustrates a three-by-three design, including three levels of treatment (optimism therapy, traditional therapy, and no therapy [control]) and three levels of depression status (*minimal, mild/moderate,* and *severe*).

In each of these examples, the dependent variable (DV) is the optimism score of the person at the end of the study. Scores on this hypothetical optimism scale can theoretically range from 0 to 100, with a value of 50 representing the average optimism for people in general.

In discussing these designs, I sometimes use shorthand terms commonly used in ANOVA. For example, when I describe an IV and its levels, it can be useful to refer to the IV using a single letter and to the levels using a single letter followed by a number designating the level. Consider the variable depression status that has three levels: 1) minimally depressed, 2) mildly/moderately depressed, and 3) severely depressed. Using shorthand, I might refer to this as factor D (for depression). Also, instead of referring to the minimally depressed group, I might simply refer to that group as D1.

7.2   Two-by-two models: Example 1

In this first hypothetical study, two IVs are included, both with two levels. The first is type of therapy, comparing optimism therapy with a control. The second IV is depression status, comparing those with low levels of depression with those with high levels of depression. This study uses the (hypothetical) simple depression scale (SDS) to assess the initial depression of participants. The scores can range from 0 to 60. To aid in the interpretation of the SDS, the following imaginary cutoffs are commonly used.

- 0–9: *minimal* depression
- 10–19: *mild* depression
- 20–29: *moderate* depression
- 30–60: *severe* depression

Those who scored in the minimal/mild depression range (0 to 19) are categorized into group 1, labeled as *minimal/mild* depression; those who scored 20 and up are categorized into group 2, labeled as *moderate/severe* depression.[1] The focus of the study is on the interaction of treatment by depression status to determine whether optimism therapy is more effective for those with lower (that is, *minimal/mild*) depression than for those with greater (that is, *moderate/severe*) depression.

Tip! Treating depression continuously

In this chapter, I will be treating depression as a categorical variable. But you can see that it actually is measured along a continuum. Rather than categorizing depression, one can find it more beneficial to leave it in its original continuous form. Chapter 8 illustrates analyses using the continuous version of this depression scale and investigating interactions between treatment group assignment and the continuous measure of depression.

The dataset for this example is used below, and the first five observations are displayed. The variable `treat` indicates the treatment assignment, coded as follows: 1 = control (`Con`), 2 = optimism therapy (`OT`). The variable `depstat` reflects the person's depression status at the start of the study and is coded 1 = *minimal/mild*, which means that the person scored in the minimal/mild depression range (0 to 19) and 2 = *moderate/severe*, which means that the person scored in the moderate or severely depressed range (20 and up). The variable `opt` is the optimism score at the end of the study.

```
. use opt-2by2, clear
. list in 1/5
```

	id	treat	depstat	opt
1.	1	Con	Min/Mild	44.0
2.	2	OT	Min/Mild	63.0
3.	3	Con	Mod/Sev	50.0
4.	4	OT	Mod/Sev	39.0
5.	5	Con	Min/Mild	32.0

---

1. Note that I use the italics typeface when showing the labels for the levels of depression, such as *minimal/mild* depression or *moderate/severe* depression. I slant the labels to call attention to the fact that these levels constructed based on cutoff scores using the SDS. In later examples, I will illustrate a group that is composed of people with *minimal* depression, scoring between 0 and 9 on the SDS scale. This group will be described as those with *minimal* depression.

The `tabulate` command is used to show the frequencies broken down by `treat` and `depstat`. This illustrates the research design, showing that the study involved a total of 200 participants. Half of them (100) had *minimal/mild* depression and half had *moderate/severe* depression. Each of these 100 people was randomly assigned (in equal numbers) to the treatment or control group.

```
. tabulate depstat treat
```

Depression status	Treatment group Con	OT	Total
Min/Mild	50	50	100
Mod/Sev	50	50	100
Total	100	100	200

The `table` command is used to display the mean optimism broken down by treatment and depression status. For example, the mean optimism for those with *minimal/mild* depression in the control group was 46.6.

```
. table treat depstat, contents(mean opt)
```

Treatment group	Depression status Min/Mild	Mod/Sev
Con	46.6	41.8
OT	56.3	45.2

The effect of optimism therapy (as compared with the control group) can be assessed by comparing the mean optimism of the optimism therapy group with the mean optimism of the control group. For example, among those with *minimal/mild* depression, the effect of optimism therapy is 56.3 minus 46.6, or 9.7. Among those with *moderate/severe* depression, the effect of optimism therapy is 45.2 minus 41.8, or 3.4. Each of these effects is called simple effects—the simple effect of treatment (`OT` versus `Con`) for those with *minimal/mild* depression is 9.7, while the simple effect of treatment for those with *moderate/severe* depression is 3.4. The difference in these simple effects (that is, 9.7 versus 3.4) yields 6.3, which estimates the size of the interaction of treatment by depression status, telling us the extent to which the effect of treatment differs by depression status.

Let's visualize this pattern of results by graphing the mean optimism by treatment and depression status, as shown in figure 7.1 (I will show you how to make such a graph later).

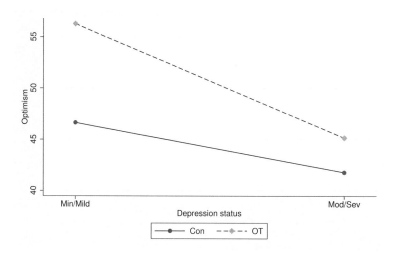

Figure 7.1: Mean optimism by treatment group and depression status

The graph shows us the effect of optimism therapy at each level of depression. The gap between the lines represents the effect of treatment (OT versus Con). We can see that the gap is greater for those with *minimal/mild* depression (56.3 minus 46.6, or 9.7) than for those with *moderate/severe* depression (45.2 minus 41.8, or 3.4).

Both the results of the table of means and the graph of the means are suggesting that optimism therapy might be more effective for those with *minimal/mild* depression than for those with *moderate/severe* depression. We can statistically assess this by running a two-by-two ANOVA and inspect the treat-by-depstat interaction.

Let's now run an analysis that predicts opt based on treatment assignment, depression status, and the interaction of these two variables. Note how the interaction is expressed by specifying depstat#treat.[2]

_____

2. Using shorthand, we could specify depstat##treat, which expands to depstat treat depstat#treat.

```
. anova opt depstat treat depstat#treat
                              Number of obs =          200    R-squared      =  0.2286
                              Root MSE      =      9.98476    Adj R-squared  =  0.2168

              Source |  Partial SS           df          MS           F      Prob>F

               Model |     5789.5             3      1929.8333      19.36    0.0000

             depstat |    3184.02             1      3184.02        31.94    0.0000
               treat |     2112.5             1      2112.5         21.19    0.0000
       depstat#treat |     492.98             1      492.98          4.94    0.0273

            Residual |   19540.32           196       99.69551

               Total |   25329.82           199      127.28553
```

The `depstat#treat` interaction is statistically significant $[F(1, 196) = 4.94, p = 0.0273]$. To begin to understand the nature of this interaction, we can compute the means broken down by treatment group and depression status by using the `margins` command.[3]

```
. margins depstat#treat, nopvalues
Adjusted predictions                              Number of obs     =         200
Expression    : Linear prediction, predict()
```

	Margin	Delta-method Std. Err.	[95% Conf. Interval]	
depstat#treat				
Min/Mild#Con	46.64	1.412059	43.85522	49.42478
Min/Mild#OT	56.28	1.412059	53.49522	59.06478
Mod/Sev#Con	41.8	1.412059	39.01522	44.58522
Mod/Sev#OT	45.16	1.412059	42.37522	47.94478

The average optimism for those with *minimal/mild* depression and in the `Con` group is 46.6, compared with 56.3 for those in the `OT` group. The effect of treatment for those with *minimal/mild* depression is 9.7 (56.3 minus 46.6).[4] The average optimism for those with *moderate/severe* depression and in the `Con` group is 41.8, compared with 45.2 for those in the `OT` group. The effect of treatment for those with *moderate/severe* depression is 3.4 (45.2 minus 41.8).

The significant interaction indicates that these effects are significantly different (that 9.7 significantly differs from 3.4). In other words, the effect of optimism therapy for

---

3. Note that the means are the same as those that we obtained from the `table` command earlier (except they are expressed to two decimal places). This is because there are no additional predictors (covariates) in the model. Had there been additional predictors in the model, the means from the `margins` command would have been adjusted for those predictors and would have differed from the means produced by the `table` command.

4. I have rounded these values to one decimal place to help relate these values to the means from the previous `table` command.

those with *minimal/mild* depression (9.7) is significantly different from the effect of optimism therapy for those with *moderate/severe* depression (3.4).

It can be easier to interpret the results by displaying them on a graph. The `marginsplot` command creates a graph of the means computed by the `margins` command. This creates a graph very similar to figure 7.1, which we saw previously.

```
. marginsplot
  Variables that uniquely identify margins: depstat treat
```

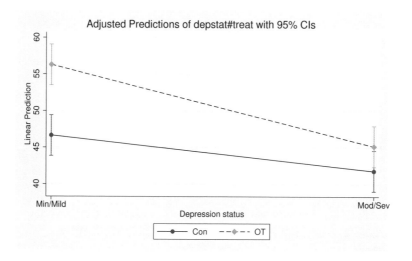

Figure 7.2: Mean optimism by treatment group and depression status

The graph in figure 7.2 illustrates how the effect of optimism therapy is greater for those with lower depression than for those with higher depression. It also helps us see the effect of optimism therapy at each level of depression status. It appears that among those with *moderate/severe* depression, the optimism therapy group may be no different from the control group. We can assess the effect of treatment at each level of depression through the use of simple effects analysis, described below.[5]

7.2.1 Simple effects

The significant interaction indicates that the effect of optimism therapy is different for those with *minimal/mild* depression versus those with *moderate/severe* depression.

---

5. It can be tempting to conclude that the difference between the optimism therapy group and control group among those with *moderate/severe* depression is not significant because the confidence intervals (CIs) for the two groups overlap. In the comparison between groups, overlapping CIs do not provide a proper test of the equality of the group means. A precise (and correct) method is to test the simple effect of treatment group assignment at the specified level of depression status (as illustrated in the following section). This issue is discussed further in section 7.6.

Each of these effects is called a "simple effect" because they reflect the effect of one variable while holding another variable constant. We can further probe the nature of this interaction by looking at the simple effect of optimism therapy separately for those with *minimal/mild* depression and for those with *moderate/severe* depression. Referring to figure 7.2, we see that these simple effects reflect the gaps between the lines at each level of depression status. We can estimate and test these simple effects using the `contrast` command. Note the @ symbol. This requests the simple effect of `treat` *at* each level of `depstat`.

```
. contrast treat@depstat
Contrasts of marginal linear predictions
Margins       : asbalanced
```

	df	F	P>F
treat@depstat			
Min/Mild	1	23.30	0.0000
Mod/Sev	1	2.83	0.0941
Joint	2	13.07	0.0000
Denominator	196		

The first test focuses on those who are in the *minimal/mild* depression group. For this group, the effect of optimism therapy is significant [$F(1, 196) = 23.30$, $p < 0.001$]. The second test shows that among those with *moderate/severe* depression, the effect of optimism therapy is not significant [$F(1, 196) = 2.83$, $p = 0.0941$].

Let's run this `contrast` command again but add the `pveffects` and `nowald` options.

```
. contrast treat@depstat, pveffects nowald
Contrasts of marginal linear predictions
Margins       : asbalanced
```

	Contrast	Std. Err.	t	P>|t|
treat@depstat				
(OT vs base) Min/Mild	9.64	1.996953	4.83	0.000
(OT vs base) Mod/Sev	3.36	1.996953	1.68	0.094

By adding the `pveffects` and `nowald` options, we see that the contrast command displays a table with the estimate of the simple effect, the standard error, and a significance test of the simple effect. This shows the effect of treatment for those with *minimal/mild* depression is 9.64, and this effect is significant [$t(196) = 4.83$, $p < 0.001$].[6] Among those with *moderate/severe* depression, the treatment effect is 3.36, and this difference is not significant [$t(196) = 1.68$, $p = 0.094$].

---

6. This significance test is reported as a t test but is equivalent to the F test we saw from the previous example.

Note! Contrast options

Combining the `pveffects` and `nowald` options provides a concise output that includes estimate of the size of the contrast and a test of its significance. Many examples in this chapter will incorporate these options with the `contrast` command because it produces a very concise output (which is good for keeping the book short). But this concise output omits things like confidence intervals, which are important. When you run these commands for yourself, you can omit these options. See section 21.2 for more about the `nowald` and `pveffects` options for customizing the output from the `contrast` command.

7.2.2 Estimating the size of the interaction

As we saw in the analysis of the simple effects, the simple effect of treatment for those with *moderate/severe* depression is 3.36 compared with 9.64 for those with *moderate/severe* depression. Taking the difference in these simple effects (3.36 minus 9.64) yields -6.28, which is an estimate of the size of the interaction. This is the same value that we obtain if we estimate the interaction effect using the `contrast` command below.

```
. contrast depstat#treat, nowald pveffects
Contrasts of marginal linear predictions
Margins       : asbalanced
```

	Contrast	Std. Err.	t	P>|t|
depstat#treat				
(Mod/Sev vs base) (OT vs base)	-6.28	2.824118	-2.22	0.027

This compares the optimism therapy group with the control group for those who have *moderate/severe* depression (group 2) versus those who have *minimal/mild* depression (group 1).

It might make more sense to compare *minimal/mild* depression (group 1) with those with *moderate/severe* depression (group 2) by applying the `r1b2` contrast operator to `depstat`.

```
. contrast r2b1.treat#r1b2.depstat, nowald pveffects
Contrasts of marginal linear predictions
Margins        : asbalanced
```

	Contrast	Std. Err.	t	P>|t|
treat#depstat (OT vs Con) (Min/Mild vs Mod/Sev)	6.28	2.824118	2.22	0.027

This provides the estimate of the size of the interaction effect, along with the standard error and test of the significance of the interaction. Note that the p-value for this test matches the p-value of the `treat#depstat` interaction from the original `anova` command.

Let's run this command again but use `cieffects` in place of the `pveffects` option, which now shows us a 95% confidence interval for the size of the interaction.

```
. contrast r2b1.treat#r1b2.depstat, nowald cieffects
Contrasts of marginal linear predictions
Margins        : asbalanced
```

	Contrast	Std. Err.	[95% Conf. Interval]	
treat#depstat (OT vs Con) (Min/Mild vs Mod/Sev)	6.28	2.824118	.7104411	11.84956

This allows us to quantify the extent to which optimism therapy is more effective for those with *minimal/mild* depression than for those with *moderate/severe* depression. We expect that optimism therapy will increase optimism among those with *minimal/mild* depression 6.3 units more than among those with *moderate/severe* depression. The 95% confidence interval for this estimate is $[0.71, 11.85]$.

7.2.3   More about interaction

Before concluding this section on two-by-two interactions, let's further explore what we mean by an interaction by considering the hypothetical pattern of results shown in figure 7.3, where there is no interaction.

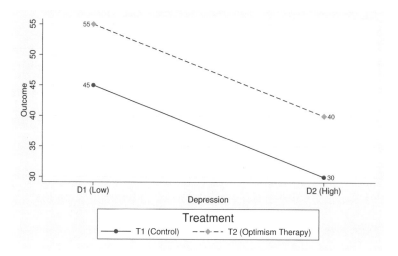

Figure 7.3: Example of a two-by-two design with no interaction

This is just one example pattern in which there is no interaction between treatment and depression status. One way we can see the absence of an interaction is by seeing that the line for T1 is parallel to the line for T2. What makes these lines parallel is the fact that the gap between T1 and T2 is the same at each level of depression. At D1, the difference between T1 and T2 is 10 (55 minus 45). At D2, the difference between T1 and T2 is also 10 (40 minus 30). The size of the interaction is the difference in these differences, 10 minus 10, or 0.

Another way to think about the absence of interaction is to think in terms of comparing the simple effect of depression for those in the control group with that for those in the optimism therapy group. The simple effect of depression (D2−D1) for those in the control group is −15 (30 minus 45). The simple effect of depression for those receiving optimism therapy is also −15 (40 minus 55). The interaction effect is the difference in these simple effects, 15 minus 15, or 0.

Video tutorial: Two-way ANOVA

See a video demonstration of a two-way ANOVA at http://www.stata.com/sbs/two-way-anova.

7.2.4 Summary

This section showed you how to perform an analysis involving a two-by-two factorial design. To help interpret the interaction, you can use the `margins` command to create a table of the DV means broken down by the IVs and the `marginsplot` command to create a graph of the means. Further, you can use the `contrast` command to test simple effects and to obtain an estimate of the size of the two-by-two interaction.

7.3 Two-by-three models

This section considers two-by-three designs where one of the IVs has two levels and the other IV has three levels. The examples from this section are an extension of the example from section 7.2.

7.3.1 Example 2

This example uses a two-by-three design with two levels of depression status (*minimal/mild* versus *moderate/severe*) and three levels of treatment (optimism therapy, traditional therapy, and no therapy [control]). In other words, this example extends the example from section 7.2 by adding traditional therapy as a third treatment group. The dataset for this example is used below.

```
. use opt-3by2
```

We use the `table` command to show the mean optimism by treatment group and depression status.

```
. table treat depstat, contents(mean opt)
```

Treatment group	Depression status Min/Mild	Mod/Sev
Con	47.1	42.4
TT	53.2	48.7
OT	56.8	45.6

Let's dive right in and analyze the data using the `anova` command with optimism as the DV and with depression status and treatment group as the IVs. The model also includes an interaction of the two IVs. The analysis shows that the `depstat#treat` interaction is significant [$F(2, 174) = 3.38$, $p = 0.0363$].

```
. anova opt depstat##treat

                          Number of obs =        180    R-squared     =  0.2701
                          Root MSE      =    8.03126    Adj R-squared =  0.2491

             Source |  Partial SS         df         MS          F      Prob>F
            --------+------------------------------------------------------------
              Model |   4152.7111          5    830.54222      12.88    0.0000

            depstat |     2080.8           1      2080.8        32.26    0.0000
              treat |   1636.0111          2    818.00556      12.68    0.0000
       depstat#treat |     435.9           2      217.95         3.38    0.0363

           Residual |    11223.2         174    64.501149
            --------+------------------------------------------------------------
              Total |   15375.911        179    85.898945
```

Let's use the **margins** command to show the mean of the DV by **treat** and **depstat**. This is followed by the **marginsplot** command to create a graph of these means, as shown in figure 7.4.

```
. margins depstat#treat, nopvalues

Adjusted predictions                          Number of obs     =        180
Expression    : Linear prediction, predict()

                           Delta-method
                  Margin    Std. Err.     [95% Conf. Interval]
             ------------+--------------------------------------------------
       depstat#treat |
        Min/Mild#Con |  47.06667    1.466301      44.17264     49.96069
         Min/Mild#TT |  53.23333    1.466301      50.33931     56.12736
         Min/Mild#OT |  56.83333    1.466301      53.93931     59.72736
         Mod/Sev#Con |  42.36667    1.466301      39.47264     45.26069
          Mod/Sev#TT |  48.73333    1.466301      45.83931     51.62736
          Mod/Sev#OT |  45.63333    1.466301      42.73931     48.52736
```

. marginsplot, noci
  Variables that uniquely identify margins: depstat treat

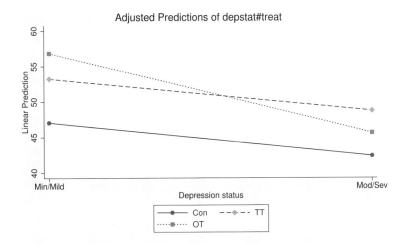

Figure 7.4: Mean optimism by treatment group and depression status

Let's inspect the graph of the means to begin to make sense of the interaction. If we focus on those with *minimal/mild* depression, it looks like traditional therapy and optimism therapy are both more effective than the control condition. Further, optimism therapy is a little bit better than traditional therapy, but this may not be significant.

Among those with *moderate/severe* depression, it looks like traditional therapy yields higher optimism scores than the control condition. The optimism scores for optimism therapy are between the scores for traditional therapy and the control condition.

Let's explore this further by looking at the effect of treatment group assignment separately for each level of depression.

Simple effects

To further understand this interaction, we can compute the simple effect of treatment at each level of depression status. This shows us whether the overall effect of treatment was significant at each level of depression status. The test of the simple effect of `treat` at each level of `depstat` is performed below using the `contrast` command.

```
. contrast treat@depstat
Contrasts of marginal linear predictions
Margins       : asbalanced
```

	df	F	P>F
treat@depstat			
Min/Mild	2	11.35	0.0000
Mod/Sev	2	4.71	0.0101
Joint	4	8.03	0.0000
Denominator	174		

Here the treatment effect is significant for those with *minimal/mild* depression $[F(2, 174) = 11.35, p < 0.001]$. The effect of treatment is also significant for those with *moderate/severe* depression $[F(2, 174) = 4.71, p = 0.0101]$. Let's further dissect these simple effects by applying contrasts to `treat`, forming simple contrasts.

Simple contrasts

We can follow up on these simple effects by performing simple contrasts on treatment at each level of depression status. For example, let's apply the `r.` contrast operator to `treat` to compare each group with the reference group (that is, comparing traditional therapy with the control and optimism therapy with the control). These comparisons are performed separately for those with *minimal/mild* depression and for those with *moderate/severe* depression.

```
. contrast r.treat@depstat, nowald pveffects
Contrasts of marginal linear predictions
Margins       : asbalanced
```

	Contrast	Std. Err.	t	P>|t|
treat@depstat				
(TT vs Con) Min/Mild	6.166667	2.073663	2.97	0.003
(TT vs Con) Mod/Sev	6.366667	2.073663	3.07	0.002
(OT vs Con) Min/Mild	9.766667	2.073663	4.71	0.000
(OT vs Con) Mod/Sev	3.266667	2.073663	1.58	0.117

The first two simple contrasts show that traditional therapy is significantly different from the control condition for those with *minimal/mild* depression $[t(174) = 2.97, p = 0.003]$ and for those with *moderate/severe* depression $[t(174) = 3.07, p = 0.002]$. The third and fourth simple contrasts show that optimism therapy is significantly different from the control condition for those with *minimal/mild* depression $[t(174) = 4.71, p < 0.001]$ but not significantly different for those with *moderate/severe* depression $[t(174) = 1.58, p = 0.117]$.

In summary, traditional therapy is significantly different from the control condition at each level of depression status. In fact, the output shows that the effect of traditional therapy (versus control) is similar across the two levels of depression—the effect is 6.17 for those with *minimal/mild* depression and is 6.37 for those with *moderate/severe* depression.

By contrast, the effect of optimism therapy (versus control) appears larger for those with *minimal/mild* depression (9.77) than for those with *moderate/severe* depression (3.27). We will investigate whether this is significant in the following section.

Partial interaction

Another way to dissect a three-by-two interaction is through a partial interaction. This applies a contrast to the three-level IV and interacts those contrasts with the two-level IV. Let's apply the `r.` contrast operator to the treatment IV. This yields two contrasts: group 2 versus 1 and group 3 versus 1 (that is, traditional therapy versus control and optimism therapy versus control). Interacting these contrasts with depression status forms two partial interactions. The first partial interaction forms a two-by-two interaction of treatment (traditional therapy versus control) by depression status (as pictured in the left panel of figure 7.5). The second partial interaction forms a two-by-two interaction of treatment group (optimism therapy versus control) by depression status (as pictured in the right panel of figure 7.5).[7]

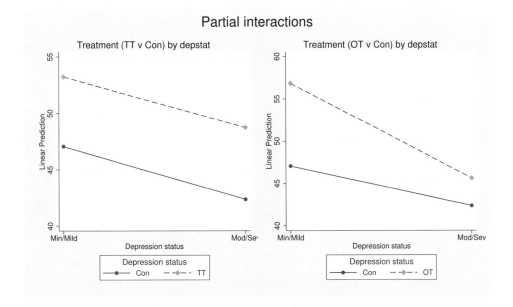

Figure 7.5: Partial interactions

---

7. Note that I manually created the graph shown in figure 7.5.

These partial interactions can be estimated by applying the `r.` contrast operator to `treat` and interacting that with `depstat`.

```
. contrast r.treat#depstat
Contrasts of marginal linear predictions
Margins     : asbalanced
```

	df	F	P>F
treat#depstat			
(TT vs Con) (joint)	1	0.00	0.9457
(OT vs Con) (joint)	1	4.91	0.0280
Joint	2	3.38	0.0363
Denominator	174		

The first partial interaction is not significant $[F(1, 174) = 0.00, p = 0.9457]$. This shows that the effect of traditional therapy (when compared with the control) is not significantly different for those with *minimal/mild* depression than for those with *moderate/severe* depression. This is illustrated in the left panel of figure 7.5.

The second partial interaction is significant $[F(1, 174) = 4.91, p = 0.0280]$. This shows that the effect of optimism therapy (when compared to the control) is significantly different for those with*minimal/mild* depression than for those with *moderate/severe* depression. This test is illustrated in the right panel of figure 7.5, which shows that optimism therapy is significantly more effective for those with *minimal/mild* depression than for those with *moderate/severe* depression.

Comparing optimism therapy with traditional therapy

The tests so far have focused on comparing each treatment (optimism therapy and traditional therapy) with the control condition. Let's say our interest was in comparing the effectiveness of optimism therapy (`OT`) with that of traditional therapy (`TT`) as a function of depression status. Let's start by comparing optimism therapy with traditional therapy at each level of depression. We specify this comparison via the `r3b2.treat` contrast, which compares group 3 of treatment (`OT`) with group 2 of treatment (`TT`).

```
. contrast r3b2.treat@depstat, nowald pveffects
Contrasts of marginal linear predictions
Margins     : asbalanced
```

	Contrast	Std. Err.	t	P>|t|
treat@depstat				
(OT vs TT) Min/Mild	3.6	2.073663	1.74	0.084
(OT vs TT) Mod/Sev	-3.1	2.073663	-1.49	0.137

The mean optimism is 3.6 units higher for the optimism therapy group versus the traditional therapy group for those with *minimal/mild* depression, but this difference is not significant [$t(174) = 1.74$, $p = 0.084$]. Among those with *moderate/severe* depression, mean optimism for optimism therapy versus traditional therapy is 3.1 units lower, and this difference is not significant [$t(174) = -1.49$, $p = 0.137$]. This pattern is visualized via the crossing lines for TT and OT in figure 7.6.

Let's further explore this by testing a partial interaction that interacts the comparison of optimism therapy versus traditional therapy by depression status. This partial interaction asks whether the difference in optimism for these two therapies is the same by depression status. This comparison is illustrated in figure 7.6.

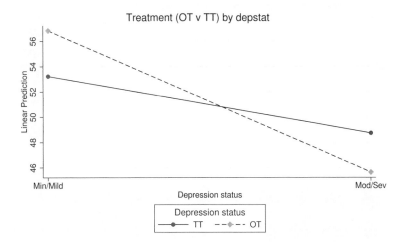

Figure 7.6: Partial interaction contrasting OT versus TT by depression status

We can perform the partial interaction visualized in figure 7.6 by specifying a contrast between optimism therapy and traditional therapy (r3b2.treat) interacted with depression status, as shown below.

```
. contrast r3b2.treat#depstat
Contrasts of marginal linear predictions
Margins       : asbalanced
```

	df	F	P>F
treat#depstat	1	5.22	0.0235
Denominator	174		

This partial interaction is significant [$F(1, 174) = 5.22$, $p = 0.0235$]. The effect of optimism therapy (versus traditional therapy) differs when depression is *minimal/mild* versus *moderate/severe* (that is, 3.6 versus −3.1). In other words, when we compare those with *minimal/mild* depression with those with *moderate/severe* depression, the effect of optimism therapy (versus traditional therapy) is significantly different.

7.3.2 Example 3

Referring back to the example from section 7.2, we see that there were two treatment groups (optimism therapy and control) and two levels of depression status (*minimal/mild* and *moderate/severe*). Suppose we instead use three categories for depression status: minimal depression (*minimal*), mild and moderate depression (*mild/moderate*), and severe depression (*severe*). This will allow us to better understand the effect of optimism therapy (versus the control) for those with *mild/moderate* depression compared with those with *severe* depression.

Let's begin by using the dataset for this example and showing a table of the mean of optimism broken down by treatment group and depression group.

```
. use opt-2by3, clear
. table treat depstat, contents(mean opt)
```

Treatment group	Depression group		
	Min	Mild/Mod	Sev
Con	49.6	43.1	40.4
OT	60.1	50.2	40.1

Let's look at the effect of optimism therapy (compared with the control) for each of the depression groups. For those with *minimal* depression, the effect of optimism therapy is 60.1 compared with 49.6 (or 10.5). Compare that with the effect of optimism therapy for those with *mild/moderate* depression (50.2 − 43.1 = 7.1) and for those with *severe* depression (40.1 − 40.4 = −0.3). This suggests that the effect of optimism therapy (compared with the control condition) may be greater for those with *minimal* depression than for those with *severe* depression. Likewise, optimism therapy may be more effective for those with *mild/moderate* depression than for those with *severe* depression. In other words, it appears that there may be a treatment group by depression group interaction.

Let's look at a graph of these means as shown in figure 7.7. This is another way to see that optimism therapy appears to be more effective for those with *minimal* versus *severe* depression and more effective for those with *mild/moderate* versus *severe* depression.

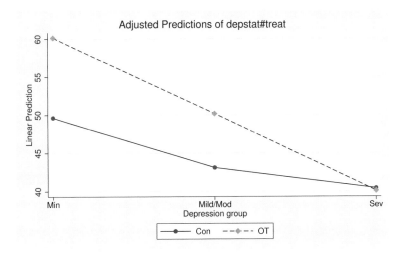

Figure 7.7: Mean optimism by treatment group and depression status

Let's now perform a two-by-three ANOVA with optimism as the DV and treatment group as a two-level IV and depression status as a three-level IV.

```
. anova opt depstat##treat
```

		Number of obs =	180	R-squared	=	0.4439
		Root MSE =	7.99438	Adj R-squared =		0.4279

Source	Partial SS	df	MS	F	Prob>F
Model	8877.3833	5	1775.4767	27.78	0.0000
depstat	6472.9	2	3236.45	50.64	0.0000
treat	1484.9389	1	1484.9389	23.23	0.0000
depstat#treat	919.54444	2	459.77222	7.19	0.0010
Residual	11120.367	174	63.910153		
Total	19997.75	179	111.71927		

As expected, the `depstat#treat` interaction is significant $[F(2, 174) = 7.19$, $p = 0.001]$. We can compute the mean optimism as a function of depression status and treatment group using the `margins` command below. Had there been additional predictors in the model, the `margins` command would have produced adjusted means, adjusting for the other predictors in the model.

```
. margins depstat#treat, nopvalues
Adjusted predictions                              Number of obs    =       180
Expression   : Linear prediction, predict()
```

	Margin	Delta-method Std. Err.	[95% Conf. Interval]	
depstat#treat				
Min#Con	49.63333	1.459568	46.7526	52.51407
Min#OT	60.13333	1.459568	57.2526	63.01407
Mild/Mod#Con	43.1	1.459568	40.21926	45.98074
Mild/Mod#OT	50.16667	1.459568	47.28593	53.0474
Sev#Con	40.4	1.459568	37.51926	43.28074
Sev#OT	40.06667	1.459568	37.18593	42.9474

The `marginsplot` command can be used to create a graph of the means created by the `margins` command. This produces the same graph that we saw in figure 7.7.

```
. marginsplot, noci
  (output omitted)
```

Now that we know the interaction is significant, we can say that the effect of optimism therapy (compared with the control) does differ as a function of depression status. One way to further understand this interaction is through tests of simple effects.

Simple effects

We can ask whether the effect of optimism therapy is significant at each level of depression status. In other words, we can test the simple effect of `treat` at each level of `depstat`. We test this using the `contrast` command.

```
. contrast treat@depstat, nowald pveffects
Contrasts of marginal linear predictions
Margins       : asbalanced
```

	Contrast	Std. Err.	t	P>|t|
treat@depstat				
(OT vs base) Min	10.5	2.064141	5.09	0.000
(OT vs base) Mild/Mod	7.066667	2.064141	3.42	0.001
(OT vs base) Sev	-.3333333	2.064141	-0.16	0.872

Optimism therapy is superior to the control condition for those with *minimal* depression $[t(174) = 5.09, p < 0.001]$ and for those with *mild/moderate* depression $[t(174) = 3.42, p = 0.001]$. For those with *severe* depression, the effect of optimism therapy (versus the control) is not significant $(t = -0.16, p = 0.872)$. Optimism therapy raises optimism (versus the control condition) by 10.5 units for those with *minimal* depression and by 7.1 units for those with *mild/moderate* depression.

This test of simple effects tells us that optimism therapy is significantly better than the control condition for those with *minimal* depression and with *mild/moderate* depression. For those with *severe* depression, optimism therapy is not significantly different from the control condition.

Partial interactions

Another way to dissect a two-by-three interaction is through partial interactions. In this example, a partial interaction can be constructed by applying a contrast operator to `depstat` and interacting that with `treat`. For example, applying the `r.` contrast operator to `depstat` yields two contrasts that compare groups: 2 versus 1 and 3 versus 1. Substantively, these contrasts compare *minimal* versus *mild/moderate* depression and *minimal* versus *severe* depression. Interacting `r.depstat` with `treat` forms two partial interactions. I used the means from the `margins` command to manually create graphs that illustrate these two partial interactions, shown in figure 7.8.

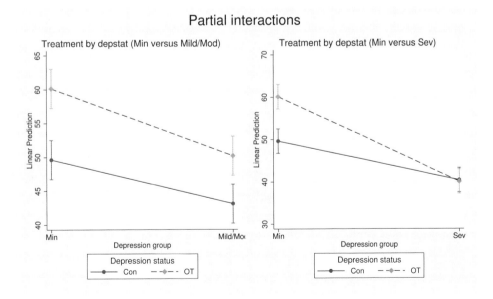

Figure 7.8: Partial interactions

The first partial interaction forms a two-by-two interaction of treatment by depression groups (*minimal* versus *mild/moderate*), as pictured in the left panel of figure 7.8. This tests whether the treatment effect is the same for those with *minimal* versus *mild/moderate* depression. The second partial interaction forms a two-by-two interaction of treatment group by depression groups (*minimal* versus *severe*), as pictured in the right panel of figure 7.8. This tests whether the treatment effect is the same for those with *minimal* depression versus those with *severe* depression. We can test these partial interactions using the `contrast` command.

```
. contrast r.depstat#treat
Contrasts of marginal linear predictions
Margins      : asbalanced
```

	df	F	P>F
depstat#treat			
(Mild/Mod vs Min) (joint)	1	1.38	0.2411
(Sev vs Min) (joint)	1	13.77	0.0003
Joint	2	7.19	0.0010
Denominator	174		

The first partial interaction is not significant [$F(1, 174) = 1.38$, $p = 0.2411$]. The effect of optimism therapy (compared with the control condition) is not significantly different for those with *minimal* depression versus *mild/moderate* depression. Referring to the left panel of figure 7.8, we can see that the effect of optimism therapy (compared with the control) is very similar for those with *minimal* depression and those with *mild/moderate* depression.

The second partial interaction is significant [$F(1, 174) = 13.77$, $p = 0.0003$]. The effect of optimism therapy (compared with the control) is significantly different when comparing those with *severe* depression versus those with *minimal* depression. Referring to the graph of the means in the right panel of figure 7.8, we can see that the effect of optimism therapy (compared with the control) is greater for those with *minimal* depression than for those with *severe* depression.

Let's investigate one further partial interaction that tests the interaction of treatment by depression group (*mild/moderate* versus *severe*). This will help us understand whether the effectiveness of optimism therapy differs between those with *mild/moderate* depression and those with *severe* depression. This partial interaction is tested below by applying the `r2b3` contrast operator to `depstat`[8] and interacting that with `treat`.

---

8. This compares groups 2 and 3, comparing *mild/moderate* depression with *severe* depression.

```
. contrast r2b3.depstat#treat
Contrasts of marginal linear predictions
Margins      : asbalanced
```

	df	F	P>F
depstat#treat	1	6.43	0.0121
Denominator	174		

This test is significant $[F(1, 174) = 6.43, p = 0.0121]$. The effectiveness of optimism therapy is significantly different for those with *mild/moderate* depression than for those with *severe* depression. We can estimate the magnitude of the difference with the `margins` command.

```
. margins r2b3.depstat#treat, contrast(cieffects nowald)
Contrasts of adjusted predictions
Expression   : Linear prediction, predict()
```

	Contrast	Delta-method Std. Err.	[95% Conf. Interval]	
depstat#treat (Mild/Mod vs Sev) (OT vs base)	7.4	2.919136	1.638527	13.16147

The increase in optimism for those with *mild/moderate* depression is 7.4 units greater than for those with *severe* depression (95% CI $= [1.6, 13.2]$).

7.3.3 Summary

This section has used two examples to illustrate interactions of two IVs where one variable has two levels and the other has three levels. The `margins` command creates a table of means broken down by the IVs, and then the `marginsplot` command can graph the means. The interaction can be further explored through tests of simple effects that look at the effect of one IV at each of the levels of the other IV. We can further explore the simple effects by performing simple contrasts that apply contrasts to one IV at each of the levels of the other IV. Finally, we saw that we can perform partial interaction tests by applying contrasts to the three-level IV and interacting those contrasts with the two-level IV.

7.4   Three-by-three models: Example 4

Let's now consider an example that illustrates a three-by-three design. Like the previous examples, the two IVs are the treatment group and depression group, and the DV is optimism. In this example, there are three levels of treatment (optimism therapy, traditional therapy, or no therapy [control]) and three levels of depression (*minimal, mild/moderate,* and *severe*).

The dataset for this analysis is used below, followed by a tabulation of the frequencies of participants by `depstat` and `treat`.

```
. use opt-3by3, clear
. tabulate depstat treat
Depression |      Treatment group
     group |     Con        TT        OT |    Total
-----------+---------------------------------+----------
       Min |      30        30        30 |       90
  Mild/Mod |      30        30        30 |       90
       Sev |      30        30        30 |       90
-----------+---------------------------------+----------
     Total |      90        90        90 |      270
```

There were a total of 270 participants in the study, of whom 90 had *minimal* depression before the start of the study, 90 had *mild/moderate* depression, and 90 had *severe* depression. Each of these 90 participants was randomly assigned (in equal numbers) to one of three treatments: optimism therapy, traditional therapy, or no therapy (control).

Let's use the `anova` command to perform a three-by-three ANOVA with `treat` and `depstat` as the IVs and `opt` as the DV.

```
. anova opt depstat##treat
                        Number of obs =      270    R-squared     =  0.3801
                        Root MSE      =  8.48124    Adj R-squared =  0.3611

          Source |    Partial SS        df         MS          F    Prob>F
-----------------+------------------------------------------------------------
           Model |    11511.341          8   1438.9176      20.00   0.0000
                 |
          depstat |    7880.9852          2   3940.4926      54.78   0.0000
            treat |    2294.0963          2   1147.0481      15.95   0.0000
    depstat#treat |    1336.2593          4   334.06481       4.64   0.0012
                 |
        Residual |     18774.1         261   71.931418
-----------------+------------------------------------------------------------
           Total |    30285.441        269   112.58528
```

The `depstat#treat` interaction is significant $[F(4, 261) = 4.64, p = 0.0012]$. Let's use the `margins` command to compute the mean optimism by `treat` and `depstat` and then use the `marginsplot` command to graph the means computed by the `margins` command (see figure 7.9). We can use this table and graph to help us interpret the `depstat#treat` interaction.

```
. margins depstat#treat, nopvalues
Adjusted predictions                              Number of obs    =        270
Expression     : Linear prediction, predict()
```

	Margin	Delta-method Std. Err.	[95% Conf. Interval]	
depstat#treat				
Min#Con	49.7	1.548455	46.65094	52.74906
Min#TT	55.1	1.548455	52.05094	58.14906
Min#OT	60.5	1.548455	57.45094	63.54906
Mild/Mod#Con	43.23333	1.548455	40.18428	46.28239
Mild/Mod#TT	50.06667	1.548455	47.01761	53.11572
Mild/Mod#OT	52.43333	1.548455	49.38428	55.48239
Sev#Con	40.33333	1.548455	37.28428	43.38239
Sev#TT	45.23333	1.548455	42.18428	48.28239
Sev#OT	40.03333	1.548455	36.98428	43.08239

```
. marginsplot, noci
    Variables that uniquely identify margins: depstat treat
```

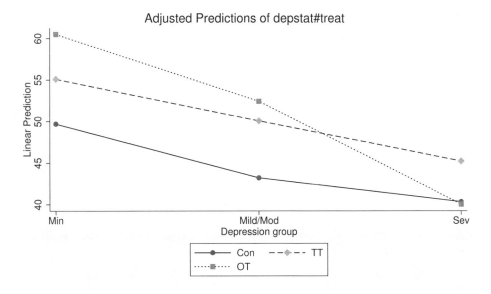

Figure 7.9: Mean optimism by treatment group and depression status

Let's focus on the comparison of the treatment groups among those with *minimal* depression. Optimism therapy yields the greatest optimism, followed by traditional therapy and then the control. A similar pattern of results is found for those with *mild/moderate* depression; however, the advantage of optimism therapy over traditional therapy appears to be reduced. The effect of treatment is very different among those with severe depression—traditional therapy has higher optimism scores than optimism therapy and the control condition, both of which have similar levels of optimism.

We can statistically dissect this interaction in four different ways, using simple effects (see section 7.4.1), simple contrasts (see section 7.4.2), partial interactions (see section 7.4.3), and interaction contrasts (see section 7.4.4). Each of these techniques is illustrated below.

7.4.1   Simple effects

One way to dissect the interaction is by looking at the effect of treatment at each level of depression status. This is performed using the `contrast` command.

```
. contrast treat@depstat
Contrasts of marginal linear predictions
Margins        : asbalanced
```

	df	F	P>F
treat@depstat			
Min	2	12.16	0.0000
Mild/Mod	2	9.52	0.0001
Sev	2	3.55	0.0300
Joint	6	8.41	0.0000
Denominator	261		

Three tests are performed showing the effect of `treat` at each of the three levels of `despstat`. These results show that the effect of `treat` is significant at each level of `depstat`. We can further dissect the simple effects by applying contrasts to `treat`, yielding simple contrasts.

7.4.2   Simple contrasts

Let's repeat the previous `contrast` command but apply the `r.` contrast operator to `treat`. This yields a comparison of each treatment group with the reference group (that is, group 1, the control group) at each level of depression status. To keep the output simple, we will focus on those with *minimal* depression (by specifying `1.depstat`).

```
. contrast r.treat@1.depstat, nowald pveffects
Contrasts of marginal linear predictions
Margins      : asbalanced
```

	Contrast	Std. Err.	t	P>|t|
treat@depstat				
(TT vs Con) Min	5.4	2.189847	2.47	0.014
(OT vs Con) Min	10.8	2.189847	4.93	0.000

For those with *minimal* depression, the comparison of traditional therapy with control is significant $[t(261) = 2.47, p = 0.014]$. The mean optimism for those receiving traditional therapy is 5.4 units higher than that for those in the control group. The comparison of optimism therapy versus control is also significant $[t(261) = 4.93, p < 0.001]$. The mean optimism score is 10.8 units higher for those receiving optimism therapy than for those in the control group.

Let's now perform these simple contrasts but for those with *mild/moderate* depression (by specifying 2.depstat).

```
. contrast r.treat@2.depstat, nowald pveffects
Contrasts of marginal linear predictions
Margins      : asbalanced
```

	Contrast	Std. Err.	t	P>|t|
treat@depstat				
(TT vs Con) Mild/Mod	6.833333	2.189847	3.12	0.002
(OT vs Con) Mild/Mod	9.2	2.189847	4.20	0.000

Both of these contrasts are significant. The mean optimism is 6.8 units higher for those in the traditional therapy group compared with those in the control group $[t(261) = 3.12, p = 0.002]$. Also, the mean optimism is 9.2 units higher for the comparison of optimism therapy with the control $[t(261) = 4.20, p < 0.001]$.

Let's now perform these simple contrasts but for those with *severe* depression (by specifying 3.depstat).

```
. contrast r.treat@3.depstat, nowald pveffects
Contrasts of marginal linear predictions
Margins      : asbalanced
```

	Contrast	Std. Err.	t	P>|t|
treat@depstat				
(TT vs Con) Sev	4.9	2.189847	2.24	0.026
(OT vs Con) Sev	-.3	2.189847	-0.14	0.891

The mean optimism for those in the traditional therapy group is 4.9 units higher than that for those in the control group $[t(261) = 2.24, p = 0.026]$. The optimism therapy group shows no significant difference in optimism compared with the control group $[t(261) = -0.14, p = 0.891]$.

Let's now consider the use of partial interactions as a means of further understanding the `treat#depstat` interaction.

7.4.3 Partial interaction

The three-by-three interaction can also be further understood through a partial interaction. A contrast is applied to one of the IVs, and those contrasts are interacted with the other three-level IVs.

For example, applying the `r.` contrast operator to `treat` yields two contrasts: group 2 versus 1 (traditional therapy versus control) and group 3 versus 1 (optimism therapy versus control). We can then interact these contrasts with depression status. This decomposes the overall three-by-three interaction into two three-by-two interactions. I used the means from the `margins` command to create a visual depiction of these two partial interactions, shown in figure 7.9. The left panel of figure 7.9 illustrates the comparison of treatment group 2 versus 1 (`TT` versus `Con`) interacted with depression status. The right panel illustrates the comparison of treatment group 3 versus 1 (`OT` versus `Con`) interacted with depression status.

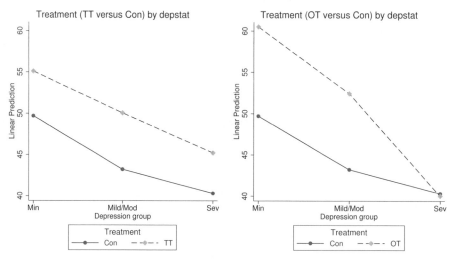

Figure 7.10: Partial interactions

The `contrast` command below tests the two partial interactions pictured in figure 7.10.

```
. contrast r.treat#depstat
Contrasts of marginal linear predictions
Margins        : asbalanced
```

	df	F	P>F
treat#depstat			
(TT vs Con) (joint)	2	0.21	0.8107
(OT vs Con) (joint)	2	7.51	0.0007
Joint	4	4.64	0.0012
Denominator	261		

The first partial interaction is not significant $[F(2, 261) = 0.21, p = 0.8107]$. The difference in optimism between the traditional therapy group and the control group does not differ among the levels of depression status. We can see this in the left panel of figure 7.10. The effect of traditional therapy (versus control) is very similar at each level of depression.

The second partial interaction is significant $[F(2, 261) = 7.51, p = 0.0007]$. The difference in optimism therapy group and the control group depends on the level of depression. Looking at the right panel of figure 7.10, we see that the effect of optimism therapy (compared with the control) may be similar for those with *minimal* depression and with *mild/moderate* depression but different for those with *severe* depression. We can investigate these differences using interaction contrasts, illustrated in the next section.

7.4.4   Interaction contrasts

An interaction contrast is formed by applying contrasts to both of the IVs and then interacting the resulting contrasts. Suppose we applied the `r.` contrast operator to the treatment group and the `r.` contrast operator to depression status. This would create contrasts of each treatment against the control group (that is, 2 versus 1 and 3 versus 1) interacted with contrasts of level of depression versus those with *minimal* depression (that is, 2 versus 1 and 3 versus 1). This yields a total of four interaction contrasts. I have created a visual representation of these four interaction contrasts based on the means from the `margins` command, pictured in figure 7.11.

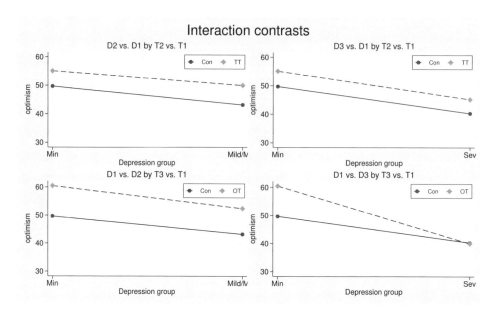

Figure 7.11: Interaction contrasts

Each of the panels of figure 7.11 forms a two-by-two interaction, selecting two levels of the treatment IV to be interacted with two levels of the depression IV. For example, the top left panel interacts the comparison of traditional therapy with the control by *mild/moderate* depression versus *minimal* depression. In the case of a three-by-three design, an interaction contrast dissects the interaction into four pieces. You can select the contrast operators for each IV to form comparisons that are of interest to you. In this example, we are applying the `r.` contrast operator to the treatment IV and the `r.` contrast operator to the depression status IV. The `contrast` command below tests each of these interaction contrasts.

```
. contrast r.treat#r.depstat, nowald pveffects
Contrasts of marginal linear predictions
Margins       : asbalanced
```

	Contrast	Std. Err.	t	P>|t|
treat#depstat				
(TT vs Con) (Mild/Mod vs Min)	1.433333	3.096911	0.46	0.644
(TT vs Con) (Sev vs Min)	-.5	3.096911	-0.16	0.872
(OT vs Con) (Mild/Mod vs Min)	-1.6	3.096911	-0.52	0.606
(OT vs Con) (Sev vs Min)	-11.1	3.096911	-3.58	0.000

The four tests created by the `contrast` command correspond to the top left, top right, bottom left, and bottom right panels of figure 7.11 (that is, the panels are read left to right, then top to bottom).

Let's start by interpreting the fourth interaction contrast, the only one that was significant $[t(261) = -3.58, p < 0.001]$. This contrast, pictured in the bottom right panel of figure 7.11, is just like a two-by-two interaction where treatment has two levels (control versus optimism therapy) and depression status has two levels (*severe* versus *minimal* depression). The fact that this interaction contrast is significant indicates that the effect of optimism therapy (compared with the control) is different for those with *minimal* versus *severe* depression. On the basis of the pattern of means we see in figure 7.11, we can say that optimism therapy is more effective for those with *minimal* depression than for those with *severe* depression.

Now let's return to interpreting the first, second, and third contrasts. The first contrast shows that the effect of traditional therapy (versus control) is not significantly different for those with *mild/moderate* depression versus those with *minimal* depression $[t(261) = 0.46, p = 0.644]$. This is consistent with the pattern of means shown in the top left panel of figure 7.11.

The second contrast (which is depicted in the top right panel) focuses on the comparison of traditional therapy with the control interacted with the comparison of *severe* depression versus *minimal* depression. This interaction contrast is not significant $[t(261) = -0.16, p = 0.872]$. The effect of traditional therapy (versus control) does not differ when comparing those with *severe* versus *minimal* depression.

The third contrast (depicted in the bottom left panel) shows that the effect of optimism therapy (versus the control) for those with *minimal* versus *mild/moderate* depression. This contrast is not significant $[t(261) = -0.52, p = 0.606]$. The effect of optimism therapy versus the control is not significantly different when comparing those with *mild/moderate* depression with those with *minimal* depression.

7.4.5   Summary

This section has illustrated how you can interpret interactions that involve two IVs that both have three levels. Actually, the principles illustrated here can be applied to designs where each of the IVs has three or more levels. The interpretation of the interaction begins with using the `margins` command to compute the means as a function of the IVs followed by the `marginsplot` command to create a graph of the means. Then, four different techniques can be used for dissecting the interaction: simple effects, simple contrasts, partial interactions, and interaction contrasts.

7.5   Unbalanced designs

The examples presented in this chapter have illustrated balanced designs. That is, the number of observations was the same for all the cells. Even in the context of a randomized experiment, it is unusual to have exactly the same number of observations in each cell. This section presents an example of an unbalanced design. This will allow us to consider two different strategies that can be used for estimating adjusted means, the "as observed" strategy and the "as balanced" strategy. As we will see, the `asbalanced` option can be used with the `margins` command to estimate margins as though the design were balanced, even if the actual design is not balanced.

For this example, let's use the General Social Survey dataset with the variable `happy7` as the DV. This Likert variable contains the respondent's rating of his or her happiness, where 1 represents Completely unhappy and 7 represents Completely happy. Let's predict happiness based on whether the respondent is married (`married`), whether the respondent is a college graduate (`cograd`), and the interaction of these two variables. Because `married` has two levels and `cograd` has two levels, this yields a two-by-two design. Let's assess whether this is a balanced design by displaying the cell frequencies of `married` by `cograd` when `happy7` is not missing.

```
. use gss2012_sbs
. tabulate married cograd if !missing(happy7), row
```

Key
frequency
row percentage

marital: married=1, unmarried= 0 (recoded)	College graudate (1=yes, 0=no) Not CO Gr	CO Grad	Total
Unmarried	527 74.54	180 25.46	707 100.00
Married	378 65.51	199 34.49	577 100.00
Total	905 70.48	379 29.52	1,284 100.00

This tabulation clearly shows that this is not a balanced design. The number of observations is not equal in each of the cells. For example, there are 527 respondents who are unmarried and not college graduates, compared with 180 respondents who are unmarried and are college graduates. Note that this table also shows us that overall, 29.52% of the respondents were college graduates, and 70.48% were not college graduates. I will refer to these percentages when manually computing adjusted means later in this section.

Before performing the analysis, let's compute the mean of `happy7` by `married` and `cograd` using the `tabulate` command.

```
. tabulate married cograd, sum(happy7)

  Means, Standard Deviations and Frequencies of how happy R is (recoded)
```

marital: married=1, unmarried= 0 (recoded)	College graduate (1=yes, 0=no)		Total
	Not CO Gr	CO Grad	
Unmarried	5.2903226	5.4388889	5.3281471
	1.0596117	.90414248	1.0237062
	527	180	707
Married	5.6984127	5.7437186	5.7140381
	.97938844	.79114304	.91838503
	378	199	577
Total	5.4607735	5.5989446	5.5015576
	1.0458819	.85918563	.99609404
	905	379	1284

This shows the mean of happy7 for each cell of the two-by-two design. It also shows the row mean of happy7 for each level of marital. I will refer to these means to illustrate how adjusted means are computed in an unbalanced design.

Let's now perform an analysis that predicts happy7 from married, cograd, and the interaction of these two variables. This is performed using the anova command below.

```
. anova happy7 married##cograd
```

		Number of obs =	1,284	R-squared	=	0.0397
		Root MSE =	.977264	Adj R-squared =		0.0375

Source	Partial SS	df	MS	F	Prob>F
Model	50.539766	3	16.846589	17.64	0.0000
married	33.606465	1	33.606465	35.19	0.0000
cograd	2.4852641	1	2.4852641	2.60	0.1070
married#cograd	.70503257	1	.70503257	0.74	0.3904
Residual	1222.4571	1,280	.95504462		
Total	1272.9969	1,283	.99220334		

We can see that the married#cograd interaction is not significant ($p = 0.3904$). For the sake of this example, let's assume that we want to retain this interaction term, even though it is not significant. Let's now turn our attention to married, which is significant ($p < 0.001$). To understand this significant result, we will use the margins command to compute the adjusted means by the levels of married.

```
. margins married, nopvalues
Predictive margins                              Number of obs   =     1,284
Expression   : Linear prediction, predict()
```

		Delta-method		
	Margin	Std. Err.	[95% Conf. Interval]
married				
Unmarried	5.334175	.0369129	5.261759	5.406591
Married	5.711786	.040906	5.631536	5.792036

Note how the adjusted means from the `margins` command are similar to, but not exactly the same as, the row means from the `tabulate` command. Consider the adjusted mean for those who are not married, 5.334175. We can think of this adjusted mean as being computed by taking each cell mean of `happy7` among those who are not married and weighting it by the corresponding proportion of those who are college graduates and noncollege graduates, as illustrated below.

```
. display  5.2903226*.7048 + 5.4388889*.2952
5.3341794
```

The cell mean for unmarried noncollege graduates (5.29) is multiplied by the overall proportion of respondents who were noncollege graduates (0.7048). The cell mean for unmarried college graduates (5.44) is multiplied by the overall proportion of college graduates (0.2952). When these weighted means are added together, we obtain 5.3341794, the adjusted mean for those who are not married.

We can likewise compute the adjusted mean for those who are married using the same strategy, as shown below.

```
. display  5.6984127*.7048 + 5.7437186*.2952
5.711787
```

The key point is that the adjusted means are computed by creating a weighted average of cell means that is weighted by the observed proportions of observations in the data (in this case, the observed proportions of `cograd`). Stata calls this the "as observed" strategy. This is the default strategy, unless we specify otherwise. We can explicitly request this strategy by adding the `asobserved` option to the `margins` command, as shown below. This yields the same adjusted means we saw in the previous `margins` command.

```
. margins married, asobserved nopvalues
Predictive margins                              Number of obs    =      1,284
Expression    : Linear prediction, predict()
```

		Delta-method		
	Margin	Std. Err.	[95% Conf. Interval]	
married				
Unmarried	5.334175	.0369129	5.261759	5.406591
Married	5.711786	.040906	5.631536	5.792036

In some cases, we might want the adjusted means to be computed using an equal weighting of the cell means, as though the design had been balanced. In the context of this example, it would mean weighting college graduates and noncollege graduates equally.[9] This can be accomplished using the asbalanced option. In the example below, the adjusted means are computed as though the design were balanced.

```
. margins married, asbalanced nopvalues
Adjusted predictions                            Number of obs    =      1,284
Expression    : Linear prediction, predict()
at            : married                (asbalanced)
                cograd                 (asbalanced)
```

		Delta-method		
	Margin	Std. Err.	[95% Conf. Interval]	
married				
Unmarried	5.364606	.0421842	5.281848	5.447364
Married	5.721066	.0427954	5.637109	5.805023

These adjusted means reflect an equal weighting of college graduates and noncollege graduates. Let's illustrate this by manually computing the adjusted mean for those who are not married. The asbalanced adjusted mean for those who are unmarried can be computed by multiplying the cell mean for unmarried noncollege graduates by 0.5 and the cell mean of unmarried college graduates by 0.5. These equally weighted cell means are added together. This is illustrated using the display command below. This yields the asbalanced adjusted mean for those who are not married.

```
. display  5.2903226*.5 + 5.4388889*.5
5.3646057
```

---

9. Such a strategy may be desirable when analyzing the results from a designed experiment. For example, imagine that we have a variable that reflects experimental group assignment (for example, treatment versus control). In computing adjusted means, the experimenter may want the treatment and control groups to be weighted equally, even if there were differing numbers of participants in each group.

Likewise, we can manually compute the `asbalanced` adjusted mean for those who are married, as shown below.

```
. display  5.6984127*.5 + 5.7437186*.5
5.7210656
```

This only touches on the ways in which you can use the `asbalanced` and `asobserved` options. For example, you can specify that certain variables should be treated with `asbalanced`, while other variables should be treated with `asobserved`. For more details regarding these options, see `help margins`.

7.6   Interpreting confidence intervals

The `marginsplot` command displays adjusted means and confidence intervals that were computed from the most recent `margins` command. The graphs produced by the `marginsplot` command, especially those created in the context of the interaction of two IVs, can tempt you into falsely believing that the confidence intervals reflect comparisons between groups. Let's consider an example where this temptation can be very compelling. Let's return to the three-by-two example we saw in section 7.3.1. Below we use the dataset for this example and run the `anova` command with `opt` as the DV and `treat` and `depstat` as the IVs.

```
. use opt-3by2
. anova opt treat##depstat
```

	Number of obs =	180	R-squared =	0.2701		
	Root MSE =	8.03126	Adj R-squared =	0.2491		
Source	Partial SS	df	MS	F		Prob>F
Model	4152.7111	5	830.54222	12.88		0.0000
treat	1636.0111	2	818.00556	12.68		0.0000
depstat	2080.8	1	2080.8	32.26		0.0000
treat#depstat	435.9	2	217.95	3.38		0.0363
Residual	11223.2	174	64.501149			
Total	15375.911	179	85.898945			

The interaction of `treat` by `depstat` is significant ($p = 0.036$). Let's compute the adjusted means of optimism by `treat` by `depstat` using the `margins` command and then graph these adjusted means using the `marginsplot` command (see figure 7.12).

```
. margins treat#depstat

Adjusted predictions                              Number of obs    =        180

Expression    : Linear prediction, predict()
```

	Margin	Delta-method Std. Err.	t	P>|t|	[95% Conf. Interval]	
treat# depstat Con #						
Min/Mild	47.06667	1.466301	32.10	0.000	44.17264	49.96069
Con#Mod/Sev	42.36667	1.466301	28.89	0.000	39.47264	45.26069
TT#Min/Mild	53.23333	1.466301	36.30	0.000	50.33931	56.12736
TT#Mod/Sev	48.73333	1.466301	33.24	0.000	45.83931	51.62736
OT#Min/Mild	56.83333	1.466301	38.76	0.000	53.93931	59.72736
OT#Mod/Sev	45.63333	1.466301	31.12	0.000	42.73931	48.52736

```
. marginsplot
    Variables that uniquely identify margins: treat depstat
```

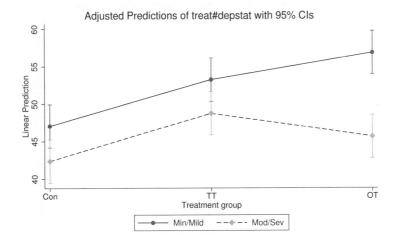

Figure 7.12: Means of optimism by treatment group and depression status

Consider the differences between the two depression status groups. Looking at the CIs in figure 7.12, you might be tempted to judge the differences between the depression groups based on the overlap (or lack of overlap) of the confidence intervals for the two groups. However, the most precise way to assess these differences is by formally testing the simple effects of depression status at each level of treatment, as shown below.

```
. contrast r.depstat@treat, nowald pveffects
Contrasts of marginal linear predictions
Margins         : asbalanced
```

	Contrast	Std. Err.	t	P>|t|
depstat@treat				
(Mod/Sev vs Min/Mild) Con	-4.7	2.073663	-2.27	0.025
(Mod/Sev vs Min/Mild) TT	-4.5	2.073663	-2.17	0.031
(Mod/Sev vs Min/Mild) OT	-11.2	2.073663	-5.40	0.000

Let's focus on the first contrast, which compares the two depression groups among the control group. This shows that the difference in optimism is significant ($p = 0.025$), even though the confidence intervals for these two groups overlap (see figure 7.12). Likewise, the test comparing the two depression groups among the traditional therapy group is also significant ($p = 0.031$), even though the confidence intervals for these two groups overlap.

In summary, the `marginsplot` command provides a graphical display of the results calculated by the `margins` command. Sometimes, the appearance of the confidence intervals of *individual* groups might tempt you to inappropriately make statistical inferences about the *comparisons* between the groups. To avoid this trap, you can directly form comparisons among the groups of interest to ascertain the significance of the group differences.

> **Video tutorial: Interactions**
>
> See a video illustrating interactions of factor variables in Stata at http://www.stata.com/sbs/interactions.

7.7   Closing thoughts

This chapter has considered three types of designs involving two IVs: two-by-two, three-by-two, and three-by-three designs. Following each design, you can use the `margins` command to obtain the means as a function of the IVs, and the `marginsplot` command can be used to display a graph of the means. You can also use a test of simple effects to assess the significance of one IV at each of the levels of the other IV.

In a two-by-three design, you can further perform simple contrasts by applying contrast coefficients to the three-level IV and examining the effect of the contrasts at each level of the two-level IV. You can also perform partial interaction tests by applying contrasts to the three-level IV and interacting that with the two-level IV. These techniques can be applied in the exact same way for four-by-two, five-by-two, and X-by-two designs.

In a three-by-three design, you can use all the techniques illustrated in the three-by-two design, namely, the use of simple effects, simple contrasts, and partial interactions. In a three-by-three design, you can also use interaction contrasts by applying contrasts to each of the three-level IVs and then interacting the contrasts.

I will reiterate a point I made at the start of the chapter. This chapter has illustrated a wide variety of methods for understanding and dissecting interactions. You do not need to apply all the methods illustrated in this chapter, and you do not need to apply them in the order in which they were illustrated. I would encourage you, before looking at your data, to describe the pattern of results that you expect and create an analysis plan (using the techniques described in this chapter) that will test for your predicted pattern of results.

8 Analysis of covariance with interactions

8.1 Chapter overview

This chapter explores analysis of covariance (ANCOVA) models that contain an interaction term between one of the independent variables (IVs) and the covariate. Let's start by considering the purpose of running an ANCOVA model with interactions by relating it to a previous example from chapter 7.

In section 7.3.2, we considered a two-by-three factorial design in which the two-level IV was treatment group assignment (optimism therapy versus control) and the three-level IV was depression status. The three levels of depression actually arose from a continuous measure of depression via the hypothetical simple depression scale (SDS). The scores on this scale can range from 0 to 60; the following imaginary cutoffs are commonly used to interpret the SDS scores.

- 0–9: minimal depression
- 10–19: mild depression
- 20–29: moderate depression
- 30–60: severe depression

In the example from section 7.3.2, the three categories of depression were formed based on the continuous SDS score using the following rules: 0 to 9 was assigned to depression group 1 (*minimal* depression); 10 to 29 was assigned to depression group 2 (*mild/moderate* depression); and 30 and higher was assigned to depression group 3 (*severely* depressed). The two-by-three analysis of variance (ANOVA) in section 7.3.2 found an interaction between treatment and depression. Exploration of the interaction found that optimism therapy significantly increased optimism (as compared with the control condition) for depression groups 1 and 2 (*minimal* and *mild/moderate* depression) but not for depression group 3 (*severe* depression). Figure 8.1 shows the means of the dependent variable (DV) (on the y axis) as a function of depression group (on the x axis) and treatment group (graphed as separate lines). This clearly shows how the effect of treatment is not significant for those with *severe* depression and how the effect of treatment is much greater for those with less depression.

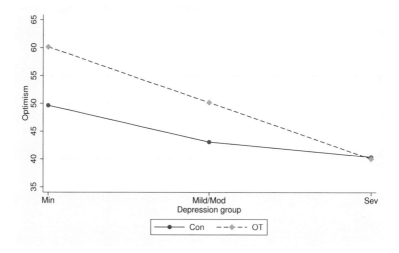

Figure 8.1: Mean optimism by treatment group and depression status

The limitations of this study are that we can compare only the treatment group with the control group for three depression groupings, even though these groupings were formed based on a continuous measure of depression.

Let's picture replacing the x axis in figure 8.1 with depression scores measured continuously. Further, imagine we fit a linear regression predicting optimism from depression for the treatment group and another such line for the control group.[1] Such a graph is shown in figure 8.2. Using the model shown in figure 8.2, we can ask whether the difference between the treatment and control groups is significant at any given level of depression.

---

1. This assumes the relationship between depression and optimism is linear for each group, an assumption we would assess using methods described in chapter 18.

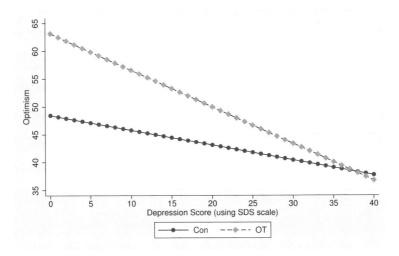

Figure 8.2: Predicted mean of optimism by depression and treatment group

The results pictured in figure 8.2 are from an ANCOVA model with an IV by covariate interaction. The DV is optimism, the IV is the treatment group assignment, and the covariate is the depression score. In this example, there is a significant interaction of the IV and the covariate (that is, treatment by depression). The significant treatment by depression interaction tells us that the effect of treatment (versus control) depends on depression status. This is evident in figure 8.2, where we can see that the effect of treatment (optimism therapy versus control) is stronger for lower levels of depression. A big advantage of this model is that we can estimate the effect of treatment at any given level of depression.

This chapter includes two examples of ANCOVA models with an interaction of the IV (treatment group assignment) and a covariate (depression status). The first example in this chapter (see section 8.2) is exactly the same as the example that I just described and portrayed in figure 8.2. In this example, optimism is the DV, treatment group assignment (optimism therapy versus control) is the IV, and the measure of depression (before the start of the study) is the covariate. The second example is similar to the first, except that the IV has three levels (optimism therapy, traditional therapy, and no therapy [control]).

Because of this emphasis on the regression aspect of the model, I will run these models using the `regress` command (which is described in part IV). Normally, we would want to check the regression model assumptions (such as checking that the relationship between the DV and the covariate is linear). For the sake of space, I won't thoroughly check such assumptions in this chapter.

> **Note: Related chapters**
>
> This chapter builds upon the ideas from the previous chapters (particularly chapters 6 and 7). In reading this chapter, you might also benefit from checking out chapter 10, which describes how to run ANOVA models via the `regress` command. Also, you might want to refer to part IV for more information on regression models, particularly chapter 10 about multiple regression and the `regress` command.

8.2   Example 1: IV has two levels

In the study shown in section 7.3.2, we found that optimism therapy increased optimism (as compared with a control condition) among those categorized as having *minimal* depression (SDS scores 0 to 9) and among those categorized as having *mild/moderate* depression (SDS scores 10 to 29). No significant difference was found for those categorized as having *severe* depression (SDS scores 30 and higher).

Professor Cheer decides that she wants to conduct a study in which the depression (SDS) scores are treated on a continuum. In this new study, she samples 200 people with a wide variety of depression levels, ensuring that the SDS depression scores represent the spectrum of depression. She randomly assigns half (100) of the people to receive optimism therapy and half (100) to the control condition. For this study, she has two focal research questions.

Question 1. Does the effect of optimism therapy (compared with the control) depend on the level of depression?

Question 2. For what range of depression scores is optimism therapy superior to the control condition.

The results of this study are contained in the dataset named `opt-ancova3.dta`. The dataset has four variables: `id`, `treat`, `depscore`, and `opt`.

```
. use opt-ancova3

. describe

Contains data from opt-ancova3.dta
  obs:            200
  vars:             4                              17 Jul 2015 15:24
  size:         3,200
```

variable name	storage type	display format	value label	variable label
id	float	%9.0g		Participant ID
treat	float	%9.0g	treat	Treatment group assignment
depscore	float	%9.0g		Depression Score (using SDS scale)
opt	float	%9.0g		Optimism score

```
Sorted by:
```

The variable **treat** reflects the treatment group assignment, showing that 100 were assigned to the control condition (group 1) and 100 assigned to optimism therapy (group 2). This is the IV for the study.

```
. fre treat

treat — Treatment group assignment
```

		Freq.	Percent	Valid	Cum.
Valid	1 Con	100	50.00	50.00	50.00
	2 OT	100	50.00	50.00	100.00
	Total	200	100.00	100.00	

The variable **depscore** is the depression score on the SDS scale measured before the study. The variable **opt** is the optimism at the end of the study. The variable **opt** is the DV and **depscore** is the covariate. Summary statistics for these two variables are shown below using the **summarize** command.

```
. summarize depscore opt
```

Variable	Obs	Mean	Std. Dev.	Min	Max
depscore	200	19.76	13.1685	0	59
opt	200	46.47	10.31168	19	78

With respect to the regression portion of this model (predicting optimism from depression), we should consider all the issues described in chapter 18 regarding the assumptions of regression. Let's focus on the issue of the linearity of the relationship between optimism and depression. Performing this check allows us to assess the assumption of linearity; it is also a stepping stone to testing the first hypothesis by visually considering the regression of depression on optimism by the two levels of treatment.

We will start by using the `graph twoway scatter` command to create a scatterplot of the relationship between depression and optimism, creating a separate scatterplot for the treatment and control groups. The `graph twoway` command creates the graph shown in figure 8.3 in which the left panel is the scatterplot for the control group and the right panel is the scatterplot for the optimism therapy group.

```
. graph twoway (scatter opt depscore), by(treat)
```

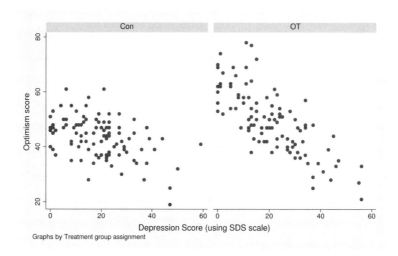

Figure 8.3: Scatterplot of optimism by depression for the control group (left panel) and optimism therapy group (right panel)

I can easily picture drawing a well-fitting regression line through the points in each of the panels shown in figure 8.3. Rather than imagining what such a linear fit would look like, we can overlay a linear fit to the scatterplots via the `lfit` command as used in the `graph` command below. This creates the graph shown in figure 8.4.

```
. graph twoway (scatter opt depscore) (lfit opt depscore), by(treat)
```

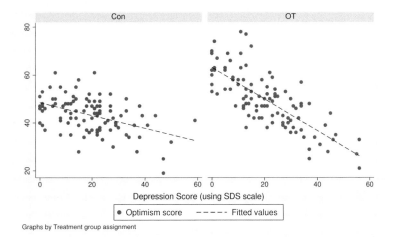

Figure 8.4: Scatterplot (with linear regression line) of optimism by depression for the control group (left panel) and optimism therapy group (right panel)

I find the graph in figure 8.4 to be very informative. The linear fit line provides a reasonable fit to the observed points.[2] Further, this shows a visual estimate of the slope for the optimism therapy group and the control group, suggesting that the slopes are not equal to each other. This suggests that there might be an interaction between treatment group assignment and depression. Let's test this by fitting an ANCOVA model with optimism as the DV, treatment group assignment as the IV, and depression status as the covariate. Further, we will include an interaction of treatment and depression status.

The following `regress` command fits the model that is described above by including `i.treat` for the effect of treatment and `c.depscore` for the covariate depression and connecting these two variables with the `##` operator to yield both the main effects and the interactions of the two variables.[3]

---

2. The regression diagnostic techniques illustrated in chapter 18 can be used to further assess the suitability of using linear regression to describe the relationship between the covariate and DV for each level of the IV.

3. If we wished to specify this model the long way, we could have run the command

    `.  regress opt i.treat c.depscore i.treat#c.depscore`

```
. regress opt i.treat##c.depscore
```

Source	SS	df	MS	Number of obs	=	200
				F(3, 196)	=	72.42
Model	11124.1773	3	3708.05909	Prob > F	=	0.0000
Residual	10035.6427	196	51.2022588	R-squared	=	0.5257
				Adj R-squared	=	0.5185
Total	21159.82	199	106.330754	Root MSE	=	7.1556

opt	Coef.	Std. Err.	t	P>|t|	[95% Conf. Interval]
treat						
OT	14.68956	1.831275	8.02	0.000	11.07803	18.3011
depscore	-.2679599	.056857	-4.71	0.000	-.38009	-.1558298
treat#						
c.depscore						
OT	-.3916125	.0774252	-5.06	0.000	-.5443059	-.2389191
_cons	48.44784	1.293407	37.46	0.000	45.89706	50.99862

Having fit this model, we will now focus on the first research question.

8.2.1   Question 1: Treatment by depression interaction

The first research question asks, Does the effect of optimism therapy (compared with the control) depend on the level of depression? We can answer this question by testing the interaction of `treat` by `depscore` using the `contrast` command, which tests the treatment by depression status interaction.

```
. contrast i.treat#c.depscore
Contrasts of marginal linear predictions
Margins      : asbalanced
```

	df	F	P>F
treat#c.depscore	1	25.58	0.0000
Denominator	196		

The interaction of `treat` by `depscore` is significant $[F(1, 196) = 25.58, p < 0.001]$. This indicates that the effect of `treat` differs based on `depscore`. To understand how the effect of treatment changes as a function of depression, let's create a graph similar to figure 8.2 that shows the adjusted mean of optimism as a function of depression score and treatment group.

Creating this graph is a two-step process. We first use the `margins` command to compute the predicted mean by `treat` when depression score is 0 and 40.[4] Then, we use the `marginsplot` command to graph the results from the `margins` command. The result is shown in figure 8.5.

```
. margins treat, at(depscore=(0 40))

Adjusted predictions                     Number of obs     =        200
Model VCE      : OLS

Expression     : Linear prediction, predict()
1._at          : depscore       =              0
2._at          : depscore       =             40
```

	Margin	Delta-method Std. Err.	t	P>|t|	[95% Conf. Interval]	
_at#treat						
1#Con	48.44784	1.293407	37.46	0.000	45.89706	50.99862
1#OT	63.1374	1.296406	48.70	0.000	60.58071	65.6941
2#Con	37.72944	1.394435	27.06	0.000	34.97942	40.47947
2#OT	36.75451	1.246886	29.48	0.000	34.29547	39.21354

```
. marginsplot, noci
  Variables that uniquely identify margins: depscore treat
```

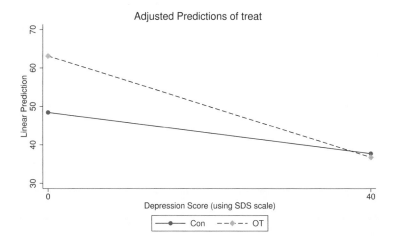

Figure 8.5: Predicted means of optimism by depression score and treatment group

---

4. Although depression scores range from 0 to 60, there are only 15 observations with depression scores greater than 40. Thus I focus on the range of scores that contains a reasonable number of observations.

The graph in figure 8.5 shows one line for those receiving optimism therapy and one line for those in the control group. Consistent with our finding of a significant `treat` by `depscore` interaction, the slopes are rather different for the optimism therapy group versus the control group. We can use the `dydx(depscore)` option with the following `margins` command to estimate the `depscore` slope for each level of `treat`.

```
. margins treat, dydx(depscore)

Average marginal effects                          Number of obs    =         200
Model VCE      : OLS

Expression     : Linear prediction, predict()
dy/dx w.r.t.   : depscore
```

	dy/dx	Delta-method Std. Err.	t	P>|t|	[95% Conf. Interval]	
depscore						
treat						
Con	-.2679599	.056857	-4.71	0.000	-.38009	-.1558298
OT	-.6595724	.0525542	-12.55	0.000	-.7632167	-.5559282

The output shows that the slope for the control group is -0.27 compared with -0.66 for the optimism therapy group. For the control group, every 1-unit increase in depression score is associated with a 0.27-unit decrease in optimism. This slope is significantly different from 0 $[t(196) = -4.71, p < 0.001]$. For the optimism therapy group, every 1-unit increase in depression is associated with a 0.66-unit decrease in optimism. This slope is also significantly different from 0 $[t(196) = -12.55, p < 0.001]$.

If we refer back to the `contrast` command, we see that test of the interaction indicated these slopes are significantly different from each other (that is, $F(1,196) = 25.58$, $p < 0.001$). We could also make such a comparison via the `margins` command by applying the `r.` contrast operator to `treat`, as shown below.

```
. margins r.treat, dydx(depscore)

Contrasts of average marginal effects
Model VCE    : OLS

Expression   : Linear prediction, predict()
dy/dx w.r.t. : depscore
```

	df	F	P>F
depscore			
treat	1	25.58	0.0000
Denominator	196		

	Contrast dy/dx	Delta-method Std. Err.	[95% Conf. Interval]	
depscore				
treat				
(OT vs Con)	-.3916125	.0774252	-.5443059	-.2389191

The output of the `margins` command yields the same F statistic that we saw from the `contrast` command, showing that the slopes of the two lines are significantly different from each other $F(1,196) = 25.58$, $p < 0.001$. The benefit of using the `margins` command is that it also provides an estimate of the difference between the two slopes (-0.39) and a 95% confidence interval (CI) for the difference in the slopes. We can be 95% confident that the difference in the slopes (for optimism therapy versus control) is between -0.54 and -0.24.[5]

Having established that the effect of optimism therapy (versus control) depends on depression, let's investigate the range of depression scores for which optimism therapy is superior to the control.

8.2.2   Question 2: When is optimism therapy superior?

Let's now consider the second research question; that is, For what range of depression scores is optimism therapy superior to the control condition? To that end, we will focus on the gap between the regression lines (shown in figure 8.5) because this reflects the effect of optimism therapy (versus the control condition). We can see that this gap is greatest for small values of `depscore`, and as depression increases, the gap decreases. Let's estimate the gap when depression status is zero by using the `margins` command. Note the use of the `at()` option to specify that we want to test the effects when `depscore` is zero.

---

5. You might have noticed that the lower portion of the output of the `contrast` command mirrors the output from the most recent `regress` command. This is because both of these portions of output represent the difference between the two slopes.

```
. margins r.treat, at(depscore=0) contrast(effects nowald)
Contrasts of adjusted predictions
Model VCE      : OLS

Expression     : Linear prediction, predict()
at             : depscore      =              0
```

	Contrast	Delta-method Std. Err.	t	P>|t|	[95% Conf. Interval]
treat (OT vs Con)	14.68956	1.831275	8.02	0.000	11.07803 18.3011

The difference between the mean of optimism for the optimism therapy group versus the control group is 14.7 points when the depression score is 0. The difference is statistically significant, $t = 8.02$, $p < 0.001$. The 95% confidence interval for this difference is 11.1 to 18.3.[6] Note that the 95% confidence interval excludes 0. Examining whether the 95% confidence interval includes or excludes 0 can be used as a method for determining whether the difference is significant at the 5% level of significance.

Let's repeat the above **margins** command using depression scores based on the commonly used cutoffs to demarcate *minimal*, *mild*, *moderate*, and *severe* depression. We include 0 as the lowest value, 9 as the top end of the *minimal* depression range, 19 as the top end of the *mild* depression range, 29 as the top end of the *moderate* depression range, and 39 as a representative value of *severe* depression. Let's test the effect of optimism therapy (versus control) at these selected values when depression scores are 0, 9, 19, 29, and 39.

---

6. It is interesting to note how the results of this **margins** command mirror the output with respect to the **treat** effect from the **regress** command. This is because the **treat** effect from the **regress** command represents the effect of **treat** when all other variables are held constant at zero (that is, when **depscore** is held constant at zero). This is the exact effect computed by the current **margins** command.

```
. margins r.treat, at(depscore=(0 9 19 29 39)) contrast(effects nowald)
> vsquish

Contrasts of adjusted predictions
Model VCE      : OLS

Expression     : Linear prediction, predict()
1._at          : depscore       =            0
2._at          : depscore       =            9
3._at          : depscore       =           19
4._at          : depscore       =           29
5._at          : depscore       =           39
```

	Contrast	Delta-method Std. Err.	t	P>|t|	[95% Conf. Interval]
treat@_at					
(OT vs Con)					
1	14.68956	1.831275	8.02	0.000	11.07803	18.3011
(OT vs Con)					
2	11.16505	1.309129	8.53	0.000	8.583264	13.74684
(OT vs Con)					
3	7.248927	1.015313	7.14	0.000	5.246585	9.251268
(OT vs Con)					
4	3.332802	1.243718	2.68	0.008	.8800135	5.78559
(OT vs Con)					
5	-.5833236	1.806029	-0.32	0.747	-4.145068	2.978421

Note! Note about the at values

The `margins` output showed the difference between the optimism therapy group and control group at 5 different levels of depression, 0, 9, 19, 29, and 39. These 5 levels of depression are labeled as 1, 2, 3, 4, and 5 in the output corresponding to the values of _at. The `margins` output includes a legend at the top showing the correspondence between depression score and each _at value (for example, showing that 1._at corresponds to a `depscore` of 0).

The above `margins` command shows the difference in the adjusted means of the DV between the optimism therapy group and the control group at the five specified levels of depression. By inspecting the p-values, we can see that the difference is significant ($p < 0.001$) when depression is 0, 9, 19. The difference is also significant when depression is 29 ($p = 0.008$). The difference is not significant when depression is 39 ($p = 0.747$). If we inspect the confidence interval for the difference, we can see that the confidence interval excludes 0 when depression is 0, 9, 19, and 29. It includes 0 when depression is 39 (95% CI $= [-4.15 \text{ to } 2.98]$).

Note! The vsquish option

The `vsquish` option vertically squishes the output by omitting extra blank lines. I frequently use this option in this book to save space and produce more compact output. You don't need to use this option when you run the examples for yourself.

From the `margins` command above, we see that optimism therapy is superior to the control condition when depression scores are 0, 9, 19, and 29 but not significantly different from the control condition when depression is 39. Furthermore, for each of these levels, we obtain an estimate of the treatment effect and a confidence interval for the treatment effect. For example, when depression is 0 before the start of the study, the estimated improvement in optimism due to optimism therapy (versus control) is 14.7 (95% CI = $[11.1, 18.3]$). When depression at the start of the study is 29, the treatment effect is much smaller (even though it is still significant). The expected treatment effect of optimism therapy is 3.33 (95% CI = $[0.88, 5.79]$).

To report these results, I could imagine nicely formatting them from the previous `margins` command. However, a graph might be more effective by helping us visualize the size of the treatment as a function of depression status. We can easily create such a graph using the `marginsplot` command. I repeat the `margins` command from above, following it with the `marginsplot` command, which yields the graph in figure 8.6.

```
. margins r.treat, at(depscore=(0 9 19 29 39)) contrast(effects nowald)
  (output omitted)
. marginsplot, yline(0)
  Variables that uniquely identify margins: depscore
```

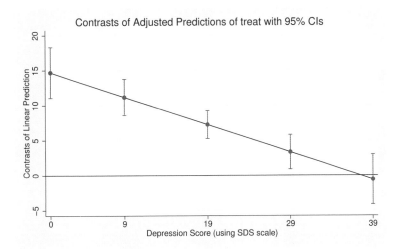

Figure 8.6: Contrasts of mean depression scores for treatment group versus control group by level of depression, with 95% confidence intervals at selected levels of depression

Because this graph is very informative, let's spend some time looking at it. It tells us about the significance of optimism therapy (versus control) at these 5 levels of depression and gives us an estimate of the size of the treatment effect (with a 95% confidence interval). We can see that the effect of optimism therapy (versus control) is significant (at the 5% level) when initial depression is 0, 9, 19, and 29 because each of the confidence intervals excludes 0. We could say that optimism therapy is significantly better than the control condition for those with *minimal* depression (0 to 9), *mild* depression (10 to 19), and *moderate* depression (20 to 29). When depression is 39, optimism therapy is no longer superior to the control condition because this confidence interval overlaps with 0.

Let's move beyond the consideration of statistical significance and consider the clinical significance of the results. Suppose that Professor Cheer considers a five-unit improvement in optimism to be clinically meaningful. Focusing on the lower confidence limit, we see that the effect of optimism therapy (versus control) exceeds this threshold when depression is 0, 9, and 19. This suggests that optimism therapy would be expected to yield a clinically meaningful improvement for those who are classified with *minimal* depression (0 to 9) and *mild* depression (10 to 19).

It is possible to explore the significance of optimism therapy versus the control at other levels of the covariate (depression). The values of 0, 9, 19, and 29 were selected because of their correspondence to agreed diagnostic cutoffs. In the absence of such, you could select values of the covariate based on other criteria, such as percentile rankings (for example, selecting the 20th, 35th, 50th, 65th, and 80th percentiles).

We want to be careful to avoid testing the effect of optimism therapy (versus control) at too many different levels of depression. A reviewer might express concern about performing too many significance tests without any form of type I error control. For example, the following `margins` command computes the effect of optimism therapy (versus control) for depression scores ranging from 0 to 40 in units of 1. (The output is omitted to save space.) Then, the `marginsplot` command is used to graph the differences (with a 95% confidence interval) by depression status, as shown in figure 8.7. The `yline(0)` option adds a reference line when optimism is 0, and the `xline()` option is used to demarcate the CIs that exclude 0 (when depression is 30 and below) versus the CIs that include 0 (when depression is 31 and above).

```
. margins r.treat, at(depscore=(0(1)40))
  (output omitted)
. marginsplot, yline(0) xline(30.5)
  Variables that uniquely identify margins: depscore
```

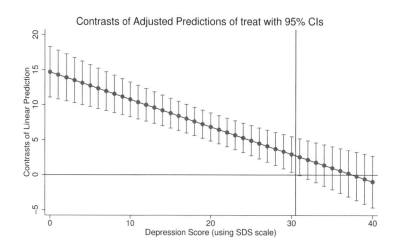

Figure 8.7: Contrasts of mean depression scores for treatment group versus control group by level of depression with 95% CIs

While I like the spirit of this graph, I think a reviewer would be rightly concerned about my having performed 41 significance tests without any adjustment to the type I error rate. Fortunately, we can use the `mcompare()` option with the `margins` command to obtain CIs (and p-values) accounting for multiple comparisons. We can specify the `mcompare(sidak)`, `mcompare(scheffe)`, or `mcompare(bonferroni)` option.

Let's repeat the `margins` and `marginsplot` commands, adding `mcompare(sidak)` to the `margins` command to obtain p-values and CIs using the Šidák adjustment for multiple comparisons. I include the `xline(27.5)` option with the `marginsplot` command to demarcate significant contrasts (27 and lower) versus nonsignificant contrasts (28 and higher). The resulting graph is shown in figure 8.8.

```
. margins r.treat, at(depscore=(0(1)40)) mcompare(sidak)
  (output omitted)
. marginsplot, yline(0) xline(27.5)
  Variables that uniquely identify margins: depscore
```

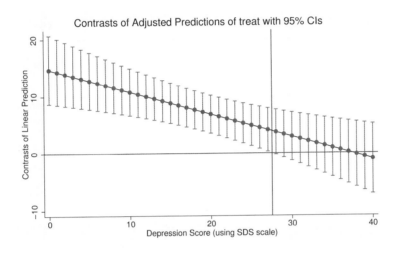

Figure 8.8: Contrasts of mean depression scores for treatment versus control group by level of depression with Šidák adjusted 95% CIs

8.2.3 Example 1: Summary

We started this chapter by examining the analysis from section 7.3.2, which used a two-by-three factorial design in which the two-level IV was treatment group assignment (optimism therapy versus control) and the three-level IV was level of depression. Depression was originally measured on a continuous scale but was categorized into three levels (1 = *minimal*, 2 = *mild/moderate*, and 3 = *severe*). I noted that we could obtain much more informative results if we treated depression as a continuous variable, thus leading to the use of ANCOVA in which the two levels of treatment were the IV and depression (measured continuously) was the covariate. Using this design, we formed two key research questions.

The first research question asked whether the effect of optimism therapy depends on the level of depression. By fitting an ANCOVA model with optimism as the DV, treatment group as the IV, and depression as the covariate, we found an interaction of treatment by depression indicating that the effect of treatment does depend on depression status. We then explored the second research question, determining the range of depression scores for which optimism therapy is superior to the control condition. When selecting depression scores based on clinical cutoffs, we found that optimism therapy was significantly better than the control condition when depression scores were 0, 9, 19, or 29. However, focusing on the clinical significance, optimism therapy yielded clinically meaningful differences from the control group when depression scores were 0, 9, and 19. When we consider the spectrum of depression scores ranging from 0 to 40 (in 1-unit increments), Šidák-adjusted CIs showed optimism therapy superior to the control condition for depression scores ranging from 0 to 27, and no significant difference for depression scores of 28 and above.

8.3   Example 2: IV has three levels

Let's extend the previous study by including a third treatment group, traditional therapy. In this study, the DV is still optimism, and the covariate is still depression measured before the start of the study. The difference is that the IV now has three levels (control = 1, traditional therapy = 2, optimism therapy = 3). The inclusion of three groups introduces the possibility of making contrasts between the groups. Professor Cheer wants to focus on a) the comparison of optimism therapy with control (as was done in the previous study) and b) the comparison of optimism therapy with traditional therapy.

The research questions are basically the same as those from section 8.2, except that now we are performing two comparisons with respect to treatment. So each research question has two parts, part "a" (focusing on the comparison of optimism therapy with control) and part "b" (focusing on the comparison of optimism therapy with traditional therapy). These questions are shown below.

- Question 1a. Does the effect of optimism therapy (versus control) depend on the level of depression?
- Question 1b. Does the effect of optimism therapy (versus traditional therapy) depend on the level of depression?
- Question 2a. For what range of depression scores is optimism therapy superior to the control condition?
- Question 2b. For what range of depression scores is optimism therapy superior to traditional therapy?

The commands we will use to answer these research questions will be very similar to those we used in section 8.2. The key difference is that when we compare optimism therapy with the control, we will apply the `r3b1` contrast operator to `treat` to obtain the comparison of group 3 with group 1 (optimism therapy versus control). Likewise, when comparing optimism therapy with traditional therapy, we will apply the `r3b2` contrast operator to `treat` to compare group 3 with group 2 (optimism therapy versus

control). (You can refer back to section 5.4 to refresh your memory about the use of
the `r.` contrast operator.)

The dataset containing the results of this study is `opt-ancova4.dta`. The dataset
has four variables: `id`, `treat`, `depscore`, and `opt`. We use this dataset and show the
variable names using the `describe` command.

```
. use opt-ancova4
. describe
Contains data from opt-ancova4.dta
  obs:           300
  vars:            4                          25 Aug 2014 01:02
  size:        4,800

              storage   display    value
variable name   type    format     label      variable label

id             float    %9.0g                 Participant ID
treat          float    %9.0g      treat      Treatment group assignment
depscore       float    %9.0g                 Depression Score (using SDS
                                                 scale)
opt            float    %6.1f                 Optimism score

Sorted by:
```

The variable `treat` reflects the treatment group assignment, showing that 100 were
assigned to the control condition (group 1), 100 were assigned to traditional therapy
(group 2), and 100 were assigned to optimism therapy (group 3). This is the IV for the
study.

```
. fre treat
treat — Treatment group assignment

                    Freq.    Percent    Valid      Cum.

Valid    1 Con        100      33.33    33.33      33.33
         2 TT         100      33.33    33.33      66.67
         3 OT         100      33.33    33.33     100.00
         Total        300     100.00   100.00
```

The variable `depscore` is the depression score (measured before the start of the
study), and the variable `opt` is the optimism at the end of the study. The summary
statistics for these two variables are shown below with the `summarize` command. The
variable `opt` is the DV and `depscore` is the covariate.

```
. summarize opt depscore
    Variable |     Obs       Mean    Std. Dev.      Min       Max

         opt |     300      47.46    9.515086        18        78
    depscore |     300      19.84   13.16608          0        60
```

As we did in section 8.2, let's create a scatterplot of optimism by depression separately for each treatment group. For each scatterplot, a linear fit line is added so that we can visually assess how well the scatterplot conforms to the linear fit line. We can also visually compare the slopes of the fit lines among the three groups. The `graph twoway` command creates the graph shown in figure 8.9.

```
. graph twoway (scatter opt depscore) (lfit opt depscore), by(treat)
```

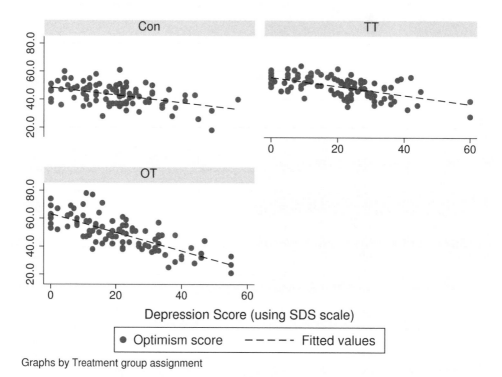

Figure 8.9: Scatterplot and linear regression fit line of optimism by depression for the control group (top left panel), traditional therapy group (top right panel), and optimism therapy group (bottom left panel)

For each level of treatment, the relationship between optimism and depression appears to be reasonably described by a linear fit. The regression diagnostic techniques illustrated in chapter 18 can (and should) be used to check the assumptions involved in regressing optimism on depression within each treatment group. For the sake of space, I have omitted showing these steps.

Let's now run the ANCOVA model with optimism as the DV, treatment group as the IV, and depression score as the covariate. The model also includes the interaction of the IV and the covariate (that is, the treatment group by depression interaction).

```
. regress opt i.treat##c.depscore
```

Source	SS	df	MS			
Model	13456.3817	5	2691.27635			
Residual	13614.1383	294	46.3065927			
Total	27070.52	299	90.5368562			

	Number of obs	=	300
	F(5, 294)	=	58.12
	Prob > F	=	0.0000
	R-squared	=	0.4971
	Adj R-squared	=	0.4885
	Root MSE	=	6.8049

opt	Coef.	Std. Err.	t	P>|t|	[95% Conf. Interval]	
treat						
TT	7.258708	1.748597	4.15	0.000	3.817354	10.70006
OT	14.60009	1.738415	8.40	0.000	11.17878	18.0214
depscore	-.2723331	.0544315	-5.00	0.000	-.3794579	-.1652083
treat#c.depscore						
TT	-.0455527	.0749492	-0.61	0.544	-.1930576	.1019521
OT	-.3860129	.0738467	-5.23	0.000	-.531348	-.2406778
_cons	48.52259	1.229369	39.47	0.000	46.10311	50.94206

Before we directly address the research questions, let's create a graph that shows optimism as a function of depression status separately for each of the treatment groups. This will give us a visual anchor to help in interpreting the results. We first use the `margins` command to compute the adjusted mean of optimism by treatment status when depression is 0 and again when depression is 40. The `marginsplot` command creates the graph shown in figure 8.10.

```
. margins treat, at(depscore=(0 40))
  (output omitted)

. marginsplot, noci
  Variables that uniquely identify margins: depscore treat
```

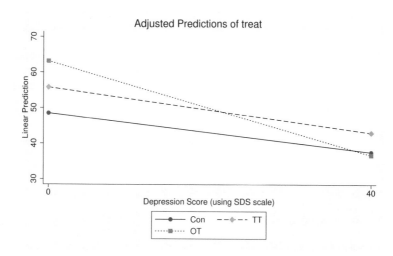

Figure 8.10: Predicted means of optimism by depression score and treatment group

Figure 8.10 illustrates the slope as a function of treatment group. The slope appears to be very similar for the control group and traditional therapy group, while the slope appears much steeper for the optimism therapy group. We can obtain the estimate of the slope for each of the treatment groups using the `margins` command combined with the `dydx(depscore)` option to estimate the `depscore` slope for each level of treatment. This shows the three slope coefficients illustrated in figure 8.10. The slopes for the control group (-0.27) and traditional therapy group (-0.32) are fairly similar. The slope for the optimism therapy group appears steeper (-0.66) than that for the other two groups.

```
. margins treat, dydx(depscore)
Average marginal effects                  Number of obs    =       300
Model VCE    : OLS
Expression   : Linear prediction, predict()
dy/dx w.r.t. : depscore
```

	dy/dx	Delta-method Std. Err.	t	P>|t|	[95% Conf. Interval]	
depscore						
treat						
Con	-.2723331	.0544315	-5.00	0.000	-.3794579	-.1652083
TT	-.3178858	.0515227	-6.17	0.000	-.4192858	-.2164858
OT	-.658346	.0499053	-13.19	0.000	-.7565629	-.560129

Let's now consider questions 1a and 1b.

8.3.1   Questions 1a and 1b

Question 1a

Let's start with question 1a, which asks whether the effect of optimism therapy versus control depends on the level of depression. With regard to figure 8.10, the effect of optimism therapy (versus control) looks greater for lower levels of depression and smaller for higher levels of depression. We test this using the `margins` command applying the `r3b1` contrast operator to `treat` to yield the comparison of group 3 with group 1 (optimism therapy versus control). We also include the `dydx(depscore)` option to specify that we are interested in comparing the `depscore` slopes.

```
. margins r3b1.treat, dydx(depscore)
Contrasts of average marginal effects
Model VCE    : OLS
Expression   : Linear prediction, predict()
dy/dx w.r.t. : depscore
```

	df	F	P>F
depscore			
treat	1	27.32	0.0000
Denominator	294		

	Contrast dy/dx	Delta-method Std. Err.	[95% Conf. Interval]	
depscore				
treat				
(OT vs Con)	-.3860129	.0738467	-.531348	-.2406778

The output shows that the effect of optimism therapy (versus control) differs as a function of depression score [$F(1, 294) = 27.32$, $p < 0.001$]. The margins command also shows the difference in the slopes for optimism therapy versus control (-0.39)[7] as well as the 95% confidence interval for the difference (95% CI $= [-0.53, -0.24]$).

Question 1b

Let's now turn to question 1b, which asks whether the effect of optimism therapy versus traditional therapy depends on the level of depression. Figure 8.10 suggests that optimism therapy yields greater optimism than traditional therapy for low depression scores but that the pattern is reversed for high depression scores (with traditional therapy yielding higher optimism scores). Let's test whether the effect of optimism therapy versus traditional therapy depends on depression status by repeating the previous margins command but instead applying the r3b2 contrast operator to treat to obtain the comparison of group 3 with group 2 (optimism therapy versus traditional therapy).

```
. margins r3b2.treat, dydx(depscore)

Contrasts of average marginal effects
Model VCE     : OLS

Expression    : Linear prediction, predict()
dy/dx w.r.t.  : depscore
```

	df	F	P>F
depscore			
treat	1	22.53	0.0000
Denominator	294		

	Contrast dy/dx	Delta-method Std. Err.	[95% Conf. Interval]	
depscore				
treat				
(OT vs TT)	-.3404602	.0717295	-.4816286	-.1992917

The output shows that the effect of optimism therapy (versus traditional therapy) differs as a function of depression score [$F(1, 294) = 22.53$, $p < 0.001$]. The slope for the optimism therapy group minus the slope for the traditional therapy group is -0.34 (95% CI $= [-0.48, -0.20]$).

---

7. Note how this corresponds to the difference in the slopes from the previous margins command, -0.658346 minus -0.2723331. Also, note how this corresponds to the estimate of the coefficient for OT in the output from the regress command. This reflects that the OT coefficient represents the comparison of the slope for group 3 (OT) with group 1 (Con, the reference group).

We have established that the effect of optimism therapy versus control depends on depression (via question 1a) and that the effect of optimism therapy versus traditional therapy depends on depression (via question 1b). Let's now explore questions 2a and 2b.

8.3.2 Questions 2a and 2b

Question 2a

Question 2a asks about the range of values for which the difference in optimism therapy (versus control) is significant. Let's address this question as we did in section 8.2.2 but applying the r3b1 operator to treat to focus on the contrast of optimism therapy versus control. The margins command below tests the effects of optimism therapy (versus control) when depression is 0, 9, 19, 29, and 39.

```
. margins r3b1.treat, at(depscore=(0 9 19 29 39)) contrast(effects nowald)
> vsquish

Contrasts of adjusted predictions
Model VCE     : OLS

Expression    : Linear prediction, predict()
1._at         : depscore        =           0
2._at         : depscore        =           9
3._at         : depscore        =          19
4._at         : depscore        =          29
5._at         : depscore        =          39
```

	Contrast	Delta-method Std. Err.	t	P>|t|	[95% Conf. Interval]
treat@_at					
(OT vs Con)					
1	14.60009	1.738415	8.40	0.000	11.17878 18.0214
(OT vs Con)					
2	11.12597	1.241453	8.96	0.000	8.682713 13.56924
(OT vs Con)					
3	7.265846	.9653587	7.53	0.000	5.365956 9.165735
(OT vs Con)					
4	3.405717	1.188821	2.86	0.004	1.06604 5.745394
(OT vs Con)					
5	-.454412	1.727813	-0.26	0.793	-3.854862 2.946038

The effect of optimism therapy versus control is significant when depression is 0,[8] 9, and 19 ($ps < 0.001$). This effect is also significant when depression is 29 ($p = 0.004$). The effect of optimism therapy versus control is not significant when depression is 39 ($p = 0.793$).

---

8. This illustrates another case where the output of the margins command mirrors the output from the previous regress command. The comparison of OT with Con when depscore is held constant at zero (from the margins command) mirrors the output associated with OT from the regress command. This is because the OT effect (from the regress command) also reflects the comparison of OT with the reference group (that is, Con) when all other variables are held constant at zero (that is, when depscore is held constant at zero).

We can visualize the results from the `margins` command using the `marginsplot` command, which yields the graph shown in figure 8.11.

```
. marginsplot, yline(0)
    Variables that uniquely identify margins: depscore
```

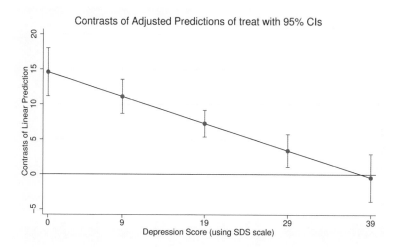

Figure 8.11: Contrasts of mean depression scores for treatment versus control group by level of depression, with 95% CIs at selected levels of depression

Question 2b

We can address question 2b in the same way we addressed question 2a, except that we change our focus to the comparison of optimism therapy with traditional therapy by using the `r3b2` contrast operator. The `margins` command below tests the effect of optimism therapy versus traditional therapy when depression is 0, 9, 19, 29, and 39.

```
. margins r3b2.treat, at(depscore=(0 9 19 29 39)) contrast(effects nowald)
> vsquish

Contrasts of adjusted predictions
Model VCE     : OLS

Expression    : Linear prediction, predict()
1._at         : depscore        =          0
2._at         : depscore        =          9
3._at         : depscore        =         19
4._at         : depscore        =         29
5._at         : depscore        =         39
```

	Contrast	Delta-method Std. Err.	t	P>|t|	[95% Conf. Interval]
treat@_at					
(OT vs TT)					
1	7.341383	1.748422	4.20	0.000	3.900373 10.78239
(OT vs TT)					
2	4.277241	1.260583	3.39	0.001	1.79633 6.758152
(OT vs TT)					
3	.8726394	.9672813	0.90	0.368	-1.031034 2.776313
(OT vs TT)					
4	-2.531962	1.145086	-2.21	0.028	-4.785566 -.2783582
(OT vs TT)					
5	-5.936564	1.647979	-3.60	0.000	-9.179895 -2.693233

When depression is 0 or 9, optimism therapy yields higher optimism scores as compared with traditional therapy ($p \leq 0.001$). The difference between the two groups is not significant when depression is 19. Optimism therapy yields lower optimism scores than traditional therapy when depression is 29 ($p = 0.028$) and when depression is 39 ($p < 0.001$).

We can visualize the effect of optimism therapy versus traditional therapy at these levels of depression by using the `marginsplot` command, which yields the graph shown in figure 8.12.

```
. marginsplot, yline(0)
  Variables that uniquely identify margins: depscore
```

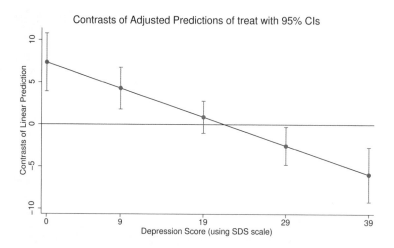

Figure 8.12: Contrasts of mean depression scores for optimism therapy versus traditional therapy group by level of depression, with 95% CIs at selected levels of depression

8.3.3 Overall interaction

Before concluding this section, let's consider one final analysis. Suppose that we were interested in testing the overall interaction of the IV by the covariate (that is, `treat` by `depstat`). That test is equivalent to testing the following null hypothesis that the three slopes are equal to each other,

$$H_0 : \beta_1 = \beta_2 = \beta_3$$

where β_1 is the slope for the control group, β_2 is the slope for the traditional therapy group, and β_3 is the slope for the optimism therapy group. The `contrast` command below tests the null hypothesis that all three slopes are equal. We can reject this null hypothesis $[F(2, 294) = 17.02, p = 0.000]$.

```
. contrast treat#c.depscore
Contrasts of marginal linear predictions
Margins        : asbalanced
```

	df	F	P>F
treat#c.depscore	2	17.02	0.0000
Denominator	294		

8.3.4   Example 2: Summary

In this section, we considered an ANCOVA with an IV by covariate interaction in which the IV has three levels. The analysis strategy was actually very similar to the case where the IV had two levels (as shown in section 8.2). The key difference is that we used contrast operators to focus on the contrasts of interest.

The analysis found that the effect of optimism therapy versus control depended on the level of depression. As shown in figure 8.11, optimism therapy yielded significantly higher optimism scores when depression was 0, 9, 19, and 29. When depression was 39, the difference in these two groups was not significant.

In addition, we found that the effect of optimism therapy versus traditional therapy depended on the level of depression before the start of the study. As illustrated in figure 8.12, optimism therapy yielded significantly higher optimism for those with *minimal* depression (ranging from 0 to 9). For those with *severe* depression (29 and higher), optimism therapy yielded a significantly lower optimism score compared with traditional therapy. For depression scores in the *mild* to *moderate* range (10 to 29), the results are mixed.

Video tutorial: ANCOVA with interactions

See a video illustrating an ANCOVA model with an interaction between the IV and the covariate at http://www.stata.com/sbs/ancova-interactions.

8.4   Closing thoughts

This chapter has shown how we can use ANCOVA models that introduce an IV by covariate interaction to understand how the effect of the IV varies as a function of the covariate. I feel that this is a technique that is underutilized in the behavioral sciences. Often, I have seen researchers convert the continuous covariate into a categorical variable so that they can use a factorial ANOVA. When the relationship between the covariate and the outcome is linear, the ANCOVA with interaction can give you a more complete understanding of how the effect of the IV varies across the entire spectrum of covariate values.

9 Three-way between-subjects analysis of variance

9.1 Chapter overview

This chapter illustrates models involving interactions of three independent variables (IVs), with an emphasis on how to interpret the interaction of the three IVs. The chapter begins with a two-by-two-by-two model (see section 9.2) followed by an example illustrating a two-by-two-by-three model (see section 9.3). The chapter then concludes with a discussion of models that have at least three levels for each IV and uses a three-by-three-by-four model as an example (see section 9.4).

Like chapter 7, this chapter illustrates the many ways in which you can dissect interactions. In the case of a three-way interaction, there are even more ways that such interactions can be dissected. The best practice would be to use your research questions to develop an analysis plan that describes how the interactions will be dissected. This plan would also consider the number of statistical tests to be performed and consider whether a strategy is required to control the overall type I error rate. By contrast, an undesirable practice would be to include an interaction because it happened to be significant, dissect it every which way possible until an unpredicted significant test is obtained, and make no adjustment for the number of unplanned statistical tests that were performed.

The examples in the chapter illustrate the possible ways to dissect interactions, so you can understand how these techniques work and can choose among them for your analysis plan. This chapter focuses on teaching these techniques; to that end, data are inspected before being analyzed, patterns of data are used to suggest interesting tests to perform, and every possible method illustrated for dissecting interactions. I think this is a useful teaching strategy but not a research strategy to emulate.

9.2 Two-by-two-by-two models

This chapter begins with the simplest example of a model with an interaction of three IVs, a two-by-two-by-two model. The research example is a variation of the one shown in chapter 7. In that chapter, the two focal IVs were treatment type and depression status; here, instead of using optimism as the outcome, we will use happiness as the main outcome. The measure of happiness ranges from 0 to 100 with an average of about 50 for the general population. Suppose that the researcher wants to further extend the work from chapter 7 by considering the role of the time of year (that is, season), thinking that happiness may be less malleable during dark winter months as compared with cheerful summer months. To investigate this, the researcher conducts a new study using a two-by-two-by-two model in which treatment has two levels (control versus optimism therapy), depression status has two levels (nondepressed versus mildly depressed), and season has two levels (winter and summer).

The results of this study are contained in the dataset named `hap-2by2by2.dta`. Let's use this dataset and show the mean of the outcome variable (happiness) by depression status, treatment, and season.

```
. use hap-2by2by2, clear
. table depstat treat season, contents(mean hap)
```

Depressio n status	Season and Treatment group			
	− Winter −		− Summer −	
	Con	OT	Con	OT
Non	45.0	60.3	52.2	65.2
Mild	39.7	51.6	45.4	65.4

The `anova` command is used below to fit a model that predicts happiness from treatment, depression status, season, all the two-way interactions of these variables, and the three-way interaction of these variables.

```
. anova hap depstat##treat##season
                        Number of obs =       240    R-squared     = 0.5702
                        Root MSE      = 8.00481    Adj R-squared = 0.5572

              Source |  Partial SS       df        MS          F     Prob>F

               Model |  19720.117        7    2817.1595      43.97   0.0000

             depstat |  1612.0167        1    1612.0167      25.16   0.0000
               treat |  13590.15         1    13590.15      212.09   0.0000
       depstat#treat |  46.816667        1    46.816667       0.73   0.3936
              season |  3728.8167        1    3728.8167      58.19   0.0000
      depstat#season |  205.35           1    205.35          3.20   0.0747
        treat#season |  126.15           1    126.15          1.97   0.1619
depstat#treat#season |  410.81667        1    410.81667       6.41   0.0120

            Residual |  14865.867      232    64.077011

               Total |  34585.983      239    144.71123
```

Note! Three-way interaction shortcut

`depstat##treat##season` is a shortcut for specifying all main effects, two-way interactions, and the three-way interaction of `depstat`, `treat`, and `season`. This both saves time and helps ensure that you include all the lower-order effects. Even if not significant, these lower-order effects should be included in the model.

The `depstat#treat#season` interaction is significant [$F(1, 232) = 6.41$, $p = 0.0120$). Let's use the `margins` command to show the mean of happiness broken down by these three IVs.

```
. margins depstat#treat#season, nopvalues
Adjusted predictions                          Number of obs     =       240
Expression    : Linear prediction, predict()
```

	Margin	Delta-method Std. Err.	[95% Conf. Interval]	
depstat#treat#season				
Non#Con#Winter	45	1.461472	42.12055	47.87945
Non#Con#Summer	52.2	1.461472	49.32055	55.07945
Non#OT#Winter	60.33333	1.461472	57.45388	63.21279
Non#OT#Summer	65.2	1.461472	62.32055	68.07945
Mild#Con#Winter	39.7	1.461472	36.82055	42.57945
Mild#Con#Summer	45.36667	1.461472	42.48721	48.24612
Mild#OT#Winter	51.56667	1.461472	48.68721	54.44612
Mild#OT#Summer	65.36667	1.461472	62.48721	68.24612

Let's then use the `marginsplot` command to make a graph of the means, with `treat` on the x axis and the different seasons in separate panels (see figure 9.1).

```
. marginsplot, xdim(treat) bydim(season) noci
  Variables that uniquely identify margins: depstat treat season
```

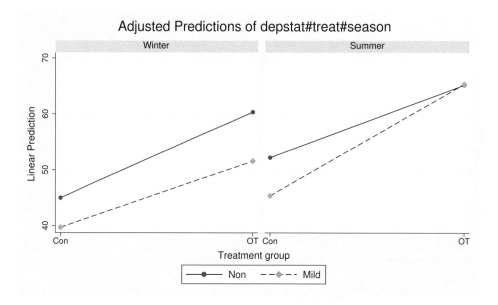

Figure 9.1: Happiness by treatment, depression status, and season

The graph in figure 9.1 illustrates the three-way interaction, showing that the size of the `treat#depstat` interaction differs for winter versus summer. It appears that there is no `treat#depstat` interaction during the winter but that there is such an interaction during the summer. In fact, we can perform such tests by looking at the simple interaction of `treat` by `depstat` at each level of `season`.

9.2.1 Simple interactions by season

One way that we can dissect the three-way interaction is by looking at the simple interactions of treatment by depression status at each season. In figure 9.1, the left panel illustrates the simple interaction of treatment by depression status for the winter, and the right panel illustrates the treatment by depression status interaction for the summer. It appears that the interaction is not significant during the winter and is significant during the summer. We test this using the `contrast` command shown below, which tests the `treat#depstat` interaction at each level of `season`.

```
. contrast treat#depstat@season

Contrasts of marginal linear predictions

Margins      : asbalanced
```

	df	F	P>F
treat#depstat@season			
Winter	1	1.41	0.2368
Summer	1	5.74	0.0174
Joint	2	3.57	0.0297
Denominator	232		

Indeed, the treatment by depression status interaction is not significant during the winter [$F(1, 232) = 1.41$, $p = 0.2368$] and is significant during the summer [$F(1, 232) = 5.74$, $p = 0.0174$]. Looking at the left panel of figure 9.1, we see that in the winter, the effect of optimism therapy (compared with the control) does not depend on depression status, hence the nonsignificant simple interaction. However, in the summer, the effect of optimism therapy does depend on depression status. Looking at the right panel of figure 9.1, we can see that during the summer months, optimism therapy is more effective for those who are mildly depressed than for those who are nondepressed.

9.2.2   Simple interactions by depression status

Another way to dissect this three-way interaction is by looking at the simple interaction of treatment by season at each level of depression status. To visualize this, let's rerun the `margins` command and then use the `marginsplot` command to graph the means, with `treat` on the x axis and separate panels for those who are nondepressed and mildly depressed (see figure 9.2).

```
. margins depstat#treat#season
  (output omitted)
. marginsplot, xdim(treat) bydim(depstat) noci
  Variables that uniquely identify margins: depstat treat season
```

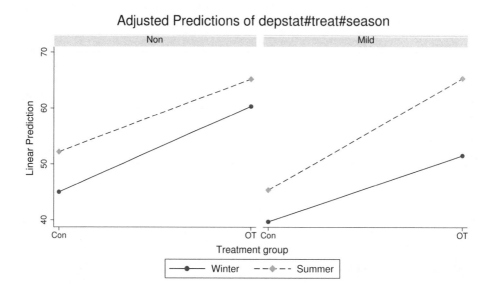

Figure 9.2: Happiness by treatment, season, and depression status

Among those who are nondepressed, it appears that optimism therapy is equally effective in the summer and winter (see the left panel of figure 9.2). By contrast, among those who are mildly depressed, it appears that the effectiveness of optimism therapy depends on the season (see the right panel of figure 9.2). In other words, it appears that there is an interaction of treatment by season for those who are mildly depressed but no such interaction for those who are not depressed. We can test the interaction of treatment by depression status at each level of season using the `contrast` command.

```
. contrast treat#season@depstat
Contrasts of marginal linear predictions
Margins       : asbalanced
```

	df	F	P>F
treat#season@depstat			
Non	1	0.64	0.4255
Mild	1	7.74	0.0058
Joint	2	4.19	0.0163
Denominator	232		

Indeed, the interaction of treatment by depression status is not significant for those who are not depressed [$F(1, 232) = 0.64$, $p = 0.4255$] but is significant for those who are mildly depressed [$F(1, 232) = 7.74$, $p = 0.0058$].

9.2.3 Simple effects

We have seen that the effect of optimism therapy depends both on season and on depression status. We might want to know whether the effect of optimism therapy is significant for each combination of season and depression status. This simple effect can be tested using the contrast command shown below. It tests the effect of treat at each combination of season and depstat.

```
. contrast treat@season#depstat, nowald pveffects
Contrasts of marginal linear predictions
Margins       : asbalanced
```

	Contrast	Std. Err.	t	P>|t|
treat@season#depstat				
(OT vs base) Winter#Non	15.33333	2.066834	7.42	0.000
(OT vs base) Winter#Mild	11.86667	2.066834	5.74	0.000
(OT vs base) Summer#Non	13	2.066834	6.29	0.000
(OT vs base) Summer#Mild	20	2.066834	9.68	0.000

The mean happiness of the optimism therapy group is always higher than the mean happiness of the control group at each level of season and depression status ($ps < 0.001$). For example, among those who are nondepressed and treated during the winter, the mean happiness is 15.3 points higher for optimism therapy versus control, a significant difference [$t(232) = 7.42$, $p < 0.001$].

9.3   Two-by-two-by-three models

Let's now consider an example with three IVs, two of which have two levels and one of which has three levels. This is an extension of the example shown in the previous section, involving treatment group, depression status, and season as IVs. In this example, the treatment variable now has three levels: (1) control, (2) traditional therapy, and (3) optimism therapy.

Let's use the dataset for this example and show the mean happiness by depression status, treatment group, and season.

```
. use hap-3by2by2, clear

. table depstat treat season, contents(mean hap)
```

Depression status	Season and Treatment group					
	— Winter (S1) —			— Summer (S2) —		
	Con	TT	OT	Con	TT	OT
Non	45.2	54.8	60.2	50.2	59.8	65.1
Mild	40.0	49.7	50.0	44.7	55.2	64.9

Let's now use the `anova` command to predict `opt` from `depstat`, `treat`, `season`, all the two-way interactions of these variables, and the three-way interaction.

```
. anova hap depstat##treat##season
```

		Number of obs =	360	R-squared	=	0.4911
		Root MSE =	8.01433	Adj R-squared =		0.4750

Source	Partial SS	df	MS	F	Prob>F
Model	21566.431	11	1960.5846	30.52	0.0000
depstat	2376.7361	1	2376.7361	37.00	0.0000
treat	13985.272	2	6992.6361	108.87	0.0000
depstat#treat	3.4388889	2	1.7194444	0.03	0.9736
season	3980.025	1	3980.025	61.97	0.0000
depstat#season	258.40278	1	258.40278	4.02	0.0457
treat#season	467.81667	2	233.90833	3.64	0.0272
depstat#treat#season	494.73889	2	247.36944	3.85	0.0222
Residual	22351.833	348	64.229406		
Total	43918.264	359	122.335		

The three-way interaction is significant $[F(2, 348) = 3.85, p = 0.0222]$. Let's use the `margins` and `marginsplot` commands to display and graph the means by each of these IVs. In creating the graph of the means, let's display separate panels for each level of depression status (see figure 9.3). This allows us to focus on the way that the `treat#season` interaction varies by depression status.

```
. margins depstat#treat#season, nopvalues
Adjusted predictions                              Number of obs    =         360
Expression   : Linear prediction, predict()
```

	Margin	Delta-method Std. Err.	[95% Conf. Interval]	
depstat#treat#season				
Non#Con#Winter (S1)	45.16667	1.463209	42.28882	48.04451
Non#Con#Summer (S2)	50.2	1.463209	47.32215	53.07785
Non#TT#Winter (S1)	54.83333	1.463209	51.95549	57.71118
Non#TT#Summer (S2)	59.8	1.463209	56.92215	62.67785
Non#OT#Winter (S1)	60.23333	1.463209	57.35549	63.11118
Non#OT#Summer (S2)	65.1	1.463209	62.22215	67.97785
Mild#Con#Winter (S1)	40.03333	1.463209	37.15549	42.91118
Mild#Con#Summer (S2)	44.7	1.463209	41.82215	47.57785
Mild#TT#Winter (S1)	49.7	1.463209	46.82215	52.57785
Mild#TT#Summer (S2)	55.2	1.463209	52.32215	58.07785
Mild#OT#Winter (S1)	50	1.463209	47.12215	52.87785
Mild#OT#Summer (S2)	64.86667	1.463209	61.98882	67.74451

```
. marginsplot, bydim(depstat) noci
  Variables that uniquely identify margins: depstat treat season
```

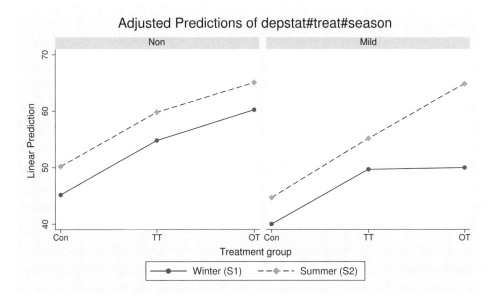

Figure 9.3: Happiness by treatment, season, and depression status

9.3.1   Simple interactions by depression status

It appears that the `treat#season` interaction might not be significant for those who are not depressed (see the left panel of figure 9.3) but significant for those who are mildly depressed (see the right panel of figure 9.3). We can explore this by assessing the `treat#season` interaction at each level of `depstat` using the `contrast` command.

```
. contrast treat#season@depstat
Contrasts of marginal linear predictions
Margins       : asbalanced
```

	df	F	P>F
treat#season@depstat			
Non	2	0.00	0.9984
Mild	2	7.49	0.0007
Joint	4	3.75	0.0053
Denominator	348		

The treatment by season interaction is not significant among those who are nondepressed $[F(2, 348) = 0.000, p = 0.9984]$. The parallel lines in the left panel of figure 9.3 illustrate the absence of an interaction. By contrast, the treatment by season interaction is significant for those who are mildly depressed $[F(2, 348) = 7.49, p = 0.0007]$ (see the right panel of figure 9.3).

Let's further dissect this simple interaction by applying contrasts to treatment through the use of simple partial interactions.

9.3.2   Simple partial interaction by depression status

If we focus on the right panel of figure 9.3, we can see that this simple interaction is really a two-by-three interaction, and we can dissect it using the tools from section 7.3 on two-by-three interactions. We can start by applying contrasts to the treatment, forming a simple partial interaction. Say that we would want to form two comparisons with respect to `treat`, comparing group 2 with 1 (traditional therapy versus control) and comparing group 3 with 2 (optimism therapy versus traditional therapy). The interaction of treatment (traditional therapy versus control) by season for those who are mildly depressed is shown in the left panel of figure 9.4, and the interaction of treatment (optimism therapy versus traditional therapy) by season for those who are mildly depressed is shown in the right panel of figure 9.4.

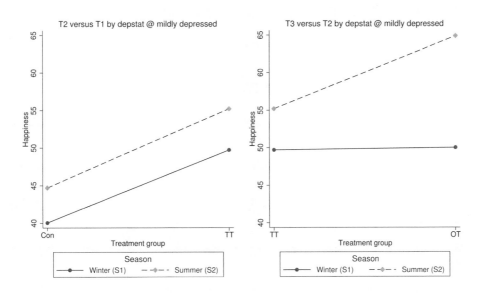

Figure 9.4: Simple partial interactions

The `contrast` command below tests the two partial interactions depicted in the left and right panels of figure 9.4 by specifying `ar.treat#season@2.depstat`. To understand this, we will break it into two parts. The first part, `ar.treat#season`, creates the interactions of treatment (2 versus 1) by season and treatment (3 versus 2) by season. The second part, `@2.depstat`, indicates the contrasts will be performed only for level 2 of `depstat`, that is, the mildly depressed group.

```
. contrast ar.treat#season@2.depstat
Contrasts of marginal linear predictions
Margins       : asbalanced
```

	df	F	P>F
treat#season@depstat			
(TT vs Con) (joint) Mild	1	0.08	0.7760
(OT vs TT) (joint) Mild	1	10.24	0.0015
Joint	2	7.49	0.0007
Denominator	348		

The first test is not significant $[F(1, 348) = 0.08, p = 0.7760]$. As depicted in the left panel of figure 9.4, the difference between traditional therapy and the control does not depend on season (with regard to those who are mildly depressed). The second test is significant $[F(1, 348) = 10.24, p = 0.0015]$. As shown in the right panel of figure 9.4, the difference between optimism therapy and traditional therapy does depend on season.

Let's now further understand this simple partial interaction through the use of simple contrasts.

9.3.3   Simple contrasts

The significant simple partial interaction shown in the right panel of figure 9.4 is really now a two-by-two analysis and can be dissected using simple contrasts. In particular, we will ask whether there is a difference between optimism therapy and traditional therapy separately for each season, focusing only on those who are mildly depressed. We can perform this test by specifying `ar3.treat@season#2.depstat` with the `contrast` command. Let's break this into two parts. The first part, `ar3.treat`, requests the comparison of treatment group 3 with 2 (optimism therapy versus traditional therapy). The second part, `@season#2.depstat`, requests that the contrasts be performed at each level of season and at level 2 of depression status (mildly depressed).

```
. contrast ar3.treat@season#2.depstat, nowald pveffects
Contrasts of marginal linear predictions
Margins      : asbalanced
```

	Contrast	Std. Err.	t	P>|t|
treat@season#depstat				
(OT vs TT) Winter (S1)#Mild	.3	2.06929	0.14	0.885
(OT vs TT) Summer (S2)#Mild	9.666667	2.06929	4.67	0.000

The results of these two tests are consistent with what we observe in the right panel of figure 9.4. The first test is not significant [$t(348) = 0.14$, $p = 0.885$]. There is no difference between optimism therapy and traditional therapy among those who are mildly depressed in the winter. The second test is significant [$t(348) = 4.67$, $p < 0.001$]. In the summer, there is a significant difference between optimism therapy and traditional therapy among those who are mildly depressed.

9.3.4   Partial interactions

Let's consider another strategy we could use to further understand the three-way interaction. Let's look back at the original graph of the results of the three-way interaction shown in figure 9.3. Say that we wanted to further understand this interaction by applying adjacent group contrasts to treatment. These contrasts compare group 2 with 1 (traditional therapy versus control) and group 3 with 2 (optimism therapy versus traditional therapy). We could interact these contrasts with season and depression status. The contrast of `treat` (traditional therapy versus control) by `depstat` by `season` is shown in figure 9.5, and the contrast of `treat` (optimism therapy versus traditional therapy) by `depstat` by `season` is shown in figure 9.6.

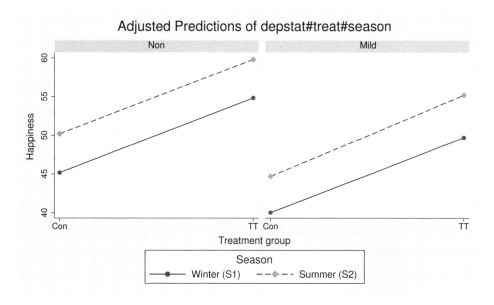

Figure 9.5: Simple partial interaction: TT versus Con by season by `depstat`

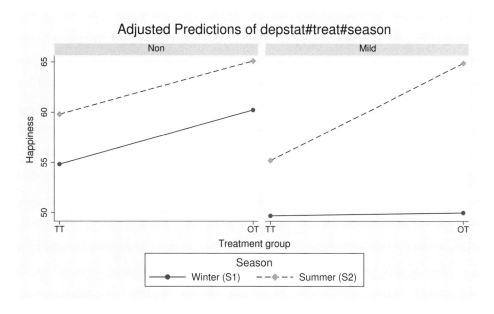

Figure 9.6: Simple partial interaction: OT versus TT by season by `depstat`

We can test each of these partial interactions using the `contrast` command.

```
. contrast ar.treat#season#depstat
Contrasts of marginal linear predictions
Margins        : asbalanced
```

	df	F	P>F
treat#season#depstat			
(TT vs Con) (joint) (joint)	1	0.05	0.8280
(OT vs TT) (joint) (joint)	1	5.23	0.0228
Joint	2	3.85	0.0222
Denominator	348		

The first partial interaction is not significant $[F(1,348) = 0.05,\ p = 0.8280]$. As we can see in figure 9.5, the `treat` (traditional therapy versus control) by `season` interaction is roughly the same for each level of depression status.

The second partial interaction is significant $[F(1,348) = 5.23,\ p = 0.0228]$. As shown in figure 9.6, the `treat` (optimism therapy versus traditional therapy) by `season` interaction differs by depression status. This partial interaction now resembles a two-by-two-by-two model, and you can further dissect it using the techniques illustrated in section 9.2.

Let's now move on to the most complex kind of three-way interactions—those that involve at least three levels for each IV.

9.4   Three-by-three-by-three models and beyond

Three-way interactions become increasingly complex as the number of levels of each IV increases. When an IV goes from having two levels to having three levels, it introduces the possibility of applying contrasts to the IV to further probe the interaction. For example, in the two-by-two-by-three model from section 9.3, we applied contrasts to the three-level `treat` IV. If we extended the `depstat` IV to also have three levels (for example, nondepressed, mildly depressed, severely depressed), then it would be possible to apply contrasts to the `depstat` IV as well. If we take this one step further and include a third level of `season`, then the model would be a three-by-three-by-three model, and it would be possible to apply contrasts to all three IVs.

Let's consider an extension of the example from section 9.3 involving the IVs `treat`, `depstat`, and `season`, except that these IVs have three, three, and four levels, respectively. The three levels of `treat` are (1) control, (2) traditional therapy, and (3) optimism therapy. The three levels of `depstat` are (1) nondepressed, (2) mildly depressed, and (3) severely depressed. The four levels of `season` are (1) winter, (2) spring, (3) summer, and (4) fall.

Let's use the dataset for this hypothetical example and show the mean of the outcome by these three IVs.

```
. use hap-3by3by4, clear
. table season treat depstat, contents(mean hap)
```

| | Depression status and Treatment group | | | | | | | | |
| | Non | | | Mild | | | Severe | | |
Season	Con	TT	OT	Con	TT	OT	Con	TT	OT
Winter	45.3	54.8	61.7	40.1	49.6	51.8	37.2	47.7	47.3
Spring	48.2	57.8	63.2	43.1	52.7	57.0	40.2	50.7	50.2
Summer	50.1	60.7	65.1	45.0	54.8	66.0	40.3	49.8	49.2
Fall	48.2	57.6	63.4	43.3	52.5	56.6	40.5	50.5	50.7

Let's analyze the data for this example using the anova command.

```
. anova hap depstat##treat##season
```

| | Number of obs = | 1,080 | R-squared | = | 0.6158 |
| | Root MSE   = | 6.00758 | Adj R-squared = | | 0.6030 |

Source	Partial SS	df	MS	F	Prob>F
Model	60402.744	35	1725.7927	47.82	0.0000
depstat	18578.052	2	9289.0259	257.38	0.0000
treat	34509.902	2	17254.951	478.10	0.0000
depstat#treat	1723.0315	4	430.75787	11.94	0.0000
season	3536.1435	3	1178.7145	32.66	0.0000
depstat#season	1001.3259	6	166.88765	4.62	0.0001
treat#season	192.03148	6	32.005247	0.89	0.5038
depstat#treat#season	862.25741	12	71.854784	1.99	0.0221
Residual	37679.033	1,044	36.091028		
Total	98081.777	1,079	90.900627		

The three-way interaction of depstat#treat#season is significant $[F(12, 1044) = 1.99, p = 0.0221]$. To begin to understand the nature of this interaction, let's first graph the interaction using the margins and marginsplot commands. The graph of the mean happiness by treatment group, depression status, and season is shown in figure 9.7.

```
. margins depstat#treat#season
  (output omitted)
. marginsplot, xdim(treat) bydim(season) noci legend(rows(1))
  Variables that uniquely identify margins: depstat treat season
```

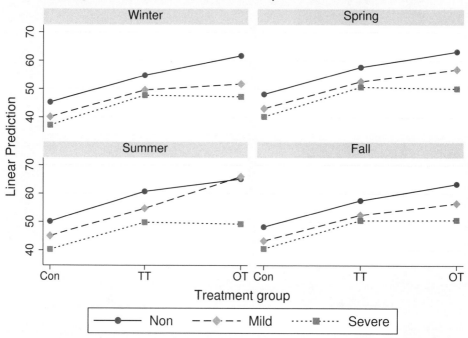

Figure 9.7: Happiness by treatment, season, and depression status

9.4.1   Partial interactions and interaction contrasts

Let's explore ways to dissect the three-way interaction of `depstat#treat#season` by applying contrasts to one or more of the IVs. Let's start by testing whether the interaction of `treat#depstat` is the same for each season compared with winter (season 1). This is performed by applying the `r.` contrast operator to season and interacting that with `treat` and `depstat`.

```
. contrast r.season#treat#depstat
Contrasts of marginal linear predictions
Margins        : asbalanced
```

	df	F	P>F
season#treat#depstat			
(Spring vs Winter) (joint) (joint)	4	0.45	0.7710
(Summer vs Winter) (joint) (joint)	4	5.23	0.0004
(Fall vs Winter) (joint) (joint)	4	0.32	0.8651
Joint	12	1.99	0.0221
Denominator	1044		

The season (spring versus winter) by treat by depstat interaction is not significant $[F(4, 1044) = 0.45, p = 0.7710]$. In figure 9.7, we can see how the two-way interaction of treat by depstat is similar for spring versus winter. Likewise, the season (fall versus winter) by treat by depstat interaction is not significant $[F(4, 1044) = 0.32, p = 0.8651]$. However, the season (summer versus winter) by treat by depstat interaction is significant $[F(4, 1044) = 5.23, p = 0.0004]$. Looking at the winter and summer panels from figure 9.7, we can see that the treat#depstat interaction differs for summer versus winter.

Let's further explore the season (summer versus winter) by treat by depstat interaction. Referring to the winter and summer panels from figure 9.7, we will focus on nondepressed and mildly depressed groups, as illustrated in figure 9.8.

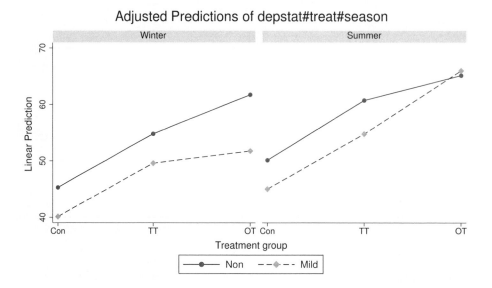

Figure 9.8: Interaction contrast of season (winter versus summer) by depression status (mild versus non) by treatment

Consider the nature of the interaction in the left panel of figure 9.8, which shows the interaction of depression status (nondepressed versus mildly depressed) by treatment during winter. Likewise, consider the interaction in the right panel of figure 9.8, which shows the same interaction, except during summer. Let's test the interaction in the right versus left panels. In other words, let's test the season (winter versus summer) by depression status (nondepressed versus mildly depressed) by treatment interaction. The `contrast` command below performs this comparison by applying the `r3.` contrast to `season` (to compare summer with winter, 3 with 1) and the `r2.` contrast to `depstat` (to compare mildly depressed with nondepressed, 2 with 1). These terms are all interacted (that is, `r3.season#r2.depstat#treat`) and yield a test of `season` (summer versus winter) by `depstat` (mild versus non) by `treat`.

```
. * Season(winter versus summer) by depstat(non versus mild) by treat
. contrast r3.season#r2.depstat#treat
Contrasts of marginal linear predictions
Margins      : asbalanced
```

	df	F	P>F
season#depstat#treat	2	8.65	0.0002
Denominator	1044		

This test is significant $[F(2, 1044) = 8.65, p = 0.0002]$. This is shown by comparing the interaction in the left panel of figure 9.8 with that in the right panel of figure 9.8. This significant test indicates that the two-way interaction formed by interacting depression status (nondepressed versus mildly depressed) by treatment differs by season (winter versus summer).

Say that we wanted to take this test and focus on the contrast of optimism therapy versus traditional therapy (group 3 versus 2).[1] This yields an interaction of season (winter versus summer) by depression status (nondepressed versus mildly depressed) by treatment (optimism therapy versus traditional). I created the graph in figure 9.9 that shows this interaction contrast. We can test this contrast using the contrast command.

```
. * season (winter versus summer) by depstat (non versus mild) by treat
> (OT versus TT)
. contrast r3.season#r2.depstat#r3b2.treat, noeffects

Contrasts of marginal linear predictions

Margins      : asbalanced
```

	df	F	P>F
season#depstat#treat	1	13.90	0.0002
Denominator	1044		

---

1. This contrast is specified with r3b2.treat, which compares group 3 with group 2. This could have also been specified as a custom contrast, that is, {treat 0 -1 1}.

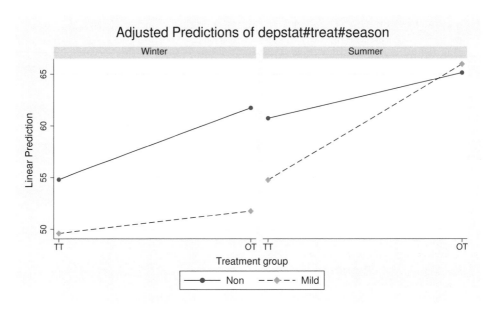

Figure 9.9: Interaction contrast of season (winter versus summer) by depression status (mild versus non) by treatment (OT versus TT)

This test is significant $[F(1, 1044) = 13.90, p = 0.0002]$. The two-way interaction of depression status (non versus mild) by treatment (optimism therapy versus traditional) significantly differs between summer and winter.

9.4.2   Simple interactions

Let's now explore a different way to dissect the three-way interaction through the use of simple interaction tests. Consider the `treat#depstat` interaction at each of the four levels of season, as shown in the four panels of figure 9.7. We can assess the `treat#depstat` interaction for each season using the `contrast` command.

```
. contrast depstat#treat@season
Contrasts of marginal linear predictions
Margins      : asbalanced
```

	df	F	P>F
depstat#treat@season			
Winter	4	3.52	0.0072
Spring	4	2.45	0.0448
Summer	4	9.84	0.0000
Fall	4	2.10	0.0790
Joint	16	4.48	0.0000
Denominator	1044		

The treat#depstat interaction is significant for winter ($p = 0.0072$), spring ($p = 0.0448$), and summer ($p < 0.0001$). The interaction is not significant for fall ($p = 0.0790$).

Let's perform this same contrast command but apply the r2. contrast to depstat, which compares those who are nondepressed with those who are mildly depressed (as illustrated in figure 9.10). This yields four partial interactions of depression status (nondepressed versus mild) by treatment at each level season. These tests are performed with the following contrast command.

```
. contrast r2.depstat#treat@season
Contrasts of marginal linear predictions
Margins      : asbalanced
```

	df	F	P>F
depstat#treat@season			
(Mild vs Non) (joint) Winter	2	3.17	0.0424
(Mild vs Non) (joint) Spring	2	0.21	0.8145
(Mild vs Non) (joint) Summer	2	5.69	0.0035
(Mild vs Non) (joint) Fall	2	0.44	0.6464
Joint	8	2.38	0.0154
Denominator	1044		

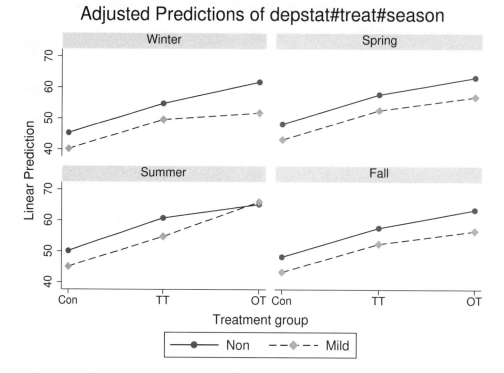

Figure 9.10: Happiness by treatment and season focusing on mildly depressed versus nondepressed

The interaction of depression status (non versus mild) by treatment is significant in winter $[F(2, 1044) = 3.17, p = 0.0424]$. In the top left panel of figure 9.10, we can see how the lines are not parallel during the winter. This test is also significant for summer $[F(2, 1044) = 5.69, p = 0.0035]$. This is consistent with the nonparallel lines displayed in the lower left panel in figure 9.10. By comparison, the lines are parallel for spring $(p = 0.8145)$ and fall $(p = 0.6464)$, which is reflected in the nonsignificant interaction of depression status (non versus mild) by treatment for those seasons.

Let's also apply the `r3b2.` contrast to `treat`, which compares optimism therapy with traditional therapy, and interact that with depression status (mild versus non) performed at each of the four seasons. These four simple interaction contrasts, tested using the `contrast` command below, are illustrated in figure 9.11.

```
. contrast r2.depstat#r3b2.treat@season, noeffects

Contrasts of marginal linear predictions

Margins        : asbalanced
```

	df	F	P>F
depstat#treat@season			
(Mild vs Non) (OT vs TT) Winter	1	4.72	0.0300
(Mild vs Non) (OT vs TT) Spring	1	0.30	0.5845
(Mild vs Non) (OT vs TT) Summer	1	9.61	0.0020
(Mild vs Non) (OT vs TT) Fall	1	0.58	0.4476
Joint	4	3.80	0.0045
Denominator	1044		

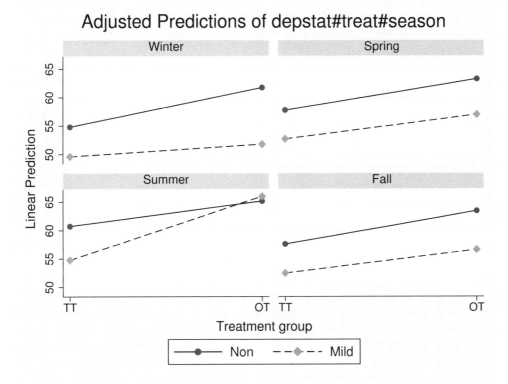

Figure 9.11: Simple interaction contrast of depression status by treatment at each season

The simple interaction contrast of depression status (non versus mild) by treatment (optimism therapy versus traditional) is significant for winter $[F(1, 1044) = 4.72, p = 0.0300]$. The top left panel of figure 9.11 shows how the effect of treatment (optimism therapy versus traditional) differs by depression status (non versus mild) during the winter. The simple interaction contrast is also significant in the summer [$F(1, 1044) =$ 9.61, $p = 0.0020$]. As we see in the lower left panel of figure 9.11, the effect of treatment (optimism therapy versus traditional) differs by depression status (non versus mild) during the summer. The simple interaction contrasts are not significant for spring ($p = 0.5845$) or fall ($p = 0.4476$). This is consistent with the parallel lines for these two seasons shown in figure 9.11.

9.4.3 Simple effects and simple contrasts

We might be interested in focusing on the simple effects of treatment across the levels of season and depression status. The following `contrast` command tests the effect of `treat` at each level of season and depression status. The output is very lengthy, so it is suppressed.

```
. contrast treat@season#depstat
(output omitted)
```

The results of this `contrast` command would show that the effect of `treat` is significant for every combination of season and depression status. We could further refine this test by focusing on the comparison of optimism therapy with traditional therapy (group 3 with 2) by applying the `r3b2.` contrast operator to `treat`, as shown below.

```
. contrast r3b2.treat@season#depstat, nowald pveffects
Contrasts of marginal linear predictions
Margins      : asbalanced
```

	Contrast	Std. Err.	t	P>|t|
treat@season#depstat				
(OT vs TT) Winter#Non	6.933333	1.551151	4.47	0.000
(OT vs TT) Winter#Mild	2.166667	1.551151	1.40	0.163
(OT vs TT) Winter#Severe	-.4666667	1.551151	-0.30	0.764
(OT vs TT) Spring#Non	5.466667	1.551151	3.52	0.000
(OT vs TT) Spring#Mild	4.266667	1.551151	2.75	0.006
(OT vs TT) Spring#Severe	-.5666667	1.551151	-0.37	0.715
(OT vs TT) Summer#Non	4.4	1.551151	2.84	0.005
(OT vs TT) Summer#Mild	11.2	1.551151	7.22	0.000
(OT vs TT) Summer#Severe	-.5666667	1.551151	-0.37	0.715
(OT vs TT) Fall#Non	5.8	1.551151	3.74	0.000
(OT vs TT) Fall#Mild	4.133333	1.551151	2.66	0.008
(OT vs TT) Fall#Severe	.1333333	1.551151	0.09	0.932

This shows that for some combinations of `season` and `depstat`, the difference between optimism therapy and traditional therapy is significant. For example, the effect of optimism therapy versus traditional therapy for nondepressed people in the winter is significant [$t(1044) = 4.47$, $p < 0.001$].

9.5   Closing thoughts

Interactions from three-way factorial designs can be very difficult to interpret. Using the `margins` and `marginsplot` commands, you can begin to visually understand the results. And as illustrated in this chapter, you can use the `contrast` command to form particular contrasts to further understand the exact nature of the interaction.

This chapter illustrated the use of these tools using three common designs with three IVs: a two-by-two-by-two design, a two-by-two-by-three design, and a three-by-three-by-four design. In each case, we saw how the `margins` and `marginsplot` commands can be used to show the interaction and how the `contrast` command can be used to dissect the interaction into its component parts, which allows you to test for specific patterns of results.

10 Supercharge your analysis of variance (via regression)

10.1 Chapter overview

When you compare analysis of variance (ANOVA) and ordinary least-squares regression, the underlying statistical models are exactly the same, as are the predicted values generated by the two different methods. One key difference is how the results are presented. ANOVA focuses on the ANOVA table, which shows sums of squares, mean squares, and F-ratios associated with each independent variable (IV). By comparison, regression models focus on regression tables, which, for each effect, present coefficients, standard errors, t-values and p-values. Dummy variables are created for all but one of the levels of the IV (the omitted group), and coefficients are presented comparing each level of the IV with the omitted group.

There are differences between ANOVA and regression that go beyond the method for displaying results. The ANOVA framework allows you to easily form tests that are difficult to perform using a regression framework. Examples include making contrasts among group means, adjusting for multiple comparisons when making contrasts in group means, and analyzing and dissecting interactions of categorical variables (which was illustrated in chapter 7 on two-way factorial ANOVA models and chapter 9 on three-way factorial ANOVA models). Let's call this bundle of techniques "the technology of ANOVA". In most cases, the technology of ANOVA is limited to just ordinary least-squares methods. But Stata has done something very innovative. Stata has brought the technology of ANOVA and applied it to virtually all of its regression-modeling commands.

This allows you to easily apply all of these techniques that normally are limited to ANOVA across the full spectrum of regression models. Let's illustrate this below by starting with an example of running an ANOVA via ordinary regression.

10.2   Performing ANOVA tests via regression

Consider the analysis we performed in chapter 5, where we looked at happiness as a function of marital status. Below we run an ANOVA with happiness as the dependent variable (DV) and marital status as the IV.

```
. use gss2012_sbs, clear
. anova happy7 marital

                            Number of obs =      1,284    R-squared     =  0.0408
                            Root MSE      =    .977102    Adj R-squared =  0.0378

      Source |   Partial SS         df         MS          F      Prob>F

       Model |    51.89968           4     12.97492      13.59   0.0000

     marital |    51.89968           4     12.97492      13.59   0.0000

    Residual |   1221.0972       1,279    .95472807

       Total |   1272.9969       1,283    .99220334
```

The ANOVA table shows the sum of squares, degrees of freedom, mean squares, F ratio, and the significance of the F ratio associated with marital status (`marital`). With respect to the test of `marital`, we could report these results by saying that there were significant differences in the average happiness among the levels of marital status $[F(4, 1279) = 13.59, p < 0.001]$. Further, we could report that marital status accounted for 3.78% of the variance in happiness (using the adjusted R-squared value).

Let's now run the same analysis but use the `regress` command in place of the `anova` command. Note that we specify `i.marital` when using the `regress` command to indicate that this is a categorical variable—otherwise, the `regress` command would treat it as a continuous variable.

```
. regress happy7 i.marital
```

Source	SS	df	MS
Model	51.8996803	4	12.9749201
Residual	1221.0972	1,279	.954728072
Total	1272.99688	1,283	.99220334

Number of obs	= 1,284
F(4, 1279)	= 13.59
Prob > F	= 0.0000
R-squared	= 0.0408
Adj R-squared	= 0.0378
Root MSE	= .9771

happy7	Coef.	Std. Err.	t	P>|t|	[95% Conf. Interval]	
marital						
widowed	-.2841316	.1028462	-2.76	0.006	-.4858973	-.0823659
divorced	-.4615629	.0798813	-5.78	0.000	-.6182756	-.3048502
separated	-.5949905	.156161	-3.81	0.000	-.9013504	-.2886306
never mar..	-.3488696	.0658518	-5.30	0.000	-.478059	-.2196801
_cons	5.714038	.0406773	140.47	0.000	5.634237	5.79384

The `regress` command creates four dummy codes for the five-level variable `marital` status. Each dummy variable provides a comparison with the omitted group (in this case, group 1—married). For each dummy variable, we see the coefficient that is the difference in the mean happiness for the current group versus the omitted group (married). For example, the average happiness for those who are widowed is 0.284 units lower than that for those who are married. The table also shows the standard error, t-value, and p-value for each coefficient. The effect of widowed (versus married) is significant ($p = 0.006$).

Note: Introduction to the regress command

This chapter discusses regression and the `regress` command, but these topics are not discussed until part IV. For a quick overview of simple and multiple regression in Stata, see chapter 14.

This output lacks an overall test of the effect of marital status on happiness (which we obtained via the `anova` command). We can obtain such a test using the `contrast` command.

```
. contrast marital
Contrasts of marginal linear predictions
Margins      : asbalanced
```

	df	F	P>F
marital	4	13.59	0.0000
Denominator	1279		

The output above corresponds exactly to the test of the main effect of `marital` we saw in the `anova` command. We could use this output to report that the effect of marital status on happiness is significant [$F(4, 1279) = 13.59$, $p < 0.001$].

The `margins`, `marginsplot`, `contrast`, and `pwcompare` commands work the same way after the `regress` command as they do after the `anova` command. For example, we can use the `margins` command to obtain the mean of the DV broken down by `marital` status.

```
. margins marital, nopvalues

Adjusted predictions                            Number of obs     =      1,284
Model VCE    : OLS

Expression   : Linear prediction, predict()
```

	Margin	Delta-method Std. Err.	[95% Conf. Interval]	
marital				
married	5.714038	.0406773	5.634237	5.79384
widowed	5.429907	.09446	5.244593	5.61522
divorced	5.252475	.0687486	5.117603	5.387348
separated	5.119048	.1507701	4.823264	5.414831
never married	5.365169	.0517863	5.263573	5.466764

We can then use the `marginsplot` command to produce the kind of graph we typically create after an ANOVA to illustrate the mean of the DV as a function of the IV. The graph in figure 10.1 shows the mean of happiness (with a 95% confidence interval) for each level of marital status.

```
. marginsplot
  Variables that uniquely identify margins: marital
```

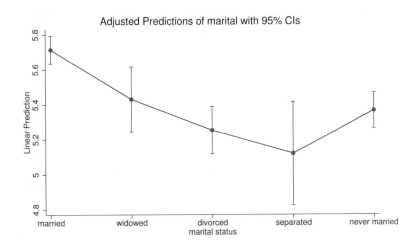

Figure 10.1: Mean happiness by marital status

Following the finding of a significant main effect from an ANOVA, we often want to test planned contrasts among the groups. We can use the `contrast` command following the `regress` command in the same way we have used it following the `anova` command. The example below compares each level of marital status with group 5 (never married).

```
. contrast rb5.marital, nowald pveffects
Contrasts of marginal linear predictions
Margins       : asbalanced
```

	Contrast	Std. Err.	t	P>|t|
marital				
(married vs never married)	.3488696	.0658518	5.30	0.000
(widowed vs never married)	.064738	.1077242	0.60	0.548
(divorced vs never married)	-.1126933	.0860709	-1.31	0.191
(separated vs never married)	-.2461209	.1594159	-1.54	0.123

After finding a significant main effect via ANOVA, we sometimes wish to test all pairwise comparisons among the levels of the IV. Further, such comparisons might be made using adjustments to the p-values to control the overall type-I error rate. We can use the `pwcompare` following the above `regress` command to make pairwise comparisons among the levels of marital status. The example below performs such comparisons and reports Bonferroni adjusted p-values.

```
. pwcompare marital, mcompare(bonferroni) pveffects
Pairwise comparisons of marginal linear predictions
Margins       : asbalanced
```

	Number of Comparisons
marital	10

	Contrast	Std. Err.	Bonferroni t	P>|t|
marital				
widowed vs married	-.2841316	.1028462	-2.76	0.058
divorced vs married	-.4615629	.0798813	-5.78	0.000
separated vs married	-.5949905	.156161	-3.81	0.001
never married vs married	-.3488696	.0658518	-5.30	0.000
divorced vs widowed	-.1774313	.1168292	-1.52	1.000
separated vs widowed	-.3108589	.1779166	-1.75	0.808
never married vs widowed	-.064738	.1077242	-0.60	1.000
separated vs divorced	-.1334276	.1657045	-0.81	1.000
never married vs divorced	.1126933	.0860709	1.31	1.000
never married vs separated	.2461209	.1594159	1.54	1.000

The `anova` and `regress` commands fit the same underlying model but display the results (initially) in different formats. The `anova` command provides the overall test of the equality of the means by default. After the `regress` command, we can use the `contrast` command to obtain a test of the overall equality of the means. Furthermore, we can use the `contrast` command to make planned comparisons, the `margins` command to obtain the estimates of the means for each level of the IV, the `marginsplot` command to graph the mean of the DV by the levels of the IV, and even the `pwcompare` command to form pairwise comparisons using adjustments for multiple comparisons. All of these tools that we normally think of as ANOVA-related tools are available after the `regress` command.

In the case of ordinary least-squares regression, you can use either the `anova` or the `regress` command, depending on your taste. However, when you use the `anova` command in Stata, you are limited to a standard ordinary least squares-style analysis. In most cases, this limitation is not a problem. However, there might be times when you want to access analytic options that go beyond ordinary least squares. This is described in the following section, which shows you how to perform some supercharged ANOVA analyses.

10.3 Supercharging your ANOVA

Let's consider the implications of what we saw in the previous section. We saw that although the `anova` and `regress` commands present results differently, the two techniques are statistically identical. Furthermore, we saw how we can use `contrast`, `margins`, `marginsplot`, and `pwcompare` after the `regress` command. Of course, Stata has many regression commands, including commands for robust regression, regression with robust standard errors, regression with standard errors adjusted for complex survey designs, median regression, logistic regression, and so forth. For all of these commands, you can follow them with the `contrast`, `margins`, `marginsplot`, and `pwcompare` commands. This means that you blend the power of these regression modeling techniques with the ANOVA technology illustrated in chapters 4 to 9 to yield some supercharged analyses. Let's consider some examples below.

10.3.1 Complex surveys

Previously, we saw the results assessing the differences in happiness between the levels of marital status (`marital`) using data from the General Social Survey (GSS). The `anova` command, illustrated below, assumes that the observations are sampled using simple random sampling.

```
. use gss2012_sbs, clear
. anova happy7 marital
```

	Number of obs =	1,284	R-squared	=	0.0408
	Root MSE =	.977102	Adj R-squared =		0.0378

Source	Partial SS	df	MS	F	Prob>F
Model	51.89968	4	12.97492	13.59	0.0000
marital	51.89968	4	12.97492	13.59	0.0000
Residual	1221.0972	1,279	.95472807		
Total	1272.9969	1,283	.99220334		

Actually, the GSS uses a complex sampling methodology that differs from simple random sampling. The analysis should account for the sampling design to obtain proper point estimates and standard errors. The `anova` command cannot be combined with complex survey data, but the `regress` command can. Using the documentation contained in "Appendix A: Sampling Design and Weighting" (see http://www3.norc.org/GSS+Website/Documentation/), GSS describes the appropriate `svyset` command that should be used to indicate the primary sampling unit, sampling weight, and stratification variable. The `svyset` command below informs Stata that the sampling design used a primary sampling unit (as indicated by the variable `sampcode`), that the sampling weight is indicated by the variable `wtss`, and that the variable `vstrat` indicates the stratification.

```
. svyset vpsu [weight=wtss], strata(vstrat)
(sampling weights assumed)
        pweight: wtss
            VCE: linearized
    Single unit: missing
       Strata 1: vstrat
          SU 1: vpsu
         FPC 1: <zero>
```

Having specified the survey design (via the `svyset` command), we can use the `svy` prefix before estimation commands (like the `regress` command) to obtain estimates that reflect the specified sampling design. As shown below, the `regress` command is used with the `svy:` prefix to perform the regression analysis accounting for the survey design.

```
. svy : regress happy7 i.marital
(running regress on estimation sample)

Survey: Linear regression

Number of strata   =        66          Number of obs     =       1,284
Number of PSUs     =       132          Population size   =  1,296.9953
                                        Design df         =          66
                                        F(   4,     63)   =        7.13
                                        Prob > F          =      0.0001
                                        R-squared         =      0.0372
```

happy7	Coef.	Linearized Std. Err.	t	P>|t|	[95% Conf. Interval]	
marital						
widowed	-.1924003	.1244397	-1.55	0.127	-.440852	.0560515
divorced	-.503711	.1193611	-4.22	0.000	-.7420231	-.2653989
separated	-.4778913	.1738385	-2.75	0.008	-.8249712	-.1308115
never mar..	-.2964566	.0750797	-3.95	0.000	-.4463579	-.1465552
_cons	5.696652	.0391279	145.59	0.000	5.618531	5.774774

We can then use the `contrast` command to obtain a test of the overall main effect of `marital` on happiness. The overall effect of marital status is significant $[F(4, 66) = 7.13, p = 0.001]$. Because of the sampling design, the denominator degrees of freedom for the F test are 66 (as compared with 1,279 from the analysis ignoring the sampling design).

```
. contrast marital
Contrasts of marginal linear predictions
                                        Design df         =        66

Margins     : asbalanced
```

	df	F	P>F
marital	4	7.13	0.0001
Design	66		

```
Note: F statistics are adjusted for the survey
      design.
```

The `margins` command can be used to compute the mean happiness by marital status, as shown below. Such means account for the sampling design to provide better estimates of the population values. When you include the `vce(unconditional)` option, the standard errors provide results that generalize better to the population.[1] Comparing these results with those from section 10.2 (ignoring the sampling design), we can see that there are small differences in the means and standard errors compared with the previous analysis.

```
. margins marital, vce(unconditional)
Adjusted predictions                    Number of obs     =      1,284
Expression   : Linear prediction, predict()
```

	Margin	Linearized Std. Err.	t	P>|t|	[95% Conf. Interval]
marital						
married	5.696652	.0391279	145.59	0.000	5.618531	5.774774
widowed	5.504252	.1174779	46.85	0.000	5.2697	5.738804
divorced	5.192941	.1089735	47.65	0.000	4.975369	5.410514
separated	5.218761	.1625046	32.11	0.000	4.89431	5.543212
never mar..	5.400196	.0597054	90.45	0.000	5.28099	5.519402

If we wished, we could follow the above `margins` command with the `marginsplot` command to obtain a graph of the means (with confidence intervals), each of which accounted for the sampling design.

We can also apply contrasts to make comparisons among groups. For example, let's compare each level of marital status with those who have never been married. Again the results differ slightly from those shown in section 10.2, where we did not account for the sampling design.

---

1. In this example, with only one predictor, the `vce(unconditional)` option is not very useful. However, if there were any additional variables in the model, the `vce(unconditional)` option would provide standard-error estimates that better generalize to the population.

```
. contrast rb5.marital, nowald pveffects
Contrasts of marginal linear predictions

                                        Design df          =          66
Margins        : asbalanced

                              │  Contrast   Std. Err.      t      P>|t|
                     marital  │
   (married vs never married) │  .2964566   .0750797     3.95    0.000
   (widowed vs never married) │  .1040563   .1302758     0.80    0.427
  (divorced vs never married) │ -.2072544   .1301141    -1.59    0.116
 (separated vs never married) │ -.1814347   .1637355    -1.11    0.272
```

Note how the output indicates the degrees of freedom (via `Design df`), emphasizing that the degrees of freedom for these tests are 66, which we would use in reporting these results. For example, we could write that the difference in happiness was significantly greater for those who are married than for those who have never married [$t(66) = 3.95$, $p < 0.001$].

We can even use the `pwcompare` command to form pairwise comparisons among all levels of marital status while accounting for the sampling design. Below, the `pwcompare` command is used with the `scheffe` option to obtain Scheffé-adjusted significance tests.[2]

Note: What's new?

Have you wondered what has been recently added to Stata. If so, you can type `help whatsnew`. This shows, in reverse chronological order, additions and fixes that have been made to Stata. If you scroll to the bottom of the output, you will find links tracing the history of updates to Stata going all the way back to Stata version 6.0 (covering the years 1999 to 2000).

---

2. You can specify `bonferroni`, `sidak`, or `scheffe` within the `mcompare()` option for results obtained using the `svy` prefix. The `tukey`, `snk`, `duncan`, and `dunnett` options, however, are not allowed.

```
. pwcompare marital, mcompare(scheffe) pveffects
Pairwise comparisons of marginal linear predictions
                                          Design df          =          66

Margins       : asbalanced
```

	Number of Comparisons
marital	10

	Contrast	Std. Err.	Scheffe t	P>|t|
marital				
widowed vs married	-.1924003	.1244397	-1.55	0.666
divorced vs married	-.503711	.1193611	-4.22	0.003
separated vs married	-.4778913	.1738385	-2.75	0.123
never married vs married	-.2964566	.0750797	-3.95	0.007
divorced vs widowed	-.3113108	.1555606	-2.00	0.413
separated vs widowed	-.2854911	.1837526	-1.55	0.661
never married vs widowed	-.1040563	.1302758	-0.80	0.958
separated vs divorced	.0258197	.1946688	0.13	1.000
never married vs divorced	.2072544	.1301141	1.59	0.640
never married vs separated	.1814347	.1637355	1.11	0.872

This is just one example of how you can apply commands like `contrast`, `margins`, `marginsplot`, and `pwcompare` after a regression-style command just as you would after the `anova` command. Let's consider another example below.

10.3.2   Homogeneity of variance

The homogeneity of variance assumption is one of the important assumptions of ANOVA, especially when the groups have highly unequal sample sizes (Wilcox 1987). For example, suppose we are studying differences in income as a function of marital status. In particular, our interest is in income differences between those who are married and those who are widowed.

We can explore this using the GSS dataset. As a preliminary analysis, we compute the mean, standard deviation, and sample size for each marital status group.

```
. use gss2012_sbs, clear

. tabstat realrinc, by(marital) stat(mean sd n)
Summary for variables: realrinc
      by categories of: marital (marital status)
        marital |     mean        sd         N
----------------+-----------------------------------
        married |  35557.09   68444.13       550
        widowed |  12659.61   12104.21        32
       divorced |  29936.33   58901.01       185
      separated |  23657.07   56378.84        35
  never married |  15370.72   32772.74       344
----------------+-----------------------------------
          Total |  27587.48   57503.99      1146
```

Consider the comparison of those who are married with those who are widowed (our focus of interest). The standard deviation is over 5 times as large for those who are married versus widowed (68,444 versus 12,104). Further, the sample size for those who are married is 550 versus only 32 for those who are widowed. These differences in standard deviations and sample sizes might yield p-values that are not reliable if we use a standard ANOVA model. Nevertheless, let's start by analyzing this using a standard ANOVA via the `anova` command.

```
. anova realrinc marital
                          Number of obs =      1,146    R-squared     =  0.0251
                          Root MSE      =    56877.7    Adj R-squared =  0.0217

        Source |  Partial SS        df          MS          F     Prob>F
    -----------+----------------------------------------------------------
         Model |   9.497e+10         4      2.374e+10      7.34   0.0000

       marital |   9.497e+10         4      2.374e+10      7.34   0.0000

      Residual |   3.691e+12     1,141      3.235e+09
    -----------+----------------------------------------------------------
         Total |   3.786e+12     1,145      3.307e+09
```

The overall test of the effect of `marital` is significant $[F(4, 1141) = 7.34, p < 0.001]$.

Now let's use the `contrast` command to compare the mean income of those who are married (group 1) with that of those who are widowed (group 2). This shows that the income of those who are married is $22,897.5 higher than that of those who are widowed and that the difference is significant $[t(1141) = 2.21, p = 0.027]$.

```
. contrast r1b2.marital, nowald pveffects
Contrasts of marginal linear predictions
Margins       : asbalanced
```

	Contrast	Std. Err.	t	P>|t|
marital (married vs widowed)	22897.48	10343.01	2.21	0.027

I do not feel confident in the previous results because of the differences in the standard deviations between those who are married and those who are widowed in combination with the substantial differences in the sample sizes for these two groups. Let's try another analysis strategy, using regression with robust standard errors. This method is robust to violations of the homogeneity of variance assumption. We can run this kind of model by using the **regress** command combined with the vce(robust) option, as shown below.[3]

```
. regress realrinc i.marital, vce(robust)
Linear regression                    Number of obs   =      1,146
                                     F(4, 1141)      =      12.99
                                     Prob > F        =     0.0000
                                     R-squared       =     0.0251
                                     Root MSE        =      56878
```

realrinc	Coef.	Robust Std. Err.	t	P>|t|	[95% Conf. Interval]	
marital						
widowed	-22897.48	3604.73	-6.35	0.000	-29970.12	-15824.83
divorced	-5620.751	5222.329	-1.08	0.282	-15867.2	4625.694
separated	-11900.02	9856.341	-1.21	0.228	-31238.6	7438.569
never mar..	-20186.37	3415.557	-5.91	0.000	-26887.85	-13484.89
_cons	35557.09	2922.195	12.17	0.000	29823.61	41290.56

Although it is not our primary interest, let's estimate the main effect of marital status via the **contrast** command. Even when using robust standard errors, we see that the effect of marital status on income remains significant [$F(4, 1141) = 12.99$, $p < 0.001$].

---

3. When running an ANOVA, we talk about the homogeneity of variance assumption. When running a regression model, we talk about the assumption of homoskedasticity. The use of these different terms makes these assumptions sound different when they both fundamentally speak to the same issue. That is, we assume that the variation in the distribution of the residuals is equal across the levels of the predictors. By using the **regress** command with the vce(robust) option, we obtain estimates of standard errors that are robust to violations of the assumption of homoskedasticity, which is the same as saying that the standard errors are robust to violations of the homogeneity of variance assumption.

```
. contrast marital
Contrasts of marginal linear predictions
Margins       : asbalanced
```

	df	F	P>F
marital	4	12.99	0.0000
Denominator	1141		

Let's now test the difference in the mean income for those who are married versus widowed using the `contrast` command below. The mean income for those who are married is significantly greater than for those who are widowed [$t(1411) = 6.35$, $p < 0.001$].

```
. contrast r1b2.marital, nowald pveffects
Contrasts of marginal linear predictions
Margins       : asbalanced
```

	Contrast	Std. Err.	t	P>|t|
marital (married vs widowed)	22897.48	3604.73	6.35	0.000

Let's compare the results of this `contrast` command (using robust standard errors) with those of the `contrast` command from the standard ANOVA analysis. Note how the estimate of the difference in the means was the same in this analysis as in the standard ANOVA ($22,897.5 in both analyses). The use of robust standard errors has no effect on the estimates of the differences in the means. The methods differ in how they compute the standard errors (which in turn influences the t-values and p-values). Using robust standard errors, we see that the t-value was 6.35 and that the associated standard error was 3,604.7. The t-value from the standard ANOVA was 2.21, and the standard error was 10,343.0. In this instance, the robustly estimated standard error was smaller than the standard error from the regular ANOVA (3,604.7 versus 10,343.0). In other instances, robustly estimated standard errors could be larger. In this case, both methods pointed to the same conclusion—that the mean income of those who are married is significantly greater than that of those who are widowed. But I would put much more faith in the results using robust standard errors.

> **Note: More on robust standard errors with ANOVA**
>
> In this section, I applied the `vce(robust)` option to address the issue of violating the homogeneity of variance assumption in ANOVA. As I noted before, this is the equivalent of violating the homoskedasticity assumption. When this assumption is violated, Stata offers `vce(robust)` for computing appropriate standard errors. You can find more information about how Stata estimates robust standard errors by typing `help vce_option` and then looking for the section about the `vce(robust` option. Within that section, you can click on the link "Obtaining robust variance estimates". That will take you to a section of the manual that describes how the `vce(robust)` option uses methods independently developed by Huber (1967) and White (1980), and this is often called a "sandwich estimator". The key is that the methods used by applying the `vce(robust)` option yield appropriate p-values even in the presence of heteroskedasticity.

10.3.3   Robust regression

Let's explore a third method that we could use for the analysis of income as a function of marital status. This method is called robust regression. Robust regression is different from the previous method, which focused on creating robust estimates of the standard errors. Robust regression is a form of weighted regression in which influential observations (often extreme observations) are downweighted to reduce their disproportionate influence. For example, consider our comparison income among levels of marital status. In the comparison of married versus widowed people, a small number of extremely large incomes could have an undue influence on our estimate of the difference in the mean income between these two groups.

In fact, if you inspect the distribution of incomes via the `tabulate` command (output omitted to save space), you will see that there is a small group of people whose incomes are coded as $341,672, with the next highest income as $68,600.

```
. tabulate realrinc
(output omitted)
```

In the example below, robust regression is used via the `rreg` command to estimate the effect of marital status on income. (The `nolog` option is used to suppress display of the iteration log to save printed space in the book.) In this analysis, observations are weighted based on how influential they are. Observations that have a greater influence on the results are weighted less.

```
. rreg realrinc i.marital, nolog
Robust regression                            Number of obs   =     1,146
                                             F(  4,    1141) =     22.20
                                             Prob > F        =    0.0000

     realrinc |     Coef.   Std. Err.      t    P>|t|     [95% Conf. Interval]

      marital |
      widowed |  -7934.065   2357.578   -3.37   0.001    -12559.74   -3308.389
     divorced |  -1049.186   1101.888   -0.95   0.341     -3211.14    1112.768
    separated |  -5225.884   2260.079   -2.31   0.021     -9660.26   -791.5072
   never mar..|   -7918.89   891.1877   -8.89   0.000     -9667.44   -6170.339

        _cons |    19066.9   552.8148   34.49   0.000     17982.25    20151.55
```

We can then use the `contrast` command to obtain an estimate of the overall effect of marital status on income. The effect of marital status on income via robust regression is significant $[F(4, 1141) = 22.20, p < 0.001]$.

```
. contrast marital
Contrasts of marginal linear predictions
Margins       : asbalanced

                      df          F        P>F

      marital |        4      22.20     0.0000

  Denominator |     1141
```

Let's now use the `contrast` command to compare the income of those who are married with that of those who are widowed. This difference is significant $[t(1141) = 3.37, p = 0.001]$.

```
. contrast r1b2.marital, nowald pveffects
Contrasts of marginal linear predictions
Margins       : asbalanced
```

	Contrast	Std. Err.	t	P>|t|
marital (married vs widowed)	7934.065	2357.578	3.37	0.001

The difference in incomes between those who are married and those who are widowed remains significant. But note that the estimate of this difference is much smaller with robust regression. With robust regression, the estimated difference in income for those who are married versus widowed is \$7,934.1 compared with \$22,897.5 from the standard ANOVA (as well as from the regression with robust standard errors).

10.3.4   Quantile regression

In light of the results from the previous analysis, it appears that extreme income values could be exerting a strong influence on the estimates of the mean income for those who are married versus widowed. Let's use this as a chance to consider another analysis strategy called median regression. In this case, median regression seeks to estimate the difference in the median incomes for those who are married versus widowed. As we know, medians are less influenced by extreme observations. We can perform a median regression using the `qreg` command shown below. (The `nolog` option is used to suppress display of the iteration log to save space.)

```
. qreg realrinc i.marital, nolog
Median regression                          Number of obs =      1,146
  Raw sum of deviations 1.18e+07 (about 13475)
  Min sum of deviations 1.14e+07           Pseudo R2      =     0.0300
```

realrinc	Coef.	Std. Err.	t	P>|t|	[95% Conf. Interval]
marital					
widowed	-10412.5	2845.084	-3.66	0.000	-15994.68 -4830.316
divorced	0	1329.739	0.00	1.000	-2609.008 2609.008
separated	-6737.5	2727.423	-2.47	0.014	-12088.83 -1386.172
never mar..	-9187.5	1075.47	-8.54	0.000	-11297.62 -7077.38
_cons	18375	667.1272	27.54	0.000	17066.07 19683.93

Let's use the `contrast` command to obtain an overall test of the effect of marital status on median income. This test is significant [$F(4, 1141) = 22.43$, $p < 0.001$].

```
. contrast marital
Warning: cannot perform check for estimable functions.
Contrasts of marginal linear predictions
Margins        : asbalanced
```

	df	F	P>F
marital	4	22.43	0.0000
Denominator	1141		

Now, we can use the **contrast** command to compare the median income for those who are married with that of those who are widowed. As we see below, the median income is significantly greater for those who are married than for those who are widowed $[t(1141) = 3.66, p < 0.001]$. In fact, the median income is estimated to be \$10,412.5 higher.

```
. contrast r1b2.marital, nowald pveffects
Warning: cannot perform check for estimable functions.
Contrasts of marginal linear predictions
Margins        : asbalanced
```

	Contrast	Std. Err.	t	P>|t|
marital (married vs widowed)	10412.5	2845.084	3.66	0.000

Before concluding this chapter, we will consider one key difference between fitting models via **anova** and fitting models via a regression strategy.

10.4   Main effects with interactions: anova versus regress

This section considers the meaning of main effects in the presence of an interaction when using the **regress** command compared with the **anova** command. The tests of coefficients reported by the **regress** command correspond to the results obtained from using dummy $(0/1)$ coding. By contrast, the effects reported by the **anova** command correspond to the results obtained using contrast $(-1/1)$ coding. In the presence of interactions, this can lead to conflicting estimates of so-called main effects for the **regress** command versus the **anova** command. This is illustrated using the example from section 7.2 in which a two-by-two model was used to predict optimism from treatment (control and optimism therapy) and depression status (*minimal/mild* depression versus *moderate/severe* depression). Let's use the dataset for this example and show the mean optimism by treatment group and depression status.

```
. use opt-2by2

. tabulate depstat treat, summarize(opt)
```
Means, Standard Deviations and Frequencies of Optimism score

Depression status	Treatment group Con	OT	Total
Min/Mild	46.6	56.3	51.5
	10.0	10.0	11.1
	50	50	100
Mod/Sev	41.8	45.2	43.5
	10.0	10.0	10.1
	50	50	100
Total	44.2	50.7	47.5
	10.2	11.4	11.3
	100	100	200

Now, let's repeat the **anova** command that was used in section 7.2 to predict optimism from treatment, depression status, and the interaction of these two variables.

```
. anova   opt treat##depstat
```

Number of obs =	200	R-squared	=	0.2286
Root MSE	=	9.98476	Adj R-squared =	0.2168

Source	Partial SS	df	MS	F	Prob>F
Model	5789.5	3	1929.8333	19.36	0.0000
treat	2112.5	1	2112.5	21.19	0.0000
depstat	3184.02	1	3184.02	31.94	0.0000
treat#depstat	492.98	1	492.98	4.94	0.0273
Residual	19540.32	196	99.69551		
Total	25329.82	199	127.28553		

Let's perform this analysis but instead use the `regress` command.

```
. regress opt treat##depstat, vsquish
```

Source	SS	df	MS		
				Number of obs	= 200
				F(3, 196)	= 19.36
Model	5789.5	3	1929.83333	Prob > F	= 0.0000
Residual	19540.32	196	99.6955102	R-squared	= 0.2286
				Adj R-squared	= 0.2168
Total	25329.82	199	127.285528	Root MSE	= 9.9848

opt	Coef.	Std. Err.	t	P>|t|	[95% Conf. Interval]	
treat						
OT	9.64	1.996953	4.83	0.000	5.701727	13.57827
depstat						
Mod/Sev	-4.84	1.996953	-2.42	0.016	-8.778273	-.9017271
treat#						
depstat						
OT#Mod/Sev	-6.28	2.824118	-2.22	0.027	-11.84956	-.7104411
_cons	46.64	1.412059	33.03	0.000	43.85522	49.42478

Let's compare the results of the `anova` command with the results of the `regress` command and focus on the significance tests. These comparisons are a bit tricky because the `anova` command reports F statistics, whereas the `regress` command reports t statistics. But we can square the t-value from the `regress` command to convert it into an equivalent of an F statistic. There will be slight differences between the squared t-value and the F-value because of rounding (the t-values are rounded to two decimal places).

We will first compare the test of the `treat#depstat` interaction, comparing the results from the `regress` command with the results from the `anova` command. If we square the t-value of -2.22 from the `regress` command, we obtain 4.93, which (aside from slight rounding) is the same as the F statistic from the `anova` command, 4.94

Using the `contrast treat#depstat` command following the `regress` command also yields the exact same results as the `anova` command. The F-value from the `contrast` command is the same as the F-value from the `anova` command, 4.94.

```
. contrast depstat#treat
Contrasts of marginal linear predictions
Margins      : asbalanced
```

	df	F	P>F
depstat#treat	1	4.94	0.0273
Denominator	196		

Let's now compare the test of `treat` from the `regress` command with the `anova` command. We square the *t*-value for the `treat` effect from the `regress` command (4.83) and obtain 23.33. This is considerably different from the *F*-value for the `treat` effect from the `anova` command, 21.19. This is not just a difference due to rounding.

Suppose we use the `contrast treat` command to test the `treat` effect. The *F*-value from the `contrast` command is 21.19, which does match the *F*-value from the `anova` command.

```
. contrast treat
Contrasts of marginal linear predictions
Margins      : asbalanced
```

	df	F	P>F
treat	1	21.19	0.0000
Denominator	196		

This might seem perplexing, but there is a perfectly logical explanation for this. The reason for these discrepancies is that the `regress` command reports tests based on dummy coding (that is, 0/1 coding), whereas the `anova` and `contrast` commands report results based on effect coding (that is, 1/-1 coding). The interpretation of the interactions is the same whether you use effect coding or dummy coding, but the meaning of the main effects differs. When you use dummy coding, the coefficient for `treat` represents the effect of `treat` when `depstat` is held constant at zero (that is, for those with *minimal/mild* depression, the reference group). Referring back to the table of means, we see that the effect of `treat` for those with *minimal/mild* depression is 56.3 minus 46.6, which is 9.7. Aside from slight rounding, this matches the coefficient for `treat` from the `regress` command (9.64). Likewise, the coefficient for `depstat` corresponds to the effect of `depstat` for the control group; that is, $41.8 - 46.6 = -4.8$. Aside from slight rounding differences, this matches the coefficient of `depstat` from the `regress` command (-4.84).

When we describe results from the `regress` command, the tests of `treat` and `depstat` might be called "main effects", but this is really a misnomer. These are really simple effects. The test of `treat` from the `regress` output is the effect of `treat` when `depstat` is held constant at the reference group. And the coefficient for `depstat` is the effect of `depstat` when `treat` is held constant at the reference group.

The `anova` and `contrast` commands report results that arise from effect coding. The main effects reported by these commands represent the classic ANOVA main effects. Referring to the table of means, we see that the main effect of `treat` concerns the differences in the column means (that is, 44.2 versus 50.7) and that the main effect of `depstat` concerns the differences in the row means (that is, 51.5 versus 43.5).

This really raises the question of whether we should even be interpreting main effects in the presence of an interaction. Regardless of whether the main effects are coded using dummy coding or effect coding, effects are not very meaningful when those effects are part of a significant interaction term.

10.5   Closing thoughts

In this chapter, we have seen how we can apply techniques that are commonly used in the tradition of ANOVA with commands other than the `anova` command. We used the `margins` command to estimate the mean of the DV as a function of the IV and the `marginsplot` command to graph the mean of the DV as a function of the IV. We used the `contrast` command to test the null hypothesis regarding the equality of the DV means across all levels of the IV, and we used the `contrast` command in conjunction with contrast operators to form specific contrasts (comparisons) among the levels of the IV. We also used the `pwcompare` command to form pairwise comparisons and the `mcompare()` option to adjust p-values for multiple comparisons.

As I mentioned in section 4.5, this suite of commands (`margins`, `marginsplot`, `contrast`, and `pwcompare`) are called postestimation commands. In this chapter, we saw how these commands can be applied not just after the `anova` command but also after other estimation commands. In fact, you can use these postestimation commands after just about any Stata estimation command. You can check this out for yourself by typing

```
. help estimation commands
```

Among that long list of commands, we see the `logistic` command for logistic regression. To learn what postestimation commands can be used following the `logistic` command, we can type

```
. help logistic postestimation
```

Stata will then show the postestimation commands that are appropriate after the `logistic` command. This list includes the `margins`, `marginsplot`, `contrast`, and `pwcompare` postestimation commands.

By typing `help estimation commands`, you can learn about the variety of estimation commands that Stata offers. Then, for any given command (say, `mixed`), you can type `help mixed postestimation`, and you will see all the postestimation commands supported after the `mixed` command.[4]

You can use these two `help` commands to explore all the possibilities for applying these ANOVA-style postestimation commands, using statistical models that go beyond ordinary ANOVA and that allow you to supercharge your ANOVA.

---

4. You would find that the `margins`, `marginsplot`, `contrast`, and `pwcompare` postestimation commands are supported by the `mixed` command.

11 Power analysis for analysis of variance and covariance

11.1 Chapter overview

It is easy to overlook the importance of a power analysis as part of planning a research study. To omit a power analysis is akin to building a house without first formulating a budget for the construction. If you do not create a budget at the outset, you might get halfway through building the house and find that you have run out of money and all you have is a half-built house. Having a half-built house is analogous to having performed a study that was underpowered from the start and did not find a significant result because of the lack of power. A power analysis can tell you whether you are going to collect far too few subjects (which will leave you with very little chance of finding a significant effect). It can also tell you whether you are going to collect far too many subjects (which means that you might waste money collecting far more subjects than you actually needed).

A power analysis can seem like a daunting task: it requires knowing information about your study that you do not know, because you have not yet performed your study. The key to performing a useful power analysis is focusing on the information that you know and coming up with useful assessments of the information that you do not know. This chapter is not merely about how to perform a power analysis in Stata; it also is about how to formulate a power analysis strategy.

The first section (11.2) shows you not only how to perform a power analysis for a two-sample t test but also how to strategically approach a power analysis. The following section (11.3) illustrates a power analysis for a one-way analysis of variance (ANOVA). Section 11.4 illustrates a power analysis for an analysis of covariance (ANCOVA). Section 11.5 illustrates a power analysis for a two-way ANOVA. The chapter concludes with section 11.6, where I share closing thoughts on power analysis.

11.2   Power analysis for a two-sample t test

A two-sample t test is an excellent example to illustrate the overall concepts involved in a power analysis. Let's first describe the unknowns involved in a power analysis for a t test:

> The alpha for your hypothesis test
> The desired power for your test
> The mean for group 1
> The mean for group 2
> The SD for group 1
> The SD for group 2
> The N for group 1
> The N for group 2

This seems to be a daunting list of unknowns. Let's start with a simple example in which our aim is to replicate a previous research study.

11.2.1   Example 1: Replicating a two-group comparison

Let's say that our goal is to replicate the study we saw in section 4.2, where we used a t test to compare the optimism of those receiving optimism therapy with that of those in a control group. The `ttest` command and output are repeated and shown below.

```
. use opt-2
. ttest opt, by(treat)
Two-sample t test with equal variances
```

Group	Obs	Mean	Std. Err.	Std. Dev.	[95% Conf. Interval]	
Con	100	44.29	1.00567	10.0567	42.29453	46.28547
OT	100	49.91	1.020209	10.20209	47.88568	51.93432
combined	200	47.1	.7417215	10.48953	45.63736	48.56264
diff		-5.62	1.43255		-8.445014	-2.794986

```
     diff = mean(Con) - mean(OT)                           t =  -3.9231
Ho: diff = 0                            degrees of freedom =      198

   Ha: diff < 0              Ha: diff != 0               Ha: diff > 0
 Pr(T < t) = 0.0001     Pr(|T| > |t|) = 0.0001          Pr(T > t) = 0.9999
```

The output from the `ttest` command gives us the information that we need to determine the sample size required to replicate the effects found in this study.

Consider the `power twomeans` command below. In that command, I have inserted the values for the means for group 1 and group 2 and the standard deviations (SDs) for groups 1 and 2. I also specified that the power is 0.80 and the alpha is 0.05. Given these values, the `power twomeans` command computes the required sample size, which is N = 106 (N = 53 per group).

```
. power twomeans 44.3 49.9, sd1(10.1) sd2(10.2) power(.80) alpha(0.05)
Performing iteration ...
Estimated sample sizes for a two-sample means test
Satterthwaite's t test assuming unequal variances
Ho: m2 = m1   versus  Ha: m2 != m1

Study parameters:
        alpha =    0.0500
        power =    0.8000
        delta =    5.6000
           m1 =   44.3000
           m2 =   49.9000
          sd1 =   10.1000
          sd2 =   10.2000

Estimated sample sizes:
            N =       106
  N per group =        53
```

The `power` command actually assumes 80% power and alpha of 5%, so if we use those values, we can omit the `power()` and `alpha()` options, as shown below. (The output is omitted to save space.)

```
. power twomeans 44.3 49.9, sd1(10.1) sd2(10.2)
  (output omitted )
```

The initial study (from section 4.2) had a total sample size of 200. To replicate that study, we might have been inclined to use the same sample size. But using the `power twomeans` command, we see that we could reduce the sample size by nearly half (only using $N = 106$) and achieve 80% power to detect the effect that was previously found.

In the next section, let's repeat this power analysis but reframe it by using standardized effect sizes. This will give us a simpler way to frame power analyses to allow us to perform power analyses for situations where we are not merely replicating previous results.

11.2.2   Example 2: Using standardized effect sizes

One way to make a power analysis simpler is to use a standardized effect size in place of needing to know the means for groups 1 and 2 and the SDs for groups 1 and 2. A common practice is to use Cohen's d, which is the difference in the means divided by the pooled SD (s_p), as shown in (11.1).

$$d = \frac{\overline{x}_1 - \overline{x}_2}{s_p} \tag{11.1}$$

The pooled SD, s_p, can be computed as shown in (11.2).

$$s_p = \sqrt{\frac{(n_1 - 1)s_1^2 + (n_2 - 1)s_2^2}{n_1 + n_2 - 2}} \tag{11.2}$$

Applying (11.2) using the SDs and sample sizes shown in the `ttest` output yields a pooled SD, s_p, of 10.13. Using the means of 44.3 and 49.9 and a pooled SD of 10.13, we can compute Cohen's d as shown below and obtain a value of -0.553.

```
. display (44.3 - 49.9)  / 10.13
-.55281343
```

We can now replicate the previous power analysis by using the following `power twomeans` command. Note how I now specify a mean of 0 for the first group and the standardized effect size of 0.553 for the second group. This is combined with the `sd(1)` option to indicate that the SD is 1 because the means are expressed using standardized effect sizes.

```
. power twomeans 0 .553, sd(1)

Performing iteration ...

Estimated sample sizes for a two-sample means test
t test assuming sd1 = sd2 = sd
Ho: m2 = m1  versus  Ha: m2 != m1

Study parameters:
        alpha =    0.0500
        power =    0.8000
        delta =    0.5530
           m1 =    0.0000
           m2 =    0.5530
           sd =    1.0000

Estimated sample sizes:
            N =       106
  N per group =        53
```

The `power twomeans` command indicates the required sample size is 106. Note the required sample size did not change when we use the simplified standardized effect size. We have reduced the number of unknowns from 4 (mean 1, mean 2, SD1, SD2) to a single value (the effect size of 0.553).

The `power twomeans` command actually assumes that we are providing standardized effect sizes (that is, that the sd is 1), so we can omit the `sd(1)` option from the previous example and obtain the same results, as shown below but with the output omitted.

```
. power twomeans 0 .553
  (output omitted)
```

The `power twomeans` command can also display its output in a table using the `table` option.

```
. power twomeans 0 .553, table

Performing iteration ...

Estimated sample sizes for a two-sample means test
t test assuming sd1 = sd2 = sd
Ho: m2 = m1  versus  Ha: m2 != m1
```

alpha	power	N	N1	N2	delta	m1	m2	sd
.05	.8	106	53	53	.553	0	.553	1

This is a space-saving format for the purposes of this book: it displays the same information using fewer lines of text. Also, as we will see in section 11.2.5, this is a useful layout when you want to display the results of power analyses associated with a range of different effect sizes.

In the example below, I use the `table()` option both to request that the output be displayed in a tabular fashion and to select which columns are displayed in the table.

```
. power twomeans 0 .553, table(diff N N1 N2 alpha power)

Performing iteration ...

Estimated sample sizes for a two-sample means test
t test assuming sd1 = sd2 = sd
Ho: m2 = m1   versus   Ha: m2 != m1
```

diff	N	N1	N2	alpha	power
.553	106	53	53	.05	.8

The table above shows the effect size, the overall sample size, the sample size for groups 1, the sample size for group 2, the alpha, and the power. This makes for an even more compact output.

11.2.3 Estimating effect sizes

I cheated a bit in the previous examples because our goal is rarely to replicate an existing study. We usually want to test a new hypothesis for which data have not been collected. Our challenge is to intelligently express the expected effect of the study in terms of Cohen's d in the absence of a previous study. To that end, I highly recommend that you read the article by Cohen (1992) titled "A power primer". In this article, Cohen (1992) gives an overview of the fundamental concepts of power analysis and provides guidelines regarding what can be considered a small, medium, or large effect size.

In terms of Cohen's d, Cohen (1992) considers a small effect to be 0.2, a medium effect to be 0.5, and a large effect to be 0.8. These are very generic guidelines that you can fall back upon in the absence of any previous data. Another way to help make these values meaningful is that an effect size of 0.2 (small) is equivalent to explaining 1% of the variance in the outcome; that an effect size of 0.5 (medium) is equivalent to explaining 5.9% of the variance in the outcome; and that a large effect size (0.8) is equivalent to explaining 13.8% of the variance in the outcome. Assuming equal Ns for the experimental and control groups, you can compute the explained variance (R^2) from Cohen's d using the formula shown in (11.3).

$$R^2 = \frac{d^2}{(d^2 + 4)} \tag{11.3}$$

So an effect size of 0.4 corresponds to 3.85% explained variance.

```
. display .4^2 / (.4^2 + 4)
.03846154
```

Tip: A concrete example of effect sizes

To make the concept of an effect size even more concrete, let's consider the height of adult men in the United States, which has a mean of 5 feet 9.3 inches and an SD of 2.92 inches (Schilling, Watkins, and Watkins 2002). Let's round this to 5 foot 9 and an SD of 3 inches. Imagine that we had a treatment to increase the height of adult men and that we wanted to determine what would constitute a small, medium, or large effect size for such a study. Using the effect-size guidelines, we see that the gain of a small effect would be 0.6 inches (0.2×3), of a medium effect would be 1.5 inches (0.5×3), and of a large effect would be 2.4 inches (0.8×3).

You can use other sources of information to create an informed estimate of the effect size for your treatment on your outcome of interest.

- **Same treatment, different outcome.** There may be previous research that shows treatment effects from previous studies using the treatment of interest but using different outcomes. You can examine this research with respect to your treatment of interest on other outcomes to obtain a sense of how potent your treatment is on affecting other outcomes.
- **Same outcome, different treatment.** You can examine previous research for the outcome of interest to determine the effect sizes that have been observed for your outcome from other treatments. Even if the studies use treatments that differ from the one you plan to implement, you can compute Cohen effect sizes using past published research to form an impression of the size of treatment effects that are found with respect to the outcome variable of interest.
- **Validated instruments.** If your outcome is a validated instrument, like the Beck Depression Inventory, it comes with a manual that describes the scoring and norms for the scale. You may be able to use that information to help you assess what would constitute a meaningful or clinically relevant treatment effect size.
- **Useful effect size.** You can frame your power analysis by asking yourself, How small can the effect be and still be useful? You can use this as the effect size for your power analysis.

Having considered different sources that you can use to estimate effect sizes, let's further explore how you can perform power analyses for a given effect size.

11.2.4   Example 3: Power for a medium effect

Let's compute the required sample size given a medium effect size (that is, a standardized effect size of 0.5) with alpha of 0.05 and power of 0.80. We can see that our required sample size is N = 128 (N = 64 per group).

```
. power twomeans 0 .5, table(diff N N1 N2 alpha power)

Performing iteration ...

Estimated sample sizes for a two-sample means test
t test assuming sd1 = sd2 = sd
Ho: m2 = m1   versus   Ha: m2 != m1
```

diff	N	N1	N2	alpha	power
.5	128	64	64	.05	.8

The `power` command allows us to specify the difference between the group means via the `diff()` option. By indicating that the control group mean is zero and `diff(0.5)`, we obtain the exact same result as shown above.

```
. power twomeans 0, diff(0.5) table(diff N N1 N2 alpha power)

Performing iteration ...

Estimated sample sizes for a two-sample means test
t test assuming sd1 = sd2 = sd
Ho: m2 = m1   versus   Ha: m2 != m1
```

diff	N	N1	N2	alpha	power
.5	128	64	64	.05	.8

We will find the `diff()` option useful in the following example, where we can specify a range of effect sizes via the `diff()` option.

11.2.5   Example 4: Power for a range of effect sizes

Let's return to the example from section 4.2, in which optimism therapy was compared with a control condition using optimism as the dependent variable (DV). Suppose that the researcher wants to perform a similar study but that the DV will be happiness.

Let's frame this power analysis by asking ourselves how big does an effect need to be to be useful, that is, how much better would optimism therapy need to be than control to be a useful treatment for increasing happiness. Our researcher feels that an effect that ranges between a small and a medium effect would make optimism therapy useful for increasing happiness. In terms of Cohen's effect size, this is a value ranging from 0.2 to 0.5. Let's determine the required sample size for effect sizes ranging from 0.2 to 0.5 in units of 0.05 by specifying `diff(.2(.05).5)` with the `power` command below.

```
. power twomeans 0, diff(.2(.05).5) table(diff N N1 N2 alpha power)

Performing iteration ...

Estimated sample sizes for a two-sample means test
t test assuming sd1 = sd2 = sd
Ho: m2 = m1   versus   Ha: m2 != m1
```

diff	N	N1	N2	alpha	power
.2	788	394	394	.05	.8
.25	506	253	253	.05	.8
.3	352	176	176	.05	.8
.35	260	130	130	.05	.8
.4	200	100	100	.05	.8
.45	158	79	79	.05	.8
.5	128	64	64	.05	.8

The output above shows a separate power calculation for each of the standardized effect sizes specified. Focusing on the smallest and largest effect sizes, we see that for a small effect size (that is, 0.2), the required sample size is $N = 788$ ($N = 394$ per group) and that for a medium effect size (that is, 0.5), the required sample size is $N = 128$ ($N = 64$ per group).

Let's repeat the above **power twomeans** command but use the **graph** option instead of the **table()** option. This creates the graph in figure 11.1, which shows the required sample size as a function of the standardized effect size.

```
. power twomeans 0, diff(.2(.05).5) graph
```

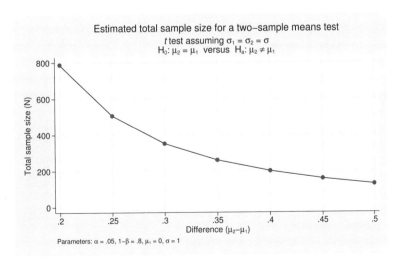

Figure 11.1: Required sample size for 80% power as a function of effect size

As figure 11.1 shows, there is not a linear relationship between the effect size (on the x axis) and the sample size (on the y axis). The additional sample size to detect an effect of 0.4 versus 0.5 (200 versus 128) is much smaller than the additional sample size to detect an effect of 0.2 versus 0.3 (788 versus 352).

We now enter the realm where power analysis becomes a balancing act. At this stage, there are no clear-cut or correct answers, only information to help us make informed decisions. Other considerations would be the cost of the entire study (which increases with greater sample size) against the size of the effect we can detect. The following example shows how we might use the power command to weigh the cost of the study against the effect size that we can detect.

11.2.6   Example 5: For a given N, compute the effect size

In examples 1, 2, 3, and 4, I have focused on computing the required N for specified effect sizes. Instead, we can determine the effect size that can be detected for a specified sample size. Continuing with the previous example, let's imagine that it costs about $1,000 per subject and that the proposed budget for this portion of the study is about $150,000. I am interested in the effect size that could be detected with this budget, that is, the effect size we could detect with a total sample size of $N = 150$. The power command below specifies power(0.80) and n(150), and it calculates the standardized effect size that can be detected given these values. A standardized effect size of 0.46 could be detected with this budget.

```
. power twomeans 0, power(.8) n(150) table(diff N N1 N2 alpha power)
Performing iteration ...

Estimated experimental-group mean for a two-sample means test
t test assuming sd1 = sd2 = sd
Ho: m2 = m1   versus  Ha: m2 != m1; m2 > m1
```

diff	N	N1	N2	alpha	power
.4605	150	75	75	.05	.8

Perhaps my budget could run from $100,000 to $200,000, which implies from 100 to 200 subjects. By specifying n(100(10)200), we can obtain the effect size that can be detected for N ranging from 100 to 200 in increments of 10.

```
. power twomeans 0, power(.8) n(100(10)200) table(diff N N1 N2 alpha power)
Performing iteration ...
Estimated experimental-group mean for a two-sample means test
t test assuming sd1 = sd2 = sd
Ho: m2 = m1   versus   Ha: m2 != m1; m2 > m1
```

diff	N	N1	N2	alpha	power
.5659	100	50	50	.05	.8
.5391	110	55	55	.05	.8
.5157	120	60	60	.05	.8
.4952	130	65	65	.05	.8
.4769	140	70	70	.05	.8
.4605	150	75	75	.05	.8
.4457	160	80	80	.05	.8
.4322	170	85	85	.05	.8
.4199	180	90	90	.05	.8
.4086	190	95	95	.05	.8
.3981	200	100	100	.05	.8

In the output above, let's focus on the columns labeled `diff` and `N`. The `diff` column is the effect size that can be detected for the specified `N`. The smallest budget ($100,000) implies an `N` of 100 and the ability to detect an effect size of 0.57, while the largest budget of $200,000 (with an `N` = 200) can detect an effect size of 0.398. This information can be used to weigh budgetary considerations against the effect size that can be detected.

11.2.7   Example 6: Compute effect sizes given unequal Ns

The results from example 5 would be accurate if the cost were the same for subjects in the treatment and control group; actually, the cost is $1,000 per subject in the treatment group but only $200 per subject in the control group. Control subjects cost less because they do not receive therapy. Let's assume that our budget is $120,000 for this example. Given this budget, we can include 100 experimental subjects (at a cost of $100,000) and 100 control subjects (at a cost of $20,000). However, if we reduced the number of experimental subjects by 10, we can include an additional 50 control subjects, yielding 90 experimental subjects and 150 control subjects. Let's compute the power assuming that we include 100, 90, 80, 70, 60, and 50 experimental subjects, which would correspond to 100, 150, 200, 250, 300, and 350 control subjects (respectively). The `power` command uses the `n1()` option to specify the number of control subjects and the `n2()` option to specify the number of experimental subjects. The `parallel` option indicates that the list of values should be processed in a parallel fashion, matching `N1` = 100 with `N2` = 100, `N1` = 150 with `N2` = 90, and so forth until `N1` = 350 and `N2` = 50. The `table()` and `graph` options are used to present the results both as a table (below) and as a graph (see figure 11.2).

```
. power twomeans 0,  alpha(0.05) power(.8)
> n1(100 150 200 250 300 350) n2(100 90 80 70 60 50) parallel
> table(diff N N1 N2 alpha power) graph

Performing iteration ...

Estimated experimental-group mean for a two-sample means test
t test assuming sd1 = sd2 = sd
Ho: m2 = m1   versus  Ha: m2 != m1; m2 > m1
```

diff	N	N1	N2	alpha	power
.3981	200	100	100	.05	.8
.3751	240	150	90	.05	.8
.3719	280	200	80	.05	.8
.38	320	250	70	.05	.8
.3973	360	300	60	.05	.8
.4246	400	350	50	.05	.8

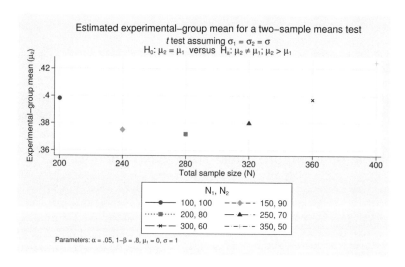

Figure 11.2: Effect size that can be detected with 80% power for specified values of N1 and N2

Looking at the table, we see that the smallest effect size that can be detected is 0.3719 and is achieved when using a total sample size of N = 280 with N1 = 200 and N2 = 80. This is the optimal allocation of subjects (given the combinations considered) because the effect size that can be detected is the smallest. We can also see this in figure 11.2, in which the point associated with the smallest effect size is when N = 280 with N1 = 200 and N2 = 80.

11.3   Power analysis for one-way ANOVA

The examples have shown you how to perform a power analysis for a comparison of two groups (which could be performed using either a *t* test or a two-group ANOVA). But what if you want to perform a power analysis for an ANOVA in which the independent variable (IV) has more than two levels? The following examples illustrate how you can perform power analyses for a three-group ANOVA, focusing on the power associated with prespecified hypotheses.

11.3.1   Overview

Referring to section 4.4, let's consider an example of a three-group ANOVA where group 1 is a control group, group 2 receives traditional therapy, and group 3 receives optimism therapy. Say that we want to assess the power to detect whether traditional therapy has a standardized effect size that is 0.2 units greater than control and whether optimism therapy has a standardized effect size that is 0.5 units greater than control. The `power oneway` command tests this, as shown below.

```
. power oneway 0 .2 .5, table(N N1 N2 N3 alpha power)
Performing iteration ...
Estimated sample size for one-way ANOVA
F test for group effect
Ho: delta = 0   versus  Ha: delta != 0
```

N	N1	N2	N3	alpha	power
234	78	78	78	.05	.8

The `power oneway` command above indicates that the sample size for each group should be 78 to achieve 80% power. That is, the probability of rejecting the following null hypothesis given this pattern of effect sizes is 80%.

$$H_0 : \mu_1 = \mu_2 = \mu_3$$

However, we rarely are interested in testing this overall null hypothesis. Instead, we are usually concerned with specific comparisons between groups. Let's consider three possible hypotheses we might be interested in.

Hypothesis 1. Traditional therapy versus control

This hypothesis is that traditional therapy (group 2) yields greater optimism than the control condition (group 1). We can use the `contrast()` option to indicate that this is the hypothesis of interest. By specifying `contrast(-1 1 0)` with the `power` command, we indicate that this is the contrast of interest.

```
. power oneway 0 .2 .5, contrast(-1 1 0) table(N N1 N2 N3 alpha power)
Performing iteration ...
Estimated sample size for one-way ANOVA
F test for contrast of means
Ho: Cm = 0   versus   Ha: Cm != 0
```

N	N1	N2	N3	alpha	power
1182	394	394	394	.05	.8

The results of the **power oneway** command show us that a sample size of 394 per group would be required to have 80% power in the test of this particular hypothesis. The sample size required for this hypothesis is much larger because it aims to detect a small effect (a standardized effect size of 0.2) in the comparison of group 2 with group 1. In fact, we saw the same required sample size ($N = 394$ per group) in the power analysis illustrated in section 11.2.5 to detect a small effect (0.2).

Hypothesis 2: Optimism therapy versus control

Another interesting hypothesis would be that optimism therapy (group 3) yields greater optimism than the control condition (group 1). We can use the **contrast()** option to indicate that this is the hypothesis of interest. By specifying **contrast(-1 0 1)**, we obtain the power analysis for the comparison of group 3 with group 1.

```
. power oneway 0 .2 .5, contrast(-1 0 1) table(alpha power N1 N2 N3)
Performing iteration ...
Estimated sample size for one-way ANOVA
F test for contrast of means
Ho: Cm = 0   versus   Ha: Cm != 0
```

alpha	power	N1	N2	N3
.05	.8	64	64	64

The results indicate that we would need a sample size of 64 per group to have 80% power in the test of this hypothesis. The sample size required for this hypothesis is much smaller than that for hypothesis 1 because it aims to detect a medium effect (a standardized effect size of 0.5) in the comparison of group 3 with group 1. In fact, we saw the same required sample size ($N = 64$ per group) in the power analysis illustrated in section 11.2.5 to detect a medium effect of 0.5.

Hypothesis 3: Optimism therapy versus traditional therapy

Let's consider a third hypothesis in which we want to test whether optimism therapy (group 3) yields greater optimism than traditional therapy (group 2). We can use the `contrast(0 -1 1)` option to request that the sample size be computed for the contrast of group 3 with group 2.

```
. power oneway 0 .2 .5, contrast(0 -1 1) table(alpha power N1 N2 N3)
Performing iteration ...
Estimated sample size for one-way ANOVA
F test for contrast of means
Ho: Cm = 0   versus   Ha: Cm != 0
```

alpha	power	N1	N2	N3
.05	.8	176	176	176

The results indicate that we need a sample size of 176 per group to have 80% power in the test of this third hypothesis. The sample size required for this hypothesis is a little bit larger than that for hypothesis 2 because it aims to detect a standardized effect size of 0.3 (0.5 minus 0.2) in the comparison of group 3 with group 2. Looking at the power analysis illustrated in section 11.2.5, we see that it also specified a required sample size of $N = 176$ per group to detect an effect size of 0.3.

Summary of hypotheses

We have seen that the required sample size for this three-group ANOVA depends on the planned hypotheses that we want to test. For hypothesis 1 (traditional therapy versus control), we would need a sample size of 394 (per group). For hypothesis 2 (optimism therapy versus control), we would need 64 per group. For hypothesis 3 (optimism therapy versus traditional therapy), we would need 176 per group.

Let's use this information to determine the sample sizes we would want to specify for each group based on the hypotheses of interest.

11.3.2   Example 7: Testing hypotheses 1 and 2

Suppose that our study hypotheses are hypothesis 1 (traditional therapy versus control) and hypothesis 2 (optimism therapy versus control). As we saw above, the test of hypothesis 1 required a sample size of 394, and the test of hypothesis 2 required a sample size of 64. Let's figure out how we can allocate subjects to each of the three groups to use the fewest subjects while still achieving 80% power for the test of these two hypotheses.

Let's formulate a power analysis that would test hypothesis 1 using 394 observations per group. The `power oneway` command specifies the sample size for groups 1, 2, and 3 as 394 via the `n1(394)`, `n2(394)`, and `n3(394)` options, respectively. This confirms we have at least 80% power (80.09%) to detect the difference between group 1 and group 2 using N1 = 394, N2 = 394, and N3 = 394.

```
. power oneway 0 .2 .5, contrast(-1 1 0) n1(394) n2(394) n3(394)
> table(alpha power N1 N2 N3)

Estimated power for one-way ANOVA
F test for contrast of means
Ho: Cm = 0   versus   Ha: Cm != 0
```

alpha	power	N1	N2	N3
.05	.8009	394	394	394

Now, let's use these same sample sizes for the test of hypothesis 2, which compares group 3 with 1 using the `contrast(-1 0 1)` option.

```
. power oneway 0 .2 .5, contrast(-1 0 1) n1(394) n2(394) n3(394)
> table(alpha power N1 N2 N3)

Estimated power for one-way ANOVA
F test for contrast of means
Ho: Cm = 0   versus   Ha: Cm != 0
```

alpha	power	N1	N2	N3
.05	1	394	394	394

The results above show that the comparison of group 1 with group 3 is actually overpowered. The power is far greater than 80% (it actually is 100%). This means that the sample size for this contrast is actually too large. We cannot reduce the sample size for group 1 (because that would reduce the power for the test of hypothesis 1); however, we could reduce the sample size for group 3. Let's try different values for N3 ranging from 30 to 100 in increments of 10 and evaluate the power of the test of the hypothesis comparing group 3 with group 1.

```
. power oneway 0 .2 .5, contrast(-1 0 1) n1(394) n2(394) n3(30(10)100)
> table(alpha power N1 N2 N3)

Estimated power for one-way ANOVA
F test for contrast of means
Ho: Cm = 0   versus   Ha: Cm != 0
```

alpha	power	N1	N2	N3
.05	.7508	394	394	30
.05	.853	394	394	40
.05	.9141	394	394	50
.05	.9499	394	394	60
.05	.9707	394	394	70
.05	.9827	394	394	80
.05	.9897	394	394	90
.05	.9938	394	394	100

The table shows that using a sample size of N3 = 40 achieves at least 80% power (85.3%).

Let's repeat the above command but instead specify that the N3 values range from 30 to 40 in increments of 1.

```
. power oneway 0 .2 .5, contrast(-1 0 1) n1(394) n2(394) n3(30(1)40)
> parallel table(alpha power N1 N2 N3)
Estimated power for one-way ANOVA
F test for contrast of means
Ho: Cm = 0  versus  Ha: Cm != 0
```

alpha	power	N1	N2	N3
.05	.7508	394	394	30
.05	.7634	394	394	31
.05	.7755	394	394	32
.05	.7869	394	394	33
.05	.7979	394	394	34
.05	.8083	394	394	35
.05	.8182	394	394	36
.05	.8276	394	394	37
.05	.8365	394	394	38
.05	.845	394	394	39
.05	.853	394	394	40

This shows that a sample size of N3 = 35 gives us at least 80% power (0.8083) in the test of the hypothesis comparing group 3 with group 1. These **power** commands have shown us that to have 80% power to test hypothesis 1 and hypothesis 2, we would want to specify N1 = 394, N2 = 394, and N3 = 35.

Let's confirm the test with respect to hypothesis 1 using these sample sizes. Indeed, these sample sizes give us at least 80% power to detect the difference specified for the comparison of group 2 with group 1 (hypothesis 1).

```
. power oneway 0 .2 .5, contrast(-1 1 0) n1(394) n2(394) n3(35)
> table(alpha power N1 N2 N3)
Estimated power for one-way ANOVA
F test for contrast of means
Ho: Cm = 0  versus  Ha: Cm != 0
```

alpha	power	N1	N2	N3
.05	.8006	394	394	35

Now, let's confirm this for hypothesis 2, the comparison of group 3 with group 1. We see that the power is 0.8083 for this test.

```
. power oneway 0 .2 .5, contrast(-1 0 1) n1(394) n2(394) n3(35)
> table(alpha power N1 N2 N3)

Estimated power for one-way ANOVA
F test for contrast of means
Ho: Cm = 0  versus  Ha: Cm != 0
```

alpha	power	N1	N2	N3
.05	.8083	394	394	35

This might seem to be a peculiar allocation of subjects, but it is the optimal sample size allocation to achieve 80% given the hypotheses of interest and the hypothesized effect sizes.

Let's consider a second example in which our goal is to test hypothesis 2 and hypothesis 3. We will see that the optimal sample size allocation for the test of those hypotheses is rather different.

11.3.3 Example 8: Testing hypotheses 2 and 3

Suppose that our study aims are to test hypothesis 2 (optimism therapy versus control) and hypothesis 3 (optimism therapy versus traditional therapy). As we saw in section 11.3.1, the test of hypothesis 2 required a sample size of 64 (per group), and the test of hypothesis 3 required a sample size of 176 (per group).

Let's formulate a power analysis that determines the sample sizes we should adopt for group 1, group 2, and group 3 and that yields at least 80% power to detect the effects specified by hypotheses 2 and 3. Let's start with the hypothesis that has the greater sample size, hypothesis 3. The `power oneway` command shows that if we specify $N = 528$ ($N = 176$ per group), we achieve 80% power for the comparison of group 3 with group 2 (hypothesis 3).

```
. power oneway 0 .2 .5, contrast(0 -1 1) n1(176) n2(176) n3(176)
> table(alpha power N N1 N2 N3)

Estimated power for one-way ANOVA
F test for contrast of means
Ho: Cm = 0  versus  Ha: Cm != 0
```

alpha	power	N	N1	N2	N3
.05	.8021	528	176	176	176

Let's then apply these same sample sizes ($n = 176$ for n1(), n2(), and n3()) for the comparison of group 3 with group 1 to test hypothesis 2.

```
. power oneway 0 .2 .5, contrast(-1 0 1) n1(176) n2(176) n3(176)
> table(alpha power N N1 N2 N3)

Estimated power for one-way ANOVA
F test for contrast of means
Ho: Cm = 0   versus   Ha: Cm != 0
```

alpha	power	N	N1	N2	N3
.05	.9968	528	176	176	176

We see that the power of this comparison well exceeds 80%. If we used $n = 176$ per group we would achieve at least 80% power for each of our hypotheses of interest (hypothesis 2 and hypothesis 3). However, we could adopt a smaller sample size for group 1 and still achieve 80% power for the test of hypothesis 2 without affecting the power for hypothesis 3. Let's try specifying sample sizes of 30 to 100 (in increments of 10) for group 1 and assess the power for the test of hypothesis 2.

```
. power oneway 0 .2 .5, contrast(-1 0 1) n1(30(10)100) n2(176) n3(176)
> table(alpha power N N1 N2 N3)

Estimated power for one-way ANOVA
F test for contrast of means
Ho: Cm = 0   versus   Ha: Cm != 0
```

alpha	power	N	N1	N2	N3
.05	.714	382	30	176	176
.05	.8126	392	40	176	176
.05	.8754	402	50	176	176
.05	.9157	412	60	176	176
.05	.9418	422	70	176	176
.05	.9591	432	80	176	176
.05	.9706	442	90	176	176
.05	.9785	452	100	176	176

This shows us that we could adopt a sample size somewhere between 30 and 40 to achieve 80% power for this comparison. Let's more precisely estimate the required sample size for group 1 by repeating the above command but specifying that the n for group 1 ranges from 30 to 40 in increments of 1. This shows that a sample size of N1 = 39 yields a power of 0.8046. This is the smallest sample size we could use for group 1 and achieve 80% power for the comparison of group 3 with group 1.

```
. power oneway 0 .2 .5, contrast(-1 0 1) n1(30(1)40) n2(176) n3(176)
> table(alpha power N N1 N2 N3)

Estimated power for one-way ANOVA
F test for contrast of means
Ho: Cm = 0  versus  Ha: Cm != 0
```

alpha	power	N	N1	N2	N3
.05	.714	382	30	176	176
.05	.7259	383	31	176	176
.05	.7374	384	32	176	176
.05	.7483	385	33	176	176
.05	.7588	386	34	176	176
.05	.7688	387	35	176	176
.05	.7784	388	36	176	176
.05	.7875	389	37	176	176
.05	.7962	390	38	176	176
.05	.8046	391	39	176	176
.05	.8126	392	40	176	176

If our goal is to have 80% power for the test of hypotheses 2 and 3, the minimum sample sizes that we would specify would be N1 = 39, N2 = 176, and N3 = 176. Let's confirm this by using the **power oneway** command to show the power for the test of hypothesis 2 and hypothesis 3 using these sample sizes.

Let's start by showing the power for the test of hypothesis 2. Indeed, the power for this test is at least 80% (0.8046).

```
. power oneway 0 .2 .5, contrast(-1 0 1) n1(39) n2(176) n3(176)
> table(alpha power N N1 N2 N3)

Estimated power for one-way ANOVA
F test for contrast of means
Ho: Cm = 0  versus  Ha: Cm != 0
```

alpha	power	N	N1	N2	N3
.05	.8046	391	39	176	176

Now, let's show the power for the test of hypothesis 3 (group 3 versus group 2). Indeed, the power for this test is at least 80% (0.8016).

```
. power oneway 0 .2 .5, contrast(0 -1 1) n1(39) n2(176) n3(176)
> table(alpha power N N1 N2 N3)

Estimated power for one-way ANOVA
F test for contrast of means
Ho: Cm = 0   versus   Ha: Cm != 0
```

alpha	power	N	N1	N2	N3
.05	.8016	391	39	176	176

11.3.4   Summary

As we have seen in this section, there is a big difference between the required sample size to reject the overall null hypothesis and the required sample size to test specific comparisons among groups. Further, the sample size can vary considerably depending on the specific comparisons (hypotheses) of interest. By focusing on the comparisons of interest, you can select the appropriate allocation of subjects to groups to have sufficient power for the tests of interest.

11.4   Power analysis for ANCOVA

11.4.1   Example 9: Using pretest as a covariate

Consider the study from section 4.3 in which two groups were studied with respect to optimism. Suppose a researcher wants to replicate this study but use happiness as the DV. Her desire is to detect a standardized effect size of 0.5. We saw such a power analysis (to detect an effect size of 0.5) in section 11.2.4. In fact, the power twomeans command from section 11.2.4 is repeated below; it shows that a total sample size of N = 128 (N = 64 per group) is needed to detect an effect size of 0.5 with 80% power.

```
. power twomeans 0 .5, table(diff N N1 N2 alpha power)

Performing iteration ...

Estimated sample sizes for a two-sample means test
t test assuming sd1 = sd2 = sd
Ho: m2 = m1   versus   Ha: m2 != m1
```

diff	N	N1	N2	alpha	power
.5	128	64	64	.05	.8

Say that it costs $1,000 per subject for this study and that our budget for subjects is $100,000. The results above show us that using a standard posttest-only design, we would need to have 64 subjects per group (128 total) to have 80% power to detect a difference of half an SD. It would cost $128,000 to have 128 subjects, which is beyond our $100,000 budget.

Imagine that we use a different design, a pretest–posttest design, in which happiness is measured both at the pretest and at the posttest. In this design, the happiness measured at the posttest would be the DV, and the happiness measured at the pretest could be used as a covariate. Given that there are individual differences in happiness (that some people tend to be happier than others), the covariate can account for some of these individual differences. If we have a pretest that is well correlated with the posttest as a covariate, we can lower the required sample size in comparison with a design that did not use the covariate.[1]

Suppose that we want to compute the required sample size using an ANCOVA design, assuming the correlation between the pretest and posttest is 0.60. We can compute the required sample size using the pretest as a covariate with the `sampsi` command.[2] In the `sampsi` command below, we specify that the mean of group 1 is 0 and group 2 is 0.5 and that the standard deviations of both groups are 1.[3] We then include the `pre(1)` option to indicate that there is one pretest and the `r01(.6)` option to indicate that the correlation between the pretest and posttest is 0.6. Finally, the `method(ancova)` option indicates we want the sample size to be computed using an ANCOVA analysis strategy.

```
. sampsi 0 .5, power(.8) sd1(1) sd2(1) pre(1) r01(.6) method(ancova)
Estimated sample size for two samples with repeated measures
Assumptions:

                                        alpha =   0.0500  (two-sided)
                                        power =   0.8000
                                           m1 =        0
                                           m2 =       .5
                                          sd1 =        1
                                          sd2 =        1
                                        n2/n1 =     1.00
            number of follow-up measurements =        1
            number of baseline measurements =         1
   correlation between baseline & follow-up =      0.600

Method: ANCOVA
    relative efficiency =     1.562
       adjustment to sd =     0.800
          adjusted sd1 =     0.800
          adjusted sd2 =     0.800

Estimated required sample sizes:
               n1 =        41
               n2 =        41
```

The `sampsi` command reports that we can use a total of $N = 82$ subjects ($N = 41$ per group) and achieve the desired power. By including the covariate, we can achieve

---

1. For example, imagine the covariate explains 25% of the total variance in the outcome, which means that after the inclusion of the covariate, the residual variance is reduced by 25%. As the covariate explains more and more variance in the DV, the more it reduces the residual variance in the DV, which means that the mean squared error (MSE) becomes smaller and smaller. As the MSE decreases, the F ratio for the treatment effect (that is, $MS_{\text{Treatment}}$ divided by MSE) gets larger, increasing our likelihood of finding a statistically significant result.

2. The `sampsi` command preceded the `power` command, but it is still useful because it allows us to compute power when using ANCOVA.

3. This specifies a standardized effect size of 0.5.

80% power but use only $N = 82$ subjects compared with the need for $N = 128$ without the covariate. Of course, this hinges on our assumption being correct about the strength of the relationship between the covariate (happiness measured at pretest) and the DV (happiness measured at posttest). You can investigate how the sample-size requirements change when you assume different correlations between the pretest and posttest. Choosing a conservative (that is, smaller) correlation can be prudent so that you do not overestimate the correlation that would lead to underestimating the required sample size.

Note: Power for ANCOVA versus repeated measures ANOVA

Another way to analyze this design would be to use a repeated measures ANOVA. A surprising fact is that the repeated measures ANOVA is not as powerful as the equivalent ANCOVA (Rausch, Maxwell, and Kelley 2003). We can see this for ourselves using `sampsi` to compute the required sample size for an equivalent repeated measures ANOVA by specifying `method(change)` instead of `method(ancova)`, as shown below.

```
. sampsi 0 .5, power(.8) sd(1) pre(1) r01(.6) method(change)
```

The output (not shown) indicates that the required sample size would be $N = 102$ ($N = 51$ per group) using a repeated measures ANOVA strategy in comparison with $N = 82$ ($N = 41$ per group) using the ANCOVA strategy.

11.4.2 Example 10: Using correlated variables as covariates

As we saw in the previous example, using the pretest as a covariate was an excellent way to increase the power of a study. However, this strategy can make us question the generalizability of our results. Answering the happiness questionnaire during the pretest phase might make one more acutely aware of one's own happiness and make happiness therapy more effective than it would have been had participants not taken the pretest. It is also possible that the fact the participants are asked about their happiness at the start of the study might influence their answers when asked about their happiness again at the end of the study. Either of these possibilities raises concerns about whether our results would generalize to a population of people who were not asked about their happiness in the first place.

Instead of using the pretest happiness as a covariate, we can use other measures (measured before the start of the study) that are related to happiness. But how could we identify potentially useful covariates, and how can we estimate how strongly such covariates would be related to the outcome (for the purposes of a power analysis)? Such covariates can be identified based on previous research or on the analysis of other datasets to identify potential covariates. In our case, we can use the General Social Survey dataset to identify variables that are predictive of happiness using the variable `happy7`. Although this single-item measure is not the same as our happiness inventory, it can act as a proxy for identifying potentially useful covariates.

The regression model below predicts happiness based on some potential covariates that I identified, namely, social class, education, age, gender, number of children, health status, family satisfaction, frequency of socializing with relatives, socializing with community, socializing with friends, and socializing at a bar. This model explains about 39% of the variance in happiness.

```
. use gss2012_sbs, clear

. regress  happy7 class educ age i.female children
> health satfam7 socrel socommun socfrend socbar
```

Source	SS	df	MS		
Model	483.772845	11	43.9793495	Number of obs	= 1,233
Residual	738.418559	1,221	.604765404	F(11, 1221) =	72.72
				Prob > F =	0.0000
				R-squared =	0.3958
Total	1222.1914	1,232	.992038477	Adj R-squared =	0.3904
				Root MSE =	.77767

happy7	Coef.	Std. Err.	t	P>|t|	[95% Conf. Interval]	
class	.1200552	.0337964	3.55	0.000	.0537497	.1863606
educ	-.0050008	.008263	-0.61	0.545	-.0212121	.0112104
age	.0011222	.0014646	0.77	0.444	-.0017512	.0039956
female						
Female	.0228167	.0453302	0.50	0.615	-.0661171	.1117505
children	.0290024	.0152172	1.91	0.057	-.0008524	.0588571
health	.1392341	.0219387	6.35	0.000	.0961924	.1822758
satfam7	-.4585798	.0207374	-22.11	0.000	-.4992648	-.4178949
socrel	.0057608	.0142016	0.41	0.685	-.0221013	.033623
socommun	.0238411	.0114285	2.09	0.037	.0014194	.0462628
socfrend	.0385213	.0153489	2.51	0.012	.0084082	.0686343
socbar	.0050099	.0148721	0.34	0.736	-.0241677	.0341876
_cons	5.505905	.185014	29.76	0.000	5.142924	5.868886

We can simplify this model down to three of the most potent predictors: satisfaction with family, health status, and frequency of socializing with friends. This model explains about 38% of the variance in happiness.

```
. use gss2012_sbs, clear
. regress  happy7 satfam7 health socfrend
```

Source	SS	df	MS		Number of obs	=	1,249
					F(3, 1245)	=	255.43
Model	471.767388	3	157.255796		Prob > F	=	0.0000
Residual	766.472804	1,245	.615640806		R-squared	=	0.3810
					Adj R-squared	=	0.3795
Total	1238.24019	1,248	.992179641		Root MSE	=	.78463

happy7	Coef.	Std. Err.	t	P>|t|	[95% Conf. Interval]	
satfam7	-.4768529	.0202565	-23.54	0.000	-.5165936	-.4371122
health	.1386011	.0209284	6.62	0.000	.0975424	.1796599
socfrend	.0393545	.0135951	2.89	0.004	.0126826	.0660263
_cons	6.000869	.1085342	55.29	0.000	5.787939	6.2138

This analysis of the General Social Survey data not only identifies potentially useful covariates but also gives us a sense of the size of the relationship between the covariates and the outcome variable. It would be overly optimistic to assume that these variables measured at baseline would explain 38% of the variance in happiness measured weeks later (at the posttest). Suppose we assume that these variables would explain 25% of the variance in our happiness measure at posttest. The square root of 0.25 is 0.5, so we insert this value into the **sampsi** command and find that under these assumptions, the required sample size would be 48.[4]

```
. sampsi 0 .5, power(.8) sd1(1) sd2(1) pre(1) r01(.5) method(ancova)
Estimated sample size for two samples with repeated measures
Assumptions:
                                    alpha =   0.0500  (two-sided)
                                    power =   0.8000
                                       m1 =        0
                                       m2 =       .5
                                      sd1 =        1
                                      sd2 =        1
                                    n2/n1 =     1.00
          number of follow-up measurements =        1
          number of baseline measurements =        1
    correlation between baseline & follow-up =    0.500

Method: ANCOVA
  relative efficiency =    1.333
     adjustment to sd =    0.866
         adjusted sd1 =    0.866
         adjusted sd2 =    0.866

Estimated required sample sizes:
                   n1 =       48
                   n2 =       48
```

---

4. This use of the **sampsi** command is slightly incorrect because it assumes we are using one covariate, but we are actually using three covariates. This simplification will usually have only a trivial impact on the sample-size estimate.

It would be prudent to evaluate a variety of scenarios with respect to the magnitude of this correlation (rather than just using a single value) to assess how the required sample size changes as a function of the correlation between the covariates and the DV.

Video tutorial: Power analysis

Take a quick video tour of power and sample size at http://www.stata.com/sbs/power.

11.5   Power analysis for two-way ANOVA

11.5.1   Example 11: Replicating a two-by-two analysis

Power analysis can be rather tricky for a two-way ANOVA. But we can apply everything that we have learned previously to help make such power analyses simpler. Let's start with a simple example where we aim to replicate the results from the two-by-two ANOVA from section 7.2. In that study, the DV was optimism; the first IV was treatment group assignment (control versus optimism therapy); and the second IV was depression status (*minimal/mild* versus *moderate/severe*). Let's start by using the dataset for that example and showing the table of means of the outcome by the two IVs.

```
. use opt-2by2, clear
. table treat depstat, contents(mean opt)
```

Treatment group	Depression status Min/Mild	Mod/Sev
Con	46.6	41.8
OT	56.3	45.2

Let's now repeat the two-way ANOVA in which `opt` was the DV and `depstat` and `treat` were the two IVs. Note that the value of the $MS_{Residual}$ is 99.70 and the root MSE is 9.98. Let's round these values to 100 and 10, respectively. The value of 100 is the estimate of the residual (error) variance, and the value of 10 is the estimate of the residual SD. We will use these values later.

```
. anova opt depstat treat depstat#treat
                    Number of obs =       200    R-squared     =  0.2286
                    Root MSE      =   9.98476    Adj R-squared =  0.2168

        Source | Partial SS       df        MS         F     Prob>F

         Model |    5789.5         3    1929.8333    19.36   0.0000

        depstat |   3184.02        1     3184.02     31.94   0.0000
         treat |    2112.5         1      2112.5     21.19   0.0000
  depstat#treat |    492.98        1      492.98      4.94   0.0273

      Residual |  19540.32       196    99.69551

         Total |  25329.82       199   127.28553
```

Let's say that we want to replicate this study with the aim of testing for the interaction effect found in the previous study. We can use the `power twoway` command to compute the required sample size as shown below. The means from the first row of the `table` command (the control group) are listed followed by a backslash that indicates the end of the row. Then, the means from the second row (optimism therapy) are listed. The `varerror()` option is used to specify that the residual (error) variance is 100. Finally, the `factor(rowcol)` option indicates we want to test the interaction effect (row-by-column effect).

```
. power twoway 46.6 41.8 \ 56.3 45.2,  varerror(100) factor(rowcol)

Performing iteration ...

Estimated sample size for two-way ANOVA
F test for row-by-column effect
Ho: delta = 0  versus  Ha: delta != 0

Study parameters:
        alpha =     0.0500
        power =     0.8000
        delta =     0.1575
          N_r =          2
          N_c =          2
        means =   <matrix>
       Var_rc =     2.4806
        Var_e =   100.0000

Estimated sample sizes:
            N =        320
    N per cell =         80
```

The output shows that the required sample size is $N = 320$ or $N = 80$ per cell.

11.5.2 Example 12: Standardized simple effects

We can simplify these computations by using standardized effects (as we did in section 11.2.2). The key difference in this factorial design is that we need to focus on the simple effects of `treat` at each level of `depstat`.[5] To that end, we will compute the standardized simple effects of treatment at each level of depression.

Let's start by computing the standardized simple effect of `treat` when depression status is *minimal/mild*. We compute this by taking the mean for the optimism therapy group (56.3) minus the mean for the control group (46.6) divided by the residual SD (10). The standardized simple effect of `treat` when depression status is *minimal/mild* is 0.97.

```
. display (56.3-46.6) / 10
.97
```

Now, let's compute the standardized simple effect of `treat` when depression status is *moderate/severe*. We compute this by taking the mean for the optimism therapy group (45.2) minus the mean for the control group (41.8) divided by the residual SD (10). The standardized simple effect of `treat` when depression status is *moderate/severe* is 0.34.

```
. display (45.2 - 41.8) / 10
.34
```

We can now run the `power twoway` command by specifying that the means with respect to the control group are 0 and that the standardized effects for the optimism therapy group are 0.34 and 0.97. We omit the `varerror()` option because we are specifying standardized effect sizes. The required sample size is exactly the same as before: N = 320 or N = 80 per cell.

```
. power twoway 0 0 \ .34 .97, factor(rowcol)
Performing iteration ...
Estimated sample size for two-way ANOVA
F test for row-by-column effect
Ho: delta = 0  versus  Ha: delta != 0
Study parameters:
        alpha =    0.0500
        power =    0.8000
        delta =    0.1575
          N_r =         2
          N_c =         2
        means =  <matrix>
      Var_rc =    0.0248
       Var_e =    1.0000
Estimated sample sizes:
            N =       320
   N per cell =        80
```

---

5. We could focus on the simple effects of `depstat` at each level of `treat`, but that feels less intuitive.

11.5.3 Example 13: Standardized interaction effect

We can extend the previous example one step further. We can take the differences in the simple effects to obtain the overall interaction effect. The standardized interaction effect for this example is 0.63.

```
. display .97 -  .34
.63
```

In the `power twoway` command below, I specify that the control group means are 0 and then specify 0 and 0.63 for the treatment group means. This computes a power analysis for a two-by-two design that has a standardized interaction effect of 0.63. It should be no surprise that the required sample size is exactly the same as before: N = 320 or N = 80 per cell.

```
. power twoway 0 0 \ 0 .63, factor(rowcol)

Performing iteration ...

Estimated sample size for two-way ANOVA
F test for row-by-column effect
Ho: delta = 0  versus  Ha: delta != 0

Study parameters:

        alpha =     0.0500
        power =     0.8000
        delta =     0.1575
          N_r =          2
          N_c =          2
        means =   <matrix>
       Var_rc =     0.0248
        Var_e =     1.0000

Estimated sample sizes:

            N =        320
   N per cell =         80
```

With respect to this two-way interaction, we now have simplified the power analysis to a single effect size, the standardized interaction effect. We accomplished this by first standardizing the simple effects and then computing the difference in the simple effects to yield 0.63.

11.5.4 Summary: Power for two-way ANOVA

These examples have just scratched the surface of the features that you can access using the `power twoway` command. But as you can see, you can apply the same concepts that we used with the `power twosample` and `power oneway` commands to the `power twoway` command. For more information, see `help power twoway`.

11.6   Closing thoughts

Power analysis is one area of statistics that is not clear cut and includes lots of judgment and educated guesses. It is an area where you can be tempted to engage in self-deception, being overly optimistic about the effect size associated with your IV or the strength of the correlation between the covariates and the DV. You will rarely have rock-solid information to guide your power analysis, so you are more likely to adapt results from studies with different IVs or different DVs, making educated guesses and investigating a variety of plausible scenarios. As you consider these scenarios, constantly challenge yourself to be skeptical about your guesses and strive to be very honest with yourself. If you are overly optimistic in your power analysis, you might be able to persuade yourself (and others, for example, agencies providing grants) that your study is properly powered only to be disappointed by nonsignificant results in the end.

At the start of the chapter, I made an analogy between performing a power analysis and creating a budget to build a house. If you create an overly optimistic budget for building a house, you may be happy at the start of construction only to be disappointed when you run out of money and cannot complete building your house. Likewise, if you deceive yourself in performing a power analysis for a prospective study, you might be happy to get the funding for the study only to be disappointed by the lack of significant results due to an insufficiently powered study.

Part III

Repeated measures and longitudinal designs

This part of the book covers two different strategies for analyzing designs with multiple observations on the same subject.

Chapter 12 covers repeated measures designs. In such designs, participants are observed at more than one time point and time, and the same time schedule is used across people. This chapter shows three examples illustrating the analysis of repeated measures designs. The first example includes a single repeated-measures independent variable (IV) (section 12.2). The second example illustrates a two-by-three between-within design where the between-subjects IV has two levels and the repeated measures IV has three levels (section 12.3). The third example illustrates a three-by-three between-within design where the repeated measures IV has three levels and the between-subjects IV also has three levels (section 12.4).

Chapter 13 covers longitudinal designs. These designs, in contrast to repeated measures designs, typically have a larger number of observations per subject, and the time gaps between the repeated measures can vary between people. This chapter includes four examples, all of which use multilevel modeling as the main analysis strategy. The first example models the dependent variable as a linear function of time (section 13.2). The second example adds a between-subjects IV, which allows us to model the linear effect of time and explore the IV by time interaction (section 13.3). This example is similar to an analysis of covariance with a treatment by covariate interaction (for instance, like the examples in chapter 8). The third example includes time as the only predictor but uses a piecewise modeling strategy for the effect of time (section 13.4). The fourth example adds a between-subjects IV to the third example, modeling the interaction of the IV with the piecewise effects of time (section 13.5).

12 Repeated measures designs

12.1 Chapter overview

This chapter considers repeated measures designs where participants are observed at more than one time point and time. In such designs, the repeated measurements are made on the same time schedule across people. In the analyses presented in this chapter, time is treated as a categorical variable. Chapter 13 illustrates longitudinal designs where analyses treat time as a continuous variable.

This chapter shows three examples illustrating the analysis of repeated measures designs. The first example includes a single repeated-measures independent variable (IV) (section 12.2). The second example illustrates a two-by-three mixed (between-within) design where the between-subjects IV has two levels and the repeated measures IV has three levels (section 12.3). The third example illustrates a mixed design where the repeated measures IV has three levels and the between-subjects IV also has three levels (section 12.4).

The analyses in this chapter will focus on the use of the `mixed` command to fit a model that is a hybrid of a traditional repeated measures analysis of variance (ANOVA) and a mixed model. As in the repeated measures ANOVA, there will be one fixed intercept (rather than the random intercepts that we commonly see when using mixed models). The `mixed` command is more flexible and powerful than a traditional repeated measures ANOVA. For example, a traditional repeated measures ANOVA omits any observation with missing data (that is, uses listwise deletion), whereas the `mixed` command retains subjects who might have missing data and performs the estimation on the time observations that are present. Additionally, a traditional repeated measures ANOVA is restricted to only two different types of residual covariance structures, exchangeable (that is, compound symmetry) and unstructured (that is, multivariate). There are many situations where an exchangeable structure is overly simple and untenable and the unstructured covariance is overly complicated and estimates too many superfluous parameters. By contrast, the `mixed` command permits you to choose from a variety of residual covariance structures such as banded, autoregressive, or toeplitz, as described in section 12.5. Although the output looks a little bit different, the `mixed` command combined with the `residuals()` option offers all the features from a repeated measures ANOVA, with the added benefit of retaining subjects who have missing time points and the ability to estimate additional residual covariance structures.

12.2 Example 1: One-way within-subjects designs

Consider a study regarding the number of minutes that people sleep at night. The study consisted of 100 people, and their sleep was measured once a month for 3 months (that is, month 1, month 2, and month 3). The measurements at month 1 were baseline measurements in that no specific treatment was applied. Before the measurement in month 2, participants were given a sleep medication to lengthen their sleep, and the use of this medication continued during month 3. In summary, sleep was measured at three points, the first a control (baseline) condition and the second and third measurements while taking a sleep medication. The study has two aims. The first aim is to assess the initial impact of sleep medication on duration of sleep. It is predicted the people will sleep longer in month 2 (while on the sleep medication) than in month 1 (before they started the medication). Even if the medication is effective in the second month, there might be a concern of whether it sustains its effectiveness (as compared with the baseline) or whether its effectiveness wears off. So the second aim is to assess the sustained effectiveness of the medication on sleep by comparing the amount of sleep in the third month with that in the first month.

Let's start by looking at the first six observations from this dataset and the summary statistics for the variables in the dataset.

```
. use sleep_cat3, clear

. list in 1/6, sepby(id)
```

	id	month	sleep
1.	1	1	303
2.	1	2	349
3.	1	3	382
4.	2	1	331
5.	2	2	380
6.	2	3	350

```
. summarize
```

Variable	Obs	Mean	Std. Dev.	Min	Max
id	300	50.5	28.9143	1	100
month	300	2	.8178608	1	3
sleep	300	361.4067	31.21017	276	466

The dataset for this example is in a long format, with one observation per person per month. The variable `id` identifies the person and ranges from 1 to 100, which corresponds to the 100 participants. The variable `month` identifies the month in which sleep was observed and ranges from 1 to 3. The variable `sleep` contains the number of minutes the person slept at night and ranges from 276 to 466 minutes.

Note! Wide and long datasets

Data for this kind of study might be stored with one observation per person and three variables representing the different time points. Sometimes, this is called a "multivariate" format, and Stata would call this a "wide" format. If your dataset is in that kind of form, you can use the `reshape` command to convert it to a long format as illustrated in section 22.9.1

Let's now use the `mixed` command to predict `sleep` from `month`, treating `month` as a repeated-subjects IV. Specifying `|| id:` introduces the random-effects part of the model and indicates that the observations are repeated within `id`. Next, the `nocons` option is specified in the random-effects options so that only one fixed intercept is fit. Further, the `residuals()` option is used to specify an unstructured covariance for the repeated measures factor.[1] This accounts for the nonindependence of the observations across months for each person.

---

1. The `un` option specifies an unstructured covariance matrix, and the `t(month)` option specifies that observations be repeated across months. Section 12.5 provides more information about how to select among residual covariance structures.

```
. mixed sleep i.month || id:, nocons residuals(un, t(month)) nolog
```

Mixed-effects ML regression   Number of obs   =   300
Group variable: id   Number of groups =   100

                                                     Obs per group:
                                                                 min =   3
                                                                 avg =   3.0
                                                                 max =   3

                                                     Wald chi2(2)   =   35.94
Log likelihood = -1437.3712   Prob > chi2   =   0.0000

sleep	Coef.	Std. Err.	z	P>|z|	[95% Conf. Interval]	
month						
2	21.07	3.8331	5.50	0.000	13.55726	28.58274
3	18.43	3.807288	4.84	0.000	10.96785	25.89215
_cons	348.24	3.144903	110.73	0.000	342.0761	354.4039

Random-effects Parameters	Estimate	Std. Err.	[95% Conf. Interval]	
id: (empty)				
Residual: Unstructured				
var(e1)	989.0415	139.8716	749.6118	1304.946
var(e2)	928.5545	131.3175	703.7677	1225.139
var(e3)	731.3612	103.4301	554.3114	964.9615
cov(e1,e2)	224.1654	98.41888	31.26793	417.0629
cov(e1,e3)	135.4291	86.12121	-33.3654	304.2235
cov(e2,e3)	107.7223	83.10908	-55.16847	270.6131

LR test vs. linear model: chi2(5) = 11.61   Prob > chi2 = 0.0405

Note: The reported degrees of freedom assumes the null hypothesis is not on
      the boundary of the parameter space. If this is not true, then the
      reported test is conservative.

The `mixed` command produces quite a bit of output. We will soon use the `contrast`, `margins`, and `marginsplot` commands to more fully understand these results. Nevertheless, let's briefly review the output, starting with the upper portion. In this study, there are 100 people (groups) who were each measured 3 times for a total of 300 observations. This is reported in the top of the output as 300 observations, 100 groups, and 3 observations per group. The test of the overall model is reported via a Wald chi-squared test [Wald $\chi^2(2, N = 100) = 35.94$, $p < 0.001$]. The center portion of the output is much like the output we would obtain from the `regress` command with columns for the regression coefficient, the standard error, a statistical test that the coefficient is zero, and a confidence interval (CI) for the coefficient. In this case, the coefficients are dummy-coded values of `month` comparing each month with the reference group (group 1). The lower part of the output provides the estimates of the unstructured covariance of sleep across months, showing that the residual variance at month 1 is 989.0, at month 2 is 928.6, and at month 3 is 731.4. The output also shows the covariance of the residuals between each pair of time points. For example, the covariance of the residuals at month 1 with month 2 is 224.2. This residual covariance structure is called an unstructured covariance and was requested with the `residuals(un, t(month))` option. Section 12.5 provides more information about ways to select and compare residual covariance structures.

Before we test the hypotheses regarding the differences in the amount of sleep across the months, let's use the `margins` command to compute the predicted mean of `sleep` for each level of `month`, as shown below. In month 1, the predicted mean of sleep is 348.24 (95% CI = [342.1, 354.4]); in month 2, the predicted mean is 369.31 (95% CI = [363.3, 375.3]); and in month 3, the predicted mean is 366.67 (95% CI = [361.4, 372.0]).

```
. margins month, nopvalue
Adjusted predictions                          Number of obs     =        300
Expression    : Linear prediction, fixed portion, predict()
```

	Margin	Delta-method Std. Err.	[95% Conf. Interval]	
month				
1	348.24	3.144903	342.0761	354.4039
2	369.31	3.047219	363.3376	375.2824
3	366.67	2.704369	361.3695	371.9705

We can use the `marginsplot` command to create a graph of the mean and 95% confidence interval for sleep across the three months, as shown in figure 12.1.

```
. marginsplot
  Variables that uniquely identify margins: month
```

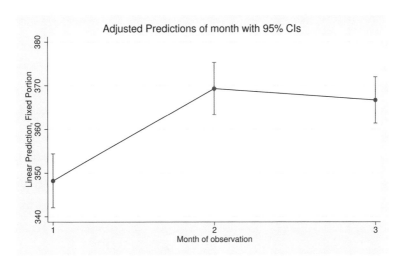

Figure 12.1: Estimated minutes of sleep at night by month

When we look at the predicted means from the `margins` and `marginsplot` commands, it looks as if sleep indeed increased from baseline (month 1) to month 2, and it appears that the amount of sleep in month 3 was also greater than the baseline value. Before we make those specific tests, let's test the overall null hypothesis that the average number of minutes of sleep is equal across all three months. This is tested using the `contrast` command below, which reports a Wald chi-squared statistic and its associated *p*-value to test the `month` effect. The effect of `month` is significant, Wald $\chi^2(2, N = 100) = 35.94$, $p < 0.001$, indicating that the average number of minutes of sleep is not equal across all the months.

```
. contrast month
Contrasts of marginal linear predictions
Margins        : asbalanced
```

	df	chi2	P>chi2
sleep			
month	2	35.94	0.0000

Now, let's use the `contrast` command to perform the specific comparisons of interest. Let's compare the amount of sleep in the second month with that of the first month to test the initial impact of the sleep medication. Let's also compare the third month with the first month to assess the extent to which any impact was sustained. These comparisons are made using the `contrast` command and applying the `r.` contrast to `month`.

```
. contrast r.month, cieffects
Contrasts of marginal linear predictions
Margins        : asbalanced
```

	df	chi2	P>chi2
sleep			
month			
(2 vs 1)	1	30.22	0.0000
(3 vs 1)	1	23.43	0.0000
Joint	2	35.94	0.0000

	Contrast	Std. Err.	[95% Conf. Interval]	
sleep				
month				
(2 vs 1)	21.07	3.8331	13.55726	28.58274
(3 vs 1)	18.43	3.807288	10.96785	25.89215

The upper portion of the output shows the significance tests using Wald chi-squared statistics, and the lower portion shows the contrast with CIs.[2] The test comparing the average sleep in month 2 with that in month 1 was significant [Wald $\chi^2(1) = 30.22$, $p < 0.001$]. In the second month, participants slept (on average) 21.1 minutes longer than they did in the first month (95% CI = $[13.56, 28.58]$). The second test comparing the third month with the first month was also significant [Wald $\chi^2(1) = 23.43$, $p < 0.001$]. Participants slept an average of 18.43 minutes longer in month 3 than they did in month 1 (95% CI = $[10.97, 25.89]$).

The example in this study had three time points; however, this could be extended to four or more time points. As illustrated in this example, the `contrast` command can be used to test the overall equality of the means at the different time points and form specific comparisons among the time points. This is a generalization of the principles illustrated in chapter 4, and you can borrow and generalize from the tools illustrated in that chapter.

12.3   Example 2: Mixed design with two groups

In the previous example, sleep was measured at three time points (month 1, month 2, and month 3). The first month was a baseline measurement, whereas participants received sleep medication during the second and third months. We found that sleep was significantly longer in the second month compared with the first. We also found that sleep was significantly longer in the third month compared with the first month.

We can augment (and improve) that design by including a control group that is measured at each of the three months but receives no treatment. This yields a two-by-three-between-within design where the between-subjects factor has two levels and the within-subjects factor has three levels. Because this is a two-by-three factorial design, we can draw upon the logic illustrated in section 7.3 of chapter 7, which shows how to interpret the results from a two-by-three design.

Using this design, we will focus on two main research questions. The first question concerns the initial effect of the sleep medication. This would be assessed by comparing the average sleep in month 2 with that in month 1 for the treatment group compared with the control group. The second question concerns the sustained effect of the medication in the third month. This would be assessed by comparing month 3 with month 1 for the treatment group compared with the control group.

Our improved study includes 100 participants in the treatment (medication) group and 100 participants in the control group. The results of this study are contained in the dataset named `sleep_catcat23.dta`. The first six observations from the dataset are listed below along with summary statistics for the variables in the dataset.

---

2. You might notice that the results of this `contrast` command mirror the upper portion of the output from the `mixed` command. This is because both the `mixed` command and the `contrast` command are performing reference group comparisons using the first group as the reference group.

```
. use sleep_catcat23, clear
. list in 1/6, sepby(id)
```

	id	month	group	sleep
1.	1	1	Control	315
2.	1	2	Control	379
3.	1	3	Control	320
4.	2	1	Control	392
5.	2	2	Control	369
6.	2	3	Control	314

```
. summarize
```

Variable	Obs	Mean	Std. Dev.	Min	Max
id	600	100.5	57.78248	1	200
month	600	2	.8171778	1	3
group	600	1.5	.5004172	1	2
sleep	600	354.49	31.36286	256	448

The variable `id` identifies the participant and ranges from 1 to 200. The variable `month` indicates the month in which sleep was measured (month 1, 2, or 3) and is the within-subjects IV. The variable `group` is the between-subjects IV and is coded 1 = control and 2 = medication. The variable `sleep` is the number of minutes of sleep and ranges from 256 to 448.

As in the previous example, the analysis is conducted using the `mixed` command and uses the `residuals()` option to specify an unstructured residual type. The model includes two IVs, `group` and `month`, which are both entered along with their interaction.

```
. mixed sleep i.group##i.month || id:, nocons residuals(un, t(month))  nolog
```

Mixed-effects ML regression   Number of obs   =   600
Group variable: id   Number of groups   =   200

                                                Obs per group:
                                                              min =   3
                                                              avg =   3.0
                                                              max =   3

                                                Wald chi2(5)   =   68.47
Log likelihood =   -2879.78   Prob > chi2   =   0.0000

sleep	Coef.	Std. Err.	z	P>|z|	[95% Conf. Interval]	
group						
Medication	-7.08	4.378767	-1.62	0.106	-15.66223	1.502225
month						
2	-3.73	3.746771	-1.00	0.319	-11.07354	3.613535
3	-.57	4.257299	-0.13	0.893	-8.914152	7.774152
group#month						
Medication #						
2	28.06	5.298734	5.30	0.000	17.67467	38.44533
Medication #						
3	24.94	6.020729	4.14	0.000	13.13959	36.74041
_cons	350.63	3.096256	113.24	0.000	344.5615	356.6985

Random-effects Parameters	Estimate	Std. Err.	[95% Conf. Interval]	
id: (empty)				
Residual: Unstructured				
var(e1)	958.68	95.86811	788.0493	1166.256
var(e2)	721.2979	72.12976	592.9179	877.475
var(e3)	980.2552	98.02552	805.7847	1192.503
cov(e1,e2)	138.0745	59.60521	21.25039	254.8985
cov(e1,e3)	63.23807	68.69312	-71.39797	197.8741
cov(e2,e3)	119.7182	60.05779	2.007147	237.4293

LR test vs. linear model: chi2(5) = 15.70   Prob > chi2 = 0.0078

Note: The reported degrees of freedom assumes the null hypothesis is not on
      the boundary of the parameter space. If this is not true, then the
      reported test is conservative.

We can estimate the mean sleep by group and month using the `margins` command below. This is followed by the `marginsplot` command to graph the means computed by the `margins` command (shown in figure 12.2).

```
. margins month#group, nopvalue
Adjusted predictions                            Number of obs    =        600
Expression    : Linear prediction, fixed portion, predict()
```

	Margin	Delta-method Std. Err.	[95% Conf.	Interval]
month#group				
1#Control	350.63	3.096256	344.5615	356.6985
1#Medication	343.55	3.096256	337.4815	349.6185
2#Control	346.9	2.685699	341.6361	352.1639
2#Medication	367.88	2.685699	362.6161	373.1439
3#Control	350.06	3.130903	343.9235	356.1965
3#Medication	367.92	3.130903	361.7835	374.0565

```
. marginsplot, noci
    Variables that uniquely identify margins: month group
```

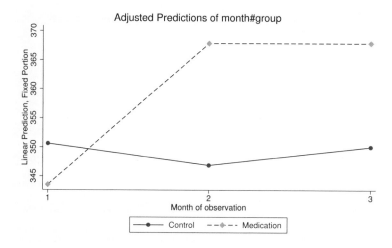

Figure 12.2: Estimated sleep by month and treatment group

The pattern of means in figure 12.2 appears to show that the medication increases sleep in month 2 compared with month 1 and that the effect is sustained in month 3. By contrast, the sleep in the control group remains rather steady at each of the three time points.

Before we test our two main questions of interest, let's assess the overall interaction of group by month using the contrast command below. The interaction of group by month is significant [Wald $\chi^2(2, N = 200) = 30.21$, $p < 0.001$].

```
. contrast group#month
Contrasts of marginal linear predictions
Margins        : asbalanced
```

	df	chi2	P>chi2
sleep			
group#month	2	30.21	0.0000

To test our questions of interest regarding the initial effect of medication and the sustained effect of medication, we can apply the r contrast operator to month and interact that with group, as shown below.

```
. contrast r.month#group, cieffects
Contrasts of marginal linear predictions
Margins        : asbalanced
```

	df	chi2	P>chi2
sleep			
month#group			
(2 vs 1) (joint)	1	28.04	0.0000
(3 vs 1) (joint)	1	17.16	0.0000
Joint	2	30.21	0.0000

	Contrast	Std. Err.	[95% Conf. Interval]	
sleep				
month#group				
(2 vs 1)				
(Medication vs base)	28.06	5.298734	17.67467	38.44533
(3 vs 1)				
(Medication vs base)	24.94	6.020729	13.13959	36.74041

The first contrast compares the sleep in month 2 with that in month 1 for the treatment group versus the control group. This test is significant [Wald $\chi^2(1, N = 200) = 28.04$, $p < 0.001$]. The average difference in sleep (month 2 versus month 1) for the medication group was 28.1 minutes more than for the control group (95% CI $= [17.7, 38.4]$).

The second contrast compares the sleep in month 3 with that in month 1 for the treatment group versus the control group. This comparison is also significant [Wald $\chi^2(1, N = 200) = 17.16$, $p < 0.001$]. The average improvement in sleep for month 3 compared with month 1 was 24.9 minutes more for the medication group than for the control group (95% CI $= [13.1, 36.7]$).

For more information about how to dissect a three-by-two interaction, see section 7.3 of chapter 7.

12.4   Example 3: Mixed design with three groups

Let's consider one final example that extends the previous example by including a third type of treatment, sleep education. The design for this example now includes three treatment groups (control, medication, and education) and three time points (month 1, month 2, month 3). Month 1 is a baseline period during which no treatment is administered to any of the groups. During months 2 and 3, the medication group receives sleep medication, and during months 2 and 3, the education group receives sleep education aimed to increase their sleep.

This study includes 300 participants, 100 assigned to each of the three treatment groups. This main variables of interest in this study are treatment group (with three levels) and month (with three levels). We can borrow from the logic and methods of interpretation that were illustrated in section 7.4 of chapter 7.

Let's use the dataset for this study and list the first six observations and show the summary statistics for the variables in the dataset.

```
. use sleep_catcat33, clear
. list in 1/6, sepby(id)
```

	id	month	group	sleep
1.	1	1	Control	305
2.	1	2	Control	313
3.	1	3	Control	351
4.	2	1	Control	295
5.	2	2	Control	289
6.	2	3	Control	374

```
. summarize
```

Variable	Obs	Mean	Std. Dev.	Min	Max
id	900	150.5	86.65021	1	300
month	900	2	.8169506	1	3
group	900	2	.8169506	1	3
sleep	900	362.8378	34.06564	260	479

The variable `id` identifies the person, and `month` indicates the month in which sleep was observed (1, 2, or 3). The variable `group` is coded $1 =$ control, $2 =$ medication, and $3 =$ education. The variable `sleep` contains the number of minutes of sleep and ranges from 260 to 479.

The `mixed` command for analyzing this example is the same as in the previous example.

```
. mixed sleep i.group##i.month || id:, nocons residuals(un, t(month))  nolog
Mixed-effects ML regression                    Number of obs     =        900
Group variable: id                             Number of groups  =        300

                                               Obs per group:
                                                             min =          3
                                                             avg =        3.0
                                                             max =          3

                                               Wald chi2(8)      =     284.84
Log likelihood = -4318.4068                    Prob > chi2       =     0.0000
```

sleep	Coef.	Std. Err.	z	P>|z|	[95% Conf. Interval]	
group						
Medication	2.6	4.24623	0.61	0.540	-5.722457	10.92246
Education	3.17	4.24623	0.75	0.455	-5.152457	11.49246
month						
2	8.06	3.71971	2.17	0.030	.7695027	15.3505
3	13.35	4.063031	3.29	0.001	5.386605	21.31339
group#month						
Medication # 2	21.88	5.260464	4.16	0.000	11.56968	32.19032
Medication # 3	18.22	5.745994	3.17	0.002	6.958059	29.48194
Education#2	7.11	5.260464	1.35	0.177	-3.20032	17.42032
Education#3	34.49	5.745994	6.00	0.000	23.22806	45.75194
_cons	344.7	3.002538	114.80	0.000	338.8151	350.5849

Random-effects Parameters	Estimate	Std. Err.	[95% Conf. Interval]	
id: (empty)				
Residual: Unstructured				
var(e1)	901.5233	73.60907	768.2041	1057.98
var(e2)	836.2546	68.27989	712.5875	981.3836
var(e3)	918.423	74.98894	782.6046	1077.812
cov(e1,e2)	177.0769	51.16172	76.80173	277.352
cov(e1,e3)	84.562	52.76138	-18.84841	187.9724
cov(e2,e3)	160.4783	51.43891	59.65987	261.2967

```
LR test vs. linear model: chi2(5) = 24.71           Prob > chi2 = 0.0002
Note: The reported degrees of freedom assumes the null hypothesis is not on
      the boundary of the parameter space.  If this is not true, then the
      reported test is conservative.
```

As with the previous example, let's use the `margins` command to compute the average sleep for each treatment group at each time point. Then, let's follow that command with the `marginsplot` command to graph the average sleep by time (on the x axis) and treatment group (using separate lines). The resulting graph is shown in figure 12.3.

```
. margins month#group, nopvalue
Adjusted predictions                        Number of obs    =      900
Expression    : Linear prediction, fixed portion, predict()
```

	Margin	Delta-method Std. Err.	[95% Conf. Interval]	
month#group				
1#Control	344.7	3.002538	338.8151	350.5849
1#Medication	347.3	3.002538	341.4151	353.1849
1#Education	347.87	3.002538	341.9851	353.7549
2#Control	352.76	2.891807	347.0922	358.4278
2#Medication	377.24	2.891807	371.5722	382.9078
2#Education	363.04	2.891807	357.3722	368.7078
3#Control	358.05	3.030549	352.1102	363.9898
3#Medication	378.87	3.030549	372.9302	384.8098
3#Education	395.71	3.030549	389.7702	401.6498

```
. marginsplot, noci
  Variables that uniquely identify margins: month group
```

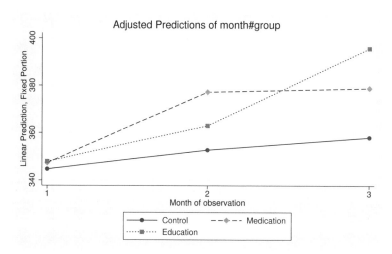

Figure 12.3: Sleep by month and treatment group

Let's now use the contrast command to test the group by month interaction. This interaction is significant [Wald $\chi^2(4, N = 300) = 62.05$, $p < 0.001$].

```
. contrast group#month
Contrasts of marginal linear predictions
Margins        : asbalanced
```

	df	chi2	P>chi2
sleep			
group#month	4	62.05	0.0000

Let's form an interaction contrast in which we apply reference group contrasts to treatment group (that is, r.group) and reference group contrasts to month (that is, r.month). The r.group contrast compares the medication group with the control group and the education group with the control group. The r.month contrast compares month 2 with 1 and month 3 with 1.

```
. contrast r.month#r.group, cieffects
Contrasts of marginal linear predictions
Margins        : asbalanced
```

	df	chi2	P>chi2
sleep			
month#group			
(2 vs 1) (Medication vs Control)	1	17.30	0.0000
(2 vs 1) (Education vs Control)	1	1.83	0.1765
(3 vs 1) (Medication vs Control)	1	10.05	0.0015
(3 vs 1) (Education vs Control)	1	36.03	0.0000
Joint	4	62.05	0.0000

	Contrast	Std. Err.	[95% Conf. Interval]	
sleep				
month#group				
(2 vs 1)				
(Medication vs Control)	21.88	5.260464	11.56968	32.19032
(2 vs 1)				
(Education vs Control)	7.11	5.260464	-3.20032	17.42032
(3 vs 1)				
(Medication vs Control)	18.22	5.745994	6.958059	29.48194
(3 vs 1)				
(Education vs Control)	34.49	5.745994	23.22806	45.75194

Consider the first two contrasts, which focus on the comparison of month 2 with 1. The first contrast shows that the gain in sleep for month 2 versus 1 for the medication group is significantly greater than that for the control group [Wald $\chi^2(1, N = 300) = 17.30$, $p < 0.001$]. From month 1 to month 2, sleep improved by 21.9 more minutes for the medication group than for the control group (95% CI = $[11.6, 32.2]$). The second contrast makes the same kind of comparison but focuses on the comparison of the education group with the control group. The gain in sleep in month 2 versus 1 for the education group versus the control group was 7.11 minutes (95% CI $= [-3.2, 17.4]$). This contrast was not significant [Wald $\chi^2(1, N = 300) = 1.83$, $p = 0.177$].

The third and fourth contrasts focus on the comparison of month 3 with 1. The third contrast shows that the average sleep in month 3 versus month 1 is significantly different for the medication group compared with the control group [Wald $\chi^2(1, N = 300) = 10.05$, $p = 0.002$]. From month 1 to month 3, sleep improved by 18.22 more minutes for the medication group than for the control group (95% CI $= [6.96, 29.48]$). The fourth contrast shows that the sleep improvement from month 1 to month 3 was 34.49 minutes greater for the education group than for the control group (95% CI $= [23.23, 45.75]$). This effect was significant [Wald $\chi^2(1, N = 300) = 36.03$, $p < 0.001$].

Let's consider another interesting pair of contrasts. In addition to comparing each treatment (education, medication) with the control group, we may also be interested in comparing education with medication. Let's test the interaction of treatment (education versus medication) by month, comparing each month with the baseline month. This is obtained by interacting the contrast of education with medication (that is, `r3b2.group`) by each month compared with month 1 (that is, `r.month`). This is tested using the `contrast` command.

```
. contrast r3b2.group#r.month, cieffects
Contrasts of marginal linear predictions
Margins        : asbalanced
```

	df	chi2	P>chi2
sleep			
group#month			
(Education vs Medication) (2 vs 1)	1	7.88	0.0050
(Education vs Medication) (3 vs 1)	1	8.02	0.0046
Joint	2	33.80	0.0000

	Contrast	Std. Err.	[95% Conf. Interval]	
sleep				
group#month				
(Education vs Medication)				
(2 vs 1)	-14.77	5.260464	-25.08032	-4.45968
(Education vs Medication)				
(3 vs 1)	16.27	5.745994	5.008059	27.53194

The first contrast is significant [Wald $\chi^2(1, N = 300) = 7.88$, $p = 0.005$]. Comparing the education with the medication group, we see that the difference in the average change in sleep from month 1 to month 2 was -14.8 minutes (95% CI $= [-25.08, -4.46]$). We can reframe this to say that the average gain in sleep from month 1 to month 2 was 14.8 minutes more for the medication group than for the education group.

The second contrast is also significant [Wald $\chi^2(1, N = 300) = 8.02$, $p = 0.0046$] but in the opposite direction. Sleep improved from month 1 to month 3 by an average of 16.3 more minutes for the education group than for the control group (95% CI = [5.01, 27.53]).

To summarize these results, we see they indicate that at the second month of the study, medication is superior to education in lengthening sleep. However, by the third month, education surpasses medication in increasing the average minutes of sleep.

You can see section 7.4 of chapter 7 for more details about how to analyze and dissect interactions from three-by-three models.

12.5   Comparing models with different residual covariance structures

The examples illustrated in this chapter all modeled the residual covariances using an unstructured covariance matrix. Using the `residuals()` option with the `mixed` command, you can choose from a variety of residual covariance structures. This section illustrates how you can choose among different residual covariance structures and assess which structure provides the best fit.

An unstructured covariance matrix is the most general choice, estimating a separate variance for every time point and a separate covariance for every pair of time points. Even though it is the most general choice, when there are only three time points, it does not estimate too many parameters; it estimates three variances (one for each time point) and three covariances (one for each pair of time points). However, as the number of time points increases, the number of variances and covariances estimated by an unstructured covariance matrix increases dramatically. For example, if you have 5 time points, an unstructured covariance estimates 5 variances and 10 covariances (a total of 15 parameters). In such cases, you might consider more parsimonious covariance structures, such as the exchangeable (compound symmetry), autoregressive, or banded residual type.[3]

This leads to the question of how to choose among models using different covariance structures. My first recommendation would be to select a residual covariance structure that is grounded in theory or suggested by previous research. However, such information may be scarce or nonexistent. In such cases, you can fit different covariance structures and compare the fit among the different structures. Among models with similar fit, simpler structures are preferable to more complex structures. This process of fitting and comparing models with different covariance structures sounds laborious, but, as illustrated below, Stata makes this process simple.

---

3. You can see `help mixed` for a list of all the available covariance structures you can choose within the `residuals()` option. Further, chapter 7 of Singer and Willett (2003) provides additional descriptions of these residual covariance structures, including information to help you choose among the different structures.

Let's use example 1 from section 12.2, which illustrated a one-way within-subjects design. Sleep time (in minutes) was measured at three time points, month 1, month 2, and month 3. Let's use the `mixed` command to analyze sleep quantity as a function of month (time) but repeat the analysis using a different residual covariance structure for each analysis. For these examples, let's use three common residual covariance structures: unstructured, exchangeable (that is, compound symmetry), and autoregressive. Using the `residuals()` option permits us to select the desired residual covariance structure. In the examples below, model 1 specifies the `un` option to use an unstructured residual covariance matrix, and model 2 specifies the `ex` option[4] to use an exchangeable (compound symmetry) residual covariance matrix. Model 3 below specifies the `ar1` option to use a first-order autoregressive residual covariance matrix. After fitting each model with the `mixed` command, we use the `estimates store` command to store the estimates from the respective model.

```
. use sleep_cat3, clear
. * Model 1. Unstructured
. mixed sleep i.month || id:, nocons residuals(un, t(month))
  (output omitted )
. estimates store m_un
. * Model 2. Exchangeable (compound symmetry)
. mixed sleep i.month || id:, nocons residuals(ex)
  (output omitted )
. estimates store m_ex
. * Model 3. Autoregressive
. mixed sleep i.month || id:, nocons residuals(ar1, t(month))
  (output omitted )
. estimates store m_ar1
```

Now, we can use `estimates stats` to summarize the results from each of the three models. Note how the `estimates stats` command below refers to the same names used in the `estimates store` commands shown above. For example, the results saved using the name m_un refer to the results from the `mixed` command using an unstructured covariance matrix. By specifying m_un, m_ex, and m_ar1 after the `estimates stats` command, we can see summary information displayed for all three of the models fit above. This creates a table of the fit indices (for example, Akaike information criterion [AIC] and Bayesian information criterion [BIC]) for each of the models and allows us to assess the quality of fit of the three models using the three different residual covariance structures.

---

4. We can specify `residuals(ex)` instead of `residuals(ex, t(month))` because the `t(month)` option is superfluous when using the `ex` option.

```
. estimates stats m_un m_ex m_ar1
Akaike's information criterion and Bayesian information criterion
```

Model	Obs	ll(null)	ll(model)	df	AIC	BIC
m_un	300	.	-1437.371	9	2892.742	2926.076
m_ex	300	.	-1438.879	5	2887.758	2906.277
m_ar1	300	.	-1439.674	5	2889.348	2907.867

```
          Note: N=Obs used in calculating BIC; see [R] BIC note.
```

Remember that when it comes to AIC and BIC, smaller is better. The `ar1` and `ex` models have lower (better) fit indices than the `un` model. These models also have the added benefit of including four fewer residual covariance parameters (two versus six). The `ex` covariance structure (that is, compound symmetry) shows the best fit in terms of both AIC and BIC (AIC = 2887.8, BIC = 2906.3). In the following section, we will show the results of analyzing this model using the `residual(ex)` option.

12.6   Example 1 revisited: Using compound symmetry

As we found in the previous section, the compound symmetry residual covariance structure fit better than the unstructured and autoregressive residual covariance structures. Let's repeat the analysis of example 1 (from section 12.2) but now use a compound symmetry residual covariance structure As shown below, this model is fit by including the `residuals(ex)` option with the `mixed` command.

> **Note: More or less**
>
> With the growth in podcasting, you might have wondered whether there is a podcast geared toward the field of statistics. In fact, there is such a podcast—it is called "More or Less: Behind the Stats". You can subscribe and find past episodes by visiting http://www.bbc.co.uk/podcasts/series/moreorless.

```
. use sleep_cat3, clear

. mixed sleep i.month || id:, nocons residuals(ex)

Obtaining starting values by EM:

Performing gradient-based optimization:

Iteration 0:    log likelihood = -1443.1779
Iteration 1:    log likelihood = -1438.8895
Iteration 2:    log likelihood =  -1438.879
Iteration 3:    log likelihood =  -1438.879

Computing standard errors:

Mixed-effects ML regression                    Number of obs       =          300
Group variable: id                             Number of groups    =          100

                                               Obs per group:
                                                              min =            3
                                                              avg =          3.0
                                                              max =            3

                                               Wald chi2(2)        =        36.24
Log likelihood =  -1438.879                    Prob > chi2         =       0.0000
```

sleep	Coef.	Std. Err.	z	P>|z|	[95% Conf. Interval]	
month						
2	21.07	3.813695	5.52	0.000	13.5953	28.5447
3	18.43	3.813695	4.83	0.000	10.9553	25.9047
_cons	348.24	2.971508	117.19	0.000	342.416	354.064

Random-effects Parameters	Estimate	Std. Err.	[95% Conf. Interval]	
id: (empty)				
Residual: Exchangeable				
var(e)	882.9858	74.3054	748.7262	1041.32
cov(e)	155.7724	61.30656	35.61377	275.9311

```
LR test vs. linear model: chi2(1) = 8.60                  Prob > chi2 = 0.0034
```

Note: The reported degrees of freedom assumes the null hypothesis is not on
    the boundary of the parameter space. If this is not true, then the
    reported test is conservative.

Let's compare the results from the `mixed` command above (using compound symmetry) with the corresponding results of the `mixed` command from section 12.2 (which used an unstructured residual covariance matrix). Note that the estimates of the coefficients are identical in both analyses (the coefficient for month 2 is 21.07; for month 3, it is 18.43). However, the standard errors are different (for example, the standard error of the coefficient for month 2 is 3.814 using compound symmetry versus 3.833 with the unstructured residual covariance matrix). This illustrates that the specification of the residual covariance structure has no impact on the estimates of the coefficients; it only affects the estimates of the standard errors. In this example, the differences in the standard errors are trivial. However, if there were more time points and the differences in the covariance structures were more striking, the choice of the residual covariance structure could have a more substantial impact on the standard errors.

Warning! Likelihood-ratio test

It is tempting to ask whether the difference in covariance structures is significantly different and to want to use a command like `lrtest` to test whether one covariance structure fits significantly better than another. A key assumption of a likelihood-ratio test is that one model is nested within another model. That is, one model can be created from the other by omitting one or more parameters. In many (or perhaps most) cases, the models formed by comparing two different residual covariance structures are not nested within each other, and the likelihood-ratio test is not valid. However, the AIC and BIC indices can be used even when models are not nested within each other.

12.7   Example 1 revisited again: Using small-sample methods

The `mixed` command uses large-sample methods for performing significance tests and computing CIs, using Wald chi-squared statistics and z statistics. When using certain designs and covariance structures, the `mixed` command provides options to use small-sample methods for significance tests and CIs. That is, in lieu of Wald chi-squared and z statistics, the `mixed` command can use F and t statistics.

Let's again consider the analysis of example 1 (introduced in section 12.2 and revisited in section 12.6). This study included $n = 100$ people whose sleep was observed at three time points, month 1, month 2, and month 3. Imagine this study actually had a smaller sample of people, say, $n = 25$ or perhaps $n = 20$ people whose sleep was observed for each of the three months. I would become concerned about using large-sample test statistics with such small sample sizes. As an analogy, imagine that I performed a paired-sample t test comparing just two of the time points (for example, time 1 versus time 2) but that I used a z distribution for judging the significance of the calculated t statistic.[5] As the sample size gets smaller and smaller (say, $n = 30$ or $n = 25$ or $n = 20$ or fewer), the differences between the z distribution and t distribution become more and more substantial. At some point, the differences are substantial enough to give us pause about using a large-sample method (that is, a z distribution) when we really should be using a small-sample method (that is, a t distribution).

---

5. For example, using a two-tailed test with alpha = 0.05 would yield a decision rule to reject the null hypothesis if the absolute value of the t statistic exceeds 1.96 (the critical value determined based on the z distribution).

The `mixed` command offers the `dfmethod()` option to fit models using small-sample methods, However, there are a limited number of situations in which these methods can be applied. The documentation for the `dfmethod()` option with the `mixed` command describes the different scenarios in which the `dfmethod()` can be used.[6] For example, the description of the `dfmethod(repeated)` option makes this sound like a useful small-sample method for analyzing the data from example 1. The documentation says that this option can be used when analyzing data from a balanced repeated-measures design with a spherical residual covariance structure. The design used in example 1 is a repeated measures design (sleep was measured at month 1, month 2, and month 3). Furthermore, the design was balanced because the number of observations was the same at each time point. The final issue is whether the residual covariance is spherical. In section 12.5, we found that among the three residual covariance structures compared, the compound symmetry residual covariance structure provided the best fit. Well, it turns out that a residual covariance matrix that has a compound symmetry structure also has a spherical structure.[7] In short, our dataset and analysis strategy from example 1 appears to meet the conditions required for using the `dfmethod(repeated)` option. Let's apply that option with the `mixed` command to analyze the data from example 1, shown below.

---

6. See `help mixed`, and then go to the section describing the `dfmethod()` option.
7. Although the reverse is not necessarily true.

```
. use sleep_cat3, clear

. mixed sleep i.month || id:, nocons residuals(ex) dfmethod(repeated)

Obtaining starting values by EM:

Performing gradient-based optimization:

Iteration 0:   log likelihood = -1443.1779
Iteration 1:   log likelihood = -1438.8895
Iteration 2:   log likelihood =  -1438.879
Iteration 3:   log likelihood =  -1438.879

Computing standard errors:

Computing degrees of freedom:
```

```
Mixed-effects ML regression                    Number of obs     =          300
Group variable: id                             Number of groups  =          100

                                               Obs per group:
                                                            min =            3
                                                            avg =          3.0
                                                            max =            3
DF method: Repeated                            DF:         min =        99.00
                                                            avg =       165.00
                                                            max =       198.00

                                               F(2,   198.00)    =        18.12
Log likelihood =  -1438.879                    Prob > F          =       0.0000
```

sleep	Coef.	Std. Err.	t	P>|t|	[95% Conf. Interval]	
month						
2	21.07	3.813695	5.52	0.000	13.54933	28.59067
3	18.43	3.813695	4.83	0.000	10.90933	25.95067
_cons	348.24	2.971508	117.19	0.000	342.3439	354.1361

Random-effects Parameters	Estimate	Std. Err.	[95% Conf. Interval]	
id: (empty)				
Residual: Exchangeable				
var(e)	882.9858	74.3054	748.7262	1041.32
cov(e)	155.7724	61.30656	35.61377	275.9311

```
LR test vs. linear model: chi2(1) = 8.60                  Prob > chi2 = 0.0034
```

Note: The reported degrees of freedom assumes the null hypothesis is not on
      the boundary of the parameter space. If this is not true, then the
      reported test is conservative.

Note how the overall test of the model is reported using an F test [that is, $F(2, 198) = 18.12$, $p < 0.001$]. Also note that the denominator degrees of freedom (d.f.) for this test is 198, which corresponds to the d.f.$_{error}$ for a univariate repeated measures ANOVA.[8] Also note how the tests of the individual coefficients are reported as t tests. For example, the test of the coefficient associated with month 2 is significant [$t(198) = 5.52$, $p <$

---

8. In that model, the d.f. for the error term is $(n-1) \times (k-1)$, where n is the number of observations per time point and k is the number of time points. Using $n = 100$ and $k = 3$ yields 99×2, which is 198.

0.001].[9] In reporting this t-value for this example, I selected d.f. = 198, based on the denominator d.f. from the overall F test.

Note! Other small-sample methods.

You are not limited to just using the `dfmethod(repeated)` option for selecting small-sample estimation methods with the `mixed` command. Using the `dfmethod()` option, you can specify `residual`, `repeated`, `anova`, `satterthwaite`, or `kroger`. However, it is important to carefully consider the conditions under which each of these options can be used. For more information, see `help mixed`, and then see the section describing the `dfmethod()` option.

When using the `contrast` command, you can specify the `df()` option to request significance tests and CIs using small-sample methods, for example, using a t distribution with the specified number of d.f. In the current example, which has balanced data and comports to using the `dfmethod(repeated)` option, we can specify `df(198)` with the `contrast` command to obtain tests and CIs that use the same small-sample methods and d.f. used in the previous `mixed` command.[10] For example, the `contrast` command below uses the `df(198)` option to use a t distribution with 198 d.f. for computing significance tests and CIs.

---

9. This compares the average sleep in month 2 with that in the omitted month, month 1.
10. Note that I selected `df(198)` based on the denominator d.f. from the F test from the `mixed` command; that is, $F(2, 198) = 18.12$. Other small-sample methods besides `dfmethod(repeated)` can give varying d.f. In those cases, it is not easy to determine the value that should be specified in the `df()` option with the `contrast` command. In fact, there may not be a single value that would be appropriate for all contrasts because the denominator d.f. may vary as a function of the effect being tested.

```
. contrast r.month, cieffects df(198)
Contrasts of marginal linear predictions
Margins       : asbalanced
```

	df	F	P>F
sleep			
month			
(2 vs 1)	1	30.52	0.0000
(3 vs 1)	1	23.35	0.0000
Joint	2	18.12	0.0000
Denominator	198		

	Contrast	Std. Err.	[95% Conf.	Interval]
sleep				
month				
(2 vs 1)	21.07	3.813695	13.54933	28.59067
(3 vs 1)	18.43	3.813695	10.90933	25.95067

The `contrast` command shows that the average minutes of sleep in month 2 were significantly different from those in those in month 1 $[F(1, 198) = 30.52, p < 0.001]$. The average minutes of sleep were 21.07 minutes longer in month 2 than in month 1 (95% CI = $[13.55, 28.59]$). The comparison of the average minutes slept in month 3 with those slept in month 1 was also significant $[F(1, 198) = 23.35, p < 0.001]$. In month 3, the average minutes slept were 18.4 minutes longer than those slept in month 1 (95% CI = $[10.91, 25.95]$).

12.8   An alternative analysis: ANCOVA

As described in section 6.2, analysis of covariance (ANCOVA) is a possible alternative to a repeated measures ANOVA. In fact, ANCOVA can be a preferred analysis (Rausch, Maxwell, and Kelley 2003). Let's look at the data from section 12.4 to show how that dataset could be analyzed using ANCOVA. Putting our ANCOVA hat on, we see that the three treatment groups in the example form the levels of the IV. The sleep score at month 1 is the covariate (in other words, a pretest). The sleep scores at months 2 and 3 are 2 separate dependent variables (DVs) (in other words, 2 posttests).

The data for the study are contained in `sleep_catcat33.dta`. As we saw in section 12.4, this dataset has one observation per person per time point. This is illustrated with the `list` command, which shows that the results associated with `id` number 1 are contained in the first 3 observations corresponding to months 1, 2, and 3.

```
. use sleep_catcat33, clear
. list in 1/6, sepby(id)
```

	id	month	group	sleep
1.	1	1	Control	305
2.	1	2	Control	313
3.	1	3	Control	351
4.	2	1	Control	295
5.	2	2	Control	289
6.	2	3	Control	374

To use ANCOVA to analyze the data from this study, we need to reshape the dataset so that the pretest and posttest scores for a given person are contained in the same row of data (that is, in the same observation). Fortunately, we can easily do this using the `reshape wide` command below (see section 22.9.2 for more details on the `reshape wide` command).

```
. reshape wide sleep, i(id) j(month)
(note: j = 1 2 3)
```

Data	long	->	wide
Number of obs.	900	->	300
Number of variables	4	->	5
j variable (3 values)	month	->	(dropped)
xij variables:			
	sleep	->	sleep1 sleep2 sleep3

Using the `list` command below, we can see the first five observations of the dataset after using the `reshape wide` command. We see that all the sleep scores for each person are contained in a single observation (row). For `id` number 1, the 3 sleep scores (305, 313, 351) are contained in the same row of data contained in the variables named `sleep1`, `sleep2`, and `sleep3`. These correspond to the sleep scores at month 1, month 2, and month 3.

```
. list in 1/5
```

	id	sleep1	sleep2	sleep3	group
1.	1	305	313	351	Control
2.	2	295	289	374	Control
3.	3	373	365	360	Control
4.	4	386	365	359	Control
5.	5	344	325	343	Control

Now that we have the data in this format, we can analyze the results using ANCOVA (see section 6.2 for additional details). We can analyze the sleep at month 2 by using sleep2 as the DV, group as the IV, and sleep1 as the covariate. Such an analysis is performed below using the anova command.

```
. anova sleep2 i.group c.sleep1
                          Number of obs =        300    R-squared     =  0.1446
                          Root MSE      =    28.5009    Adj R-squared =  0.1360

          Source |  Partial SS          df         MS         F      Prob>F

           Model |  40654.049            3     13551.35     16.68    0.0000

           group |  29058.015            2    14529.007     17.89    0.0000
          sleep1 |  10434.423            1    10434.423     12.85    0.0004

        Residual |   240441.9          296    812.30371

           Total |  281095.95          299    940.12022
```

The contrast command below tests the equality of the average month 2 sleep between the 3 groups, after adjusting for the sleep at month 1. This test is significant $[F(2, 296) = 17.89, p < 0.001]$.

```
. contrast group
Contrasts of marginal linear predictions
Margins        : asbalanced
```

	df	F	P>F
group	2	17.89	0.0000
Denominator	296		

Let's apply the r. contrast operator to group to compare each treatment group with the control group.

```
. contrast r.group, cieffects
Contrasts of marginal linear predictions
Margins       : asbalanced
```

	df	F	P>F
group			
(Medication vs Control)	1	35.32	0.0000
(Education vs Control)	1	5.73	0.0173
Joint	2	17.89	0.0000
Denominator	296		

	Contrast	Std. Err.	[95% Conf.	Interval]
group				
(Medication vs Control)	23.96931	4.03316	16.03201	31.90661
(Education vs Control)	9.65735	4.034384	1.717639	17.59706

Let's first consider the output for the first contrast, which compared the medication group with the control group. The average sleep at month 2 (adjusted for month 1 sleep) is significantly greater for the medication group compared with the control group $[F(1, 296) = 35.32, p < 0.001]$. After adjusting for the amount of sleep at month 1 (baseline), we see that the medication group slept 23.97 more minutes at month 2 than the control group (95% CI $= [16.03, 31.91]$).

Next, let's look at the output for the second contrast, which compares the education group with the control group. The difference in sleep at month 2 (adjusting for month 1) between these 2 groups is also significant $[F(1, 296) = 5.73, p = 0.0173]$.[11] The average sleep at month 2 (adjusted for month 1) for the education group was 9.66 minutes greater than that for the control group (95% CI $= [1.72, 17.60]$).

Note! Further analyses

All the analysis tools illustrated in section 6.2 can be applied to the current example. We could use the `margins` command to compute adjusted means for each treatment group (after adjusting for the sleep at month 1). The `marginsplot` command could be used to graph the adjusted means computed by the `margins` command. The `estat esize` command could be used to compute the omega-squared for the overall treatment effect. The `esizei` command could be used to compute the omega-squared for a particular contrast among treatment groups (see section 4.4.2 for more details).

---

11. Note how this test comparing the education group with the control group was significant using ANCOVA but was not significant using the repeated measures ANOVA strategy.

We can repeat the above analyses but instead use `sleep3` as the DV, `sleep1` as the covariate, and `treat` as the IV. I have omitted this analysis to save space.

The ANCOVA approach can be a preferable analysis strategy over a repeated measures analysis strategy because it has greater statistical power (as was illustrated in section 11.4.1). See Rausch, Maxwell, and Kelley (2003) for more information comparing ANCOVA with repeated measures ANOVA.

12.9 Closing thoughts

This chapter has illustrated three different examples in which time was treated as a categorical variable. These models used the `mixed` command combined with the `residuals()` option to account for the residual covariances across times. The fixed-effects portion of the `mixed` models (that is, the part before the | |) was specified as the same whether the model was longitudinal or nonlongitudinal. Further, the postestimation commands we used for interpreting and visualizing the results (that is, `margins`, `marginsplot`, and `contrast`) were exactly the same regardless of whether the model was longitudinal or nonlongitudinal. This chapter also illustrated how to select different residual covariance structures and how to compare the fit among competing structures. Further, we saw how the `dfmethod()` option can specify the use of small-sample methods, seeing the use of the `dfmethod(repeated)` option with the `mixed` command. Finally, the chapter illustrated how ANCOVA could be used as another analysis strategy for the analysis of repeated measures designs by using the pretest as a covariate and the posttest measures as the DV.

13 Longitudinal designs

13.1 Chapter overview

The previous chapter illustrated the analysis of repeated measures designs. Conventional repeated measures designs usually have a small number of repeated observations. Further, repeated measures designs assume that the repeated measurements are made on the same time schedule across people. When you use a repeated measures design, it is difficult to accommodate irregularities in the gaps between the repeated measures across people. Longitudinal designs, by contrast, typically have a larger number of observations. Further, the time gaps between the repeated measures can vary between people. Such longitudinal designs can be difficult, or even impossible, to analyze using traditional repeated measures analyses.

This chapter illustrates how you can use multilevel modeling commands in Stata to analyze data from such longitudinal designs.[1] This chapter presents four examples that illustrate how you can use the `mixed` command to analyze longitudinal designs using multilevel modeling.

The first example models the dependent variable as a linear function of time (section 13.2). The second example adds a between-subjects independent variable (IV), which allows us to model the linear effect of time as well as explore the IV by time

---

1. If you are new to using multilevel modeling for the analysis of longitudinal data, I highly recommend Singer and Willett (2003).

interaction (section 13.3). This example is similar to an analysis of covariance with a treatment by covariate interaction (for instance, like the examples in chapter 8). The third example includes time as the only predictor but uses a piecewise modeling strategy for the effect of time (section 13.4). The fourth example adds a between-subjects IV to the third example, modeling the interaction of the IV with the piecewise effects of time (section 13.5).

13.2   Example 1: Linear effect of time

Let's consider a study in which we are looking at the number of minutes people sleep at night over a seven-week period. In this hypothetical example, 75 people were enrolled, and each person's nightly sleep time (in minutes) was recorded on 8 occasions, approximately once every 7 days. The dataset for this study is organized in a long format, with one observation per person per night of observation. There are 75 people who were each observed 8 times; thus the dataset has 600 observations. Let's look at the first five observations in the dataset.

```
. use sleep_conlin
. list in 1/5
```

	id	obsday	sleep
1.	1	1	382
2.	1	6	382
3.	1	13	390
4.	1	21	378
5.	1	27	401

The day of observation is stored in the variable `obsday`. The first person slept 382 minutes on day 1. When observed again on day 6, the person again slept 382 minutes. He or she was observed again on day 13 and slept 390 minutes.

Let's now look at the summary statistics for the variables in this dataset.

```
. summarize
```

Variable	Obs	Mean	Std. Dev.	Min	Max
id	600	38	21.66677	1	75
obsday	600	23.565	14.82854	1	52
sleep	600	360.785	48.13086	175	528

The variable `id` identifies each person, and it ranges from 1 to 75, which represents the 75 people in this study. The variable `obsday` indicates the day of observation and ranges from 1 to 52. The variable `sleep` represents the number of minutes the person slept on a particular night. The average of this variable is 360.8, with a minimum of 175 and a maximum of 528.

If there were no repeated observations in the dataset (that is, the residuals were independent), we could use simple linear regression to regress `sleep` on `obsday`. Although this approach is incorrect, let's consider it below.

```
. regress sleep obsday
```

Source	SS	df	MS			
				Number of obs	=	600
				F(1, 598)	=	13.35
Model	30312.1377	1	30312.1377	Prob > F	=	0.0003
Residual	1357319.13	598	2269.76443	R-squared	=	0.0218
				Adj R-squared	=	0.0202
Total	1387631.27	599	2316.57974	Root MSE	=	47.642

sleep	Coef.	Std. Err.	t	P>|t|	[95% Conf. Interval]	
obsday	.4797296	.131274	3.65	0.000	.2219155	.7375436
_cons	349.4802	3.654108	95.64	0.000	342.3037	356.6566

Note that the regression model makes no assumption about the regularity of the time between the repeated measures across people. The days between measurements can be highly irregular, and there can even be gaps (missing weeks) between observations. The key to this analysis strategy is that it focuses on how the amount of sleep is related to the actual day of observation (that is, `obsday`). By comparison, a repeated measures analysis as illustrated in chapter 12 focuses on the measurement occasion (for example, first measurement, second measurement, third measurement). Let me pose an example to emphasize this point. Consider two participants who might have been in the study, person 1 and person 2. Because of differences in availability and scheduling issues, person 1 had his or her 3rd observation of their sleep on the 15th day, whereas person 2 had his or her 3rd observation on the 24th day. The repeated measures analysis of variance would treat both of these as the 3rd observation, ignoring the fact that person 2 was observed 9 days after person 1. The `regress` command above, on the other hand, focuses on the day of observation (that is, that person 1 was observed on day 15 and person 2 was observed on day 24).

Despite the advantages of using the `regress` command for this analysis, there is a fatal problem. It assumes that the repeated observations for the same person are independent (which is certainly not true). This yields biased standard errors and untrustworthy *p*-values. In addition, it fits a single intercept for the entire sample, which ignores the fact that people likely differ in their initial level of sleep. It is more realistic to picture separate intercepts across people, which reflects individual differences in sleep.

We can remedy both of these issues by fitting a random intercept model using the `mixed` command below. Like the `regress` command, `mixed` fits sleep as a linear function of `obsday`. Specifying `|| id:` indicates that observations are repeated within the levels of `id`, which accounts for the nonindependence of observations within a person while also allowing for different (random) intercepts across people.

```
. mixed sleep obsday || id:

Performing EM optimization:

Performing gradient-based optimization:

Iteration 0:    log likelihood = -2732.4084
Iteration 1:    log likelihood = -2732.4084

Computing standard errors:

Mixed-effects ML regression                      Number of obs      =          600
Group variable: id                               Number of groups   =           75

                                                 Obs per group:
                                                               min =            8
                                                               avg =          8.0
                                                               max =            8

                                                 Wald chi2(1)       =       103.70
Log likelihood = -2732.4084                      Prob > chi2        =       0.0000
```

sleep	Coef.	Std. Err.	z	P>|z|	[95% Conf. Interval]	
obsday	.5086386	.0499493	10.18	0.000	.4107398	.6065374
_cons	348.7989	5.268403	66.21	0.000	338.4731	359.1248

Random-effects Parameters	Estimate	Std. Err.	[95% Conf. Interval]	
id: Identity				
var(_cons)	1937.141	322.9833	1397.136	2685.862
var(Residual)	325.2438	20.07449	288.1853	367.0679

```
LR test vs. linear model: chibar2(01) = 872.36          Prob >= chibar2 = 0.0000
```

The interpretation of the obsday coefficient is straightforward. For every additional day in the study, the number of nightly minutes of sleep is predicted to increase by 0.51. Multiplying this by seven yields perhaps a simpler interpretation. For every additional week in the study, participants slept, on average, an additional 3.6 minutes per night. This coefficient describes the trajectory of sleep durations across the days of the study.

Terminology! Slopes

In this chapter, I will refer to the obsday coefficient in a couple of different ways. For example, I will call it the "obsday slope", indicating that it is the slope associated with the day of observation. Sometimes, for simplicity, I will refer to this just as the "slope".

Let's estimate the predicted mean of sleep as a function of time by using the margins command to estimate the average sleep at the 1st day of observation to the 50th day of observation in 7-day increments.

```
. margins, at(obsday=(1(7)50)) vsquish nopvalues
Adjusted predictions                              Number of obs    =        600
Expression   : Linear prediction, fixed portion, predict()
1._at        : obsday          =            1
2._at        : obsday          =            8
3._at        : obsday          =           15
4._at        : obsday          =           22
5._at        : obsday          =           29
6._at        : obsday          =           36
7._at        : obsday          =           43
8._at        : obsday          =           50
```

	Margin	Delta-method Std. Err.	[95% Conf. Interval]	
_at				
1	349.3076	5.257469	339.0031	359.612
2	352.868	5.193752	342.6885	363.0476
3	356.4285	5.153022	346.3288	366.5282
4	359.989	5.135828	349.9229	370.055
5	363.5495	5.142403	353.4705	373.6284
6	367.1099	5.172659	356.9717	377.2481
7	370.6704	5.226184	360.4273	380.9135
8	374.2309	5.302272	363.8386	384.6231

The predicted mean of sleep for the 1st day of the study (obsday = 1) is 349.3 minutes (95% CI = [339.0, 359.6]). On the 8th day of the study, the predicted mean of sleep is 352.9 minutes (95% CI = [342.7, 363.0]). (Note how the gain in sleep from the 1st day to the 8th day is 3.6 minutes, which is the slope value multiplied by 7.)

We can graph these predicted means and confidence intervals (CIs) with marginsplot (shown below). The resulting graph is shown in figure 13.1. This graph illustrates the slope of the relationship between time and sleep. It includes CIs for each of the specified observation days, providing a 95% confidence interval for the predicted mean of sleep on the specified day.

```
. marginsplot
Variables that uniquely identify margins: obsday
```

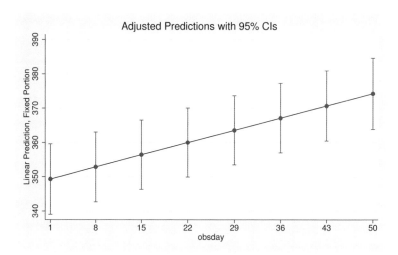

Figure 13.1: Minutes of sleep at night by time

In this model, the obsday effect is fit only in the fixed portion of the model. That is, the model fits one fixed trajectory of sleep durations across days. The model recognizes that people randomly vary in terms of the initial quality of their sleep (represented by the random intercept). But people can also vary individually in trajectory of their sleep across the weeks of the study. By adding obsday as a random coefficient (that is, a random slope), we see that the model can account for both individual differences in the initial quality of sleep (that is, a random intercept) and individual differences in the sleep trajectory across the weeks of the study (that is, a random slope for obsday). We can fit such a model using the mixed command, shown below.[2]

---

2. When we specify || id: obsday, cov(un), this model allows the intercepts to vary randomly (across people), allows the slopes to randomly vary across people, and permits the random intercepts and random slopes to be correlated. The || id: portion permits the intercepts to randomly vary across people. obsday permits this slope to also randomly vary across people. cov(un) permits the random intercept and random slope to be correlated.

```
. mixed sleep obsday || id: obsday, cov(un)

Performing EM optimization:

Performing gradient-based optimization:

Iteration 0:   log likelihood = -2488.5378
Iteration 1:   log likelihood = -2488.5378

Computing standard errors:
```

Mixed-effects ML regression				Number of obs	=	600
Group variable: id				Number of groups	=	75
				Obs per group:		
				min =		8
				avg =		8.0
				max =		8
				Wald chi2(1)	=	20.54
Log likelihood = -2488.5378				Prob > chi2	=	0.0000

sleep	Coef.	Std. Err.	z	P>|z|	[95% Conf. Interval]	
obsday	.5104496	.1126409	4.53	0.000	.2896774	.7312217
_cons	348.8205	3.308517	105.43	0.000	342.3359	355.305

Random-effects Parameters	Estimate	Std. Err.	[95% Conf. Interval]	
id: Unstructured				
var(obsday)	.8926483	.1552909	.6347458	1.255339
var(_cons)	775.8412	134.126	552.8646	1088.747
cov(obsday,_cons)	14.29639	3.556285	7.326203	21.26658
var(Residual)	101.2256	6.745405	88.83181	115.3485

```
LR test vs. linear model: chi2(3) = 1360.11          Prob > chi2 = 0.0000
Note: LR test is conservative and provided only for reference.
```

The slope of the relationship between sleep duration and observation day is significant $[z(N = 75) = 4.53, p = 0.000]$. For every additional day in the study, the participants slept an additional 0.51 minutes. (Note that this estimate is not much different from the coefficient estimated by the previous `mixed` command.)

This model added a random effect for `obsday`, which we see reflected in the estimate of `var(obsday)` in the random-effects portion of the output. This represents the degree to which the sleep trajectories vary between individuals. As described in Singer and Willett (2003), we can form a simple one-sided z test to determine whether this value is significantly greater than zero. We can compute a z statistic by dividing the estimate (that is, 0.8926483) by its standard error (that is, 0.1552909), which yields a z-value of 5.75. This value well exceeds 1.645 (the 5% cutoff for a one-sided z test), which suggests that there is significant variation in the sleep trajectories among individuals.[3]

As we did before, we can use the `margins` command to estimate the predicted mean number of minutes slept at night across the weeks of the study. We can also use the `marginsplot` command to graph the predicted means computed by the `margins` command. The `margins` and `marginsplot` commands yield results rather similar to those shown above, so I have omitted showing them to save space.

This random intercept–random slope model forms a foundation that we can use to illustrate the next example in which we extend the design to include a between-subjects IV.

Note! Linearity and other assumptions

In this model, I am assuming that the relationship between observation day and amount of sleep is linear. Naturally, this is an assumption we would want to investigate. In fact, we would want to investigate all the assumptions of this model, which would include all the kinds of assumptions we make in an ordinary linear regression model as well as the assumptions of a mixed model. See Raudenbush and Bryk (2002) and Singer and Willett (2003) for more details.

13.3   Example 2: Interacting time with a between-subjects IV

Let's consider an extension of the previous hypothetical sleep study and suppose that participants were randomly assigned to one of three different treatments: 1) control (no treatment); 2) medication (where a sleep medication is given); or 3) education (where the participants receive education about how to sleep better and longer). The treatments were applied before the first day in which sleep was observed.

---

3. Singer and Willett (2003) note that while this is a simple test to perform, there is disagreement over its appropriateness, and it is probably best treated as a quick, but imprecise, test. They provide a useful discussion of the concerns surrounding this test and provide an alternative test that may be more preferable. The most appropriate method for performing such tests appears to be an unsettled question and the topic of continued research.

In the previous example, we noted that people differed in their sleep trajectories across the weeks of the study. This study will explore whether the sleep trajectories vary as a function of treatment group assignment. In other words, we want to examine whether the slope (across time) differs across the groups. In particular, the study has two aims. The first aim is to compare the slope for the medication group with the slope for the control group. The second aim is to compare the slope for the education group with the slope of the control group.

This design involves a combination of a continuous predictor (time) and a three-level between-subjects IV (treatment group). If we think of time as a covariate, this model is much like the analysis of covariance model we saw in section 8.3. Both models include a three-level IV, a continuous covariate, and an interaction of the IV and the covariate. Although we will use a different command for the analysis of this dataset (using `mixed` instead of `regress`), the logic, interpretation, and postestimation commands for understanding the results of the model are very much the same as we saw in section 8.3.

Let's begin by using the dataset for this example, listing the first five observations, and showing summary statistics for the variables in the dataset.

```
. use sleep_cat3conlin, clear
. list in 1/5
```

	id	group	obsday	sleep
1.	1	Control	1	370
2.	1	Control	7	382
3.	1	Control	13	377
4.	1	Control	18	408
5.	1	Control	24	385

```
. summarize
```

Variable	Obs	Mean	Std. Dev.	Min	Max
id	600	38	21.66677	1	75
group	600	2	.8171778	1	3
obsday	600	20.18833	12.76994	1	46
sleep	600	358.5433	52.49169	225	504

We can see that the variable `id` identifies the person. The variable `group` reflects the treatment group assignment, coded as follows: 1 = control, 2 = medication, and 3 = education. The variable `obsday` is the day in which sleep was observed. This represents time in terms of the number of days from the start of the study. The variable `sleep` is the number of minutes slept for the particular day of observation (that is, the sleep duration). Focusing on the multilevel nature of this design, we see that the treatment group assignment is a property of the person and thus is a level 2 variable and that observation day is a level 1 variable.

We can fit a model that predicts sleep from the observation day, the group assignment, and the interaction of these two variables using the `mixed` command shown below. Note that the random-effects portion of the model specifies that `obsday` is a random effect. When we think in terms of a multilevel model, `group#obsday` is a cross-level interaction of a level 2 variable (`group`) with a level 1 variable (`obsday`).

```
. mixed sleep i.group##c.obsday || id: obsday, cov(un) nolog
Mixed-effects ML regression                     Number of obs     =        600
Group variable: id                              Number of groups  =         75

                                                Obs per group:
                                                              min =          8
                                                              avg =        8.0
                                                              max =          8

                                                Wald chi2(5)      =      66.55
Log likelihood = -2482.6017                     Prob > chi2       =     0.0000
```

sleep	Coef.	Std. Err.	z	P>|z|	[95% Conf. Interval]	
group						
Medication	35.48106	9.498675	3.74	0.000	16.86399	54.09812
Education	5.552734	9.499246	0.58	0.559	-13.06545	24.17092
obsday	-.0657762	.1798984	-0.37	0.715	-.4183707	.2868183
group#						
c.obsday						
Medication	.1815398	.2539227	0.71	0.475	-.3161395	.6792191
Education	.8307836	.2541045	3.27	0.001	.332748	1.328819
_cons	339.4458	6.717216	50.53	0.000	326.2803	352.6113

Random-effects Parameters	Estimate	Std. Err.	[95% Conf. Interval]	
id: Unstructured				
var(obsday)	.7192364	.1319566	.502	1.03048
var(_cons)	1079.505	184.1759	772.6809	1508.167
cov(obsday,_cons)	22.51981	4.232773	14.22373	30.81589
var(Residual)	108.3664	7.218497	95.10302	123.4795

```
LR test vs. linear model: chi2(3) = 1425.42              Prob > chi2 = 0.0000
Note: LR test is conservative and provided only for reference.
```

We can interpret the results of model using the same tools and techniques illustrated in section 8.3. Let's start by making a graph of the predicted means as a function of time and treatment group. We first use the `margins` command to estimate the predicted means by `group` for the 1st and 50th days of observation. We then use the `marginsplot` command to graph the predicted means computed by the `margins` command. Figure 13.2 shows the graph of the predicted means produced by the `marginsplot` command.

```
. margins group, at(obsday=(0 50)) noatlegend
(output omitted)

. marginsplot, noci
Variables that uniquely identify margins: obsday group
```

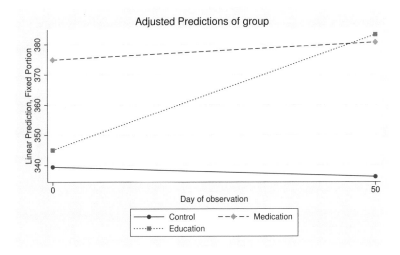

Figure 13.2: Minutes of sleep at night by time and treatment group

Let's focus on the slope of the relationship between sleep duration and day of observation. The slope appears to be slightly negative for the control group. In other words, sleep duration appears to mildly decrease as a linear function of obsday. For the medication group, the slope appears to be slightly positive. By contrast, sleep duration increases as a linear function of time for the education group. In other words, the obsday slope is positive for the education group.

Let's use the margins command to estimate the three slopes depicted in figure 13.2. The dydx(obsday) option is used to estimate the slope with respect to obsday.

```
. margins group, dydx(obsday) vsquish
Average marginal effects                        Number of obs     =        600
Expression   : Linear prediction, fixed portion, predict()
dy/dx w.r.t. : obsday
```

	dy/dx	Delta-method Std. Err.	z	P>|z|	[95% Conf. Interval]	
obsday						
group						
Control	-.0657762	.1798984	-0.37	0.715	-.4183707	.2868183
Medication	.1157636	.1792018	0.65	0.518	-.2354655	.4669927
Education	.7650074	.1794593	4.26	0.000	.4132737	1.116741

The slope for the control group is -0.07 but is not significantly different from 0 $[z(N = 75) = -0.37,\ p = 0.715]$. The slope for the medication group is 0.12 and is also not significantly different from 0 $[z(N = 75) = 0.65,\ p = 0.518]$. For the education group, the slope is 0.77 and is significantly different from 0 $[z(N = 75) = 4.26,\ p < 0.001]$. For the education group, sleep duration increased (on average) by 0.77 minutes for every additional day (or about 5.4 minutes per week). This is consistent with the positive slope for the education group illustrated in figure 13.2.

We can form comparisons of these slopes, comparing the slope for each group with group 1 (the control group). The `r.` contrast operator is applied to `group` to form reference group comparisons. The `dydx(obsday)` option is used to indicate that the comparisons are to be made on the slope with respect to `obsday`.

```
. margins r.group, dydx(obsday) contrast(pveffects nowald)

Contrasts of average marginal effects

Expression   : Linear prediction, fixed portion, predict()
dy/dx w.r.t. : obsday
```

	Contrast dy/dx	Delta-method Std. Err.	z	P>|z|
obsday				
group				
(Medication vs Control)	.1815398	.2539227	0.71	0.475
(Education vs Control)	.8307836	.2541045	3.27	0.001

The first test compares the slope with respect to `obsday` for the medication group versus the control group. The difference in these slopes is 0.18 and is not significant $[z(N = 75) = 0.71,\ p = 0.475]$. This is consistent with what we saw in figure 13.2, in which the slopes for the medication group and control group were not dissimilar.

The second test compares the slope for the education group with the slope for the control group. The difference in these slopes is 0.83 and is significant $[z(N = 75) = 3.27,\ p = 0.001]$. The slope for the education group is significantly steeper than the slope for the control group (see figure 13.2). For every additional day in the study, the education group sleeps an additional 0.83 more minutes than the control group.

Say that we wanted to compare the predicted mean of sleep duration for each group with that of the control group at different levels of time. The `margins` command below compares the average sleep for the medication to control group and the education to control group for the first day to the 50th day of observation in 7 day increments.

```
. margins r.group, at(obsday=(1(7)50)) vsquish contrast(nowald pveffects)

Contrasts of adjusted predictions
Expression    : Linear prediction, fixed portion, predict()
1._at         : obsday          =            1
2._at         : obsday          =            8
3._at         : obsday          =           15
4._at         : obsday          =           22
5._at         : obsday          =           29
6._at         : obsday          =           36
7._at         : obsday          =           43
8._at         : obsday          =           50
```

	Contrast	Delta-method Std. Err.	z	P>|z|
group@_at				
(Medication vs Control) 1	35.6626	9.675492	3.69	0.000
(Medication vs Control) 2	36.93337	10.99809	3.36	0.001
(Medication vs Control) 3	38.20415	12.43461	3.07	0.002
(Medication vs Control) 4	39.47493	13.94991	2.83	0.005
(Medication vs Control) 5	40.74571	15.52092	2.63	0.009
(Medication vs Control) 6	42.01649	17.13233	2.45	0.014
(Medication vs Control) 7	43.28727	18.77374	2.31	0.021
(Medication vs Control) 8	44.55804	20.43792	2.18	0.029
(Education vs Control) 1	6.383518	9.67595	0.66	0.509
(Education vs Control) 2	12.199	10.9981	1.11	0.267
(Education vs Control) 3	18.01449	12.43463	1.45	0.147
(Education vs Control) 4	23.82997	13.95026	1.71	0.088
(Education vs Control) 5	29.64546	15.52183	1.91	0.056
(Education vs Control) 6	35.46094	17.13395	2.07	0.038
(Education vs Control) 7	41.27643	18.77619	2.20	0.028
(Education vs Control) 8	47.09191	20.44128	2.30	0.021

The difference in sleep between the medication and control groups is significant at each of the days tested ($ps \leq 0.029$). The average sleep in the medication group is greater than that in the control group at each of the days tested.

For the comparison of education with control, the difference is not significant on the 1st, 8th, 15th, 22nd, or 29th day of observation ($ps \geq 0.056$). The average sleep for the education group is significantly greater than that for the control group at the 36th, 43rd, and 50th day of the study ($ps \leq 0.038$).

We can depict the tests from the previous `margins` command using the `marginsplot` command below. The graph created by the `marginsplot` command, shown in figure 13.3, shows the comparison of the medication group with the control group in the left panel and the comparison of the education group with the control group in the right panel. The confidence interval for each comparison is shown with a reference line at zero. Where the confidence interval excludes 0, the difference is significant at the 5% level.

```
. marginsplot, bydim(group) yline(0) xlabel(1(7)50)
  Variables that uniquely identify margins: obsday group
```

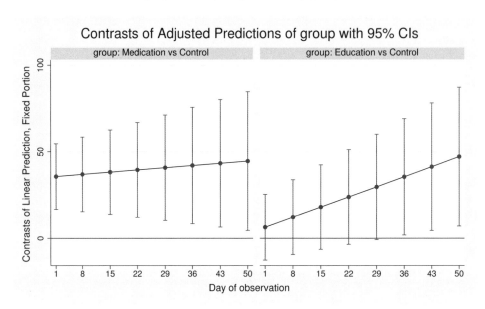

Figure 13.3: Contrasts (with 95% CIs) of medication versus control (left panel) and education versus control (right panel) by day of observation

13.4   Example 3: Piecewise modeling of time

Let's consider an example similar to the one illustrated in example 1 in section 13.2. Like that example, this one measures sleep over time with 75 participants whose sleep at night is measured approximately every 7 days for a total of 8 observations. The difference is that in this study, the first 30 days are considered a baseline period during which sleep is simply observed. On the 31st day, the treatment period begins, and participants are given a combination of sleep medication and sleep education to increase their minutes of sleep.

Time can be modeled in a piecewise fashion by breaking up the days of observation into the baseline phase (days 1 to 30) and the treatment phase (day 31 onward). Four possible patterns of results of such an analysis are illustrated in figure 13.4, which shows the amount of sleep on the y axis and the observation day on the x axis. A vertical line is drawn on day 31 to indicate the start of the treatment phase.

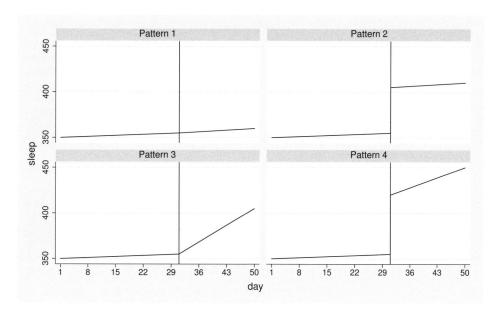

Figure 13.4: Four example patterns of results for a piecewise analysis of sleep across time

Let's compare these four patterns, starting with pattern 1 shown in figure 13.4. Sleep time very gradually improves across the days, but there is no change due to the intervention on day 31. There is no change in slope during the treatment phase compared with the baseline phase. And there is neither a sudden increase nor a decrease in sleep at the start of the treatment phase.

Let's now consider pattern 2 shown in figure 13.4. Pattern 2 is typified by a sudden increase in the amount of sleep at the start of treatment but not by a change in the slope due to the treatment. This is the kind of pattern I would expect because of the introduction of a sleep medication on day 31, which provides an instant increase in sleep on the first day of treatment.

Let's now examine pattern 3 from figure 13.4. This pattern is typified by a greater slope during the treatment phase compared with the baseline phase. However, there is no sudden increase or decrease in sleep at the start of the treatment phase. This is the kind of pattern I would expect because of the introduction of sleep education on day 31, which yields a greater increase in the quality of sleep over time but without any sudden increase in sleep quality.

Finally, let's consider pattern 4 from figure 13.4. This pattern is a blend of patterns 2 and 3. There is a sudden increase in sleep at the start of the treatment phase, and the slope during the treatment phase is greater than during the baseline phase. This is the kind of pattern I would expect because of a combination of medication (which yields an instant increase in sleep on day 31) and sleep education (which yields a greater slope during the treatment phase).

Let's now turn to the analysis of the data from this study. Let's begin by looking at the first five observations from the dataset as well as the summary statistics for the variables in the dataset.

```
. use sleep_conpw, clear
. list in 1/5, sepby(id)
```

	id	obsday	sleep
1.	1	1	285
2.	1	6	293
3.	1	10	267
4.	1	17	266
5.	1	23	288

```
. summarize
```

Variable	Obs	Mean	Std. Dev.	Min	Max
id	600	38	21.66677	1	75
obsday	600	23.81667	15.10474	1	55
sleep	600	355.095	103.3013	110	654

The variable `id` is the identifier for each person. The variable `obsday` reflects the day of observation, ranging from 1 to 55. The variable `sleep` is the duration of sleep (in minutes) for the given day of observation. The mean number of minutes of sleep is 355 and ranges from 110 to 654.

To analyze this dataset, we will use a piecewise model that allows us to model the dynamics portrayed in figure 13.4. Namely, the model will fit a slope for the baseline phase, a separate slope for the treatment phase, and a term that reflects any sudden increase (or decrease) in sleep at the start of the treatment phase.

To fit this model, we need to create some variables in advance. With respect to the slope terms, we can use the `mkspline` command to create the variables named `obsday1` and `obsday2`. Including the number 31 between these 2 variables on the `mkspline` command is our way of demarcating the end of the baseline phase and the start of the treatment phase. The variable `obsday1` will represent the slope during the baseline phase, and `obsday2` will represent the slope during the treatment phase. These 2 variables will allow us to estimate the slope during the baseline and treatment phase as well as compare these with slopes to detect changes in slope due to the treatment (as we saw in pattern 3 of figure 13.4).

```
. mkspline obsday1 31 obsday2 = obsday
```

Next, we use the `generate` and `replace` commands to create the variable `trtphase` that is coded 0 for the baseline phase (where `obsday` was 1 to 30) and 1 during the treatment phase (where `obsday` is 31 or more). This variable will capture any sudden increase (or decrease) in sleep at the start of the study (as we saw in pattern 2 of figure 13.4).

```
. generate trtphase = 0 if obsday < 31
(224 missing values generated)
. replace trtphase =  1 if obsday >= 31 & !missing(obsday)
(224 real changes made)
```

We are now ready to run a piecewise model that predicts `sleep` from `obsday1`, `obsday2`, and `trtphase`. In this example, `obsday1`, `obsday2`, and `trtphase` are also included as random effects.[4] This model is fit using the `mixed` command below. The `nolog` and `noretable` options are used to suppress the iteration log and random effects table to save printed space.

```
. mixed sleep obsday1 obsday2 trtphase || id: obsday1 obsday2 trtphase,
> cov(un) nolog noretable
Mixed-effects ML regression              Number of obs     =        600
Group variable: id                       Number of groups  =         75

                                         Obs per group:
                                                       min =          8
                                                       avg =        8.0
                                                       max =          8

                                         Wald chi2(3)      =      70.48
Log likelihood = -2601.8889              Prob > chi2       =     0.0000
```

sleep	Coef.	Std. Err.	z	P>|z|	[95% Conf. Interval]	
obsday1	-.005795	.1193361	-0.05	0.961	-.2396894	.2280995
obsday2	.537627	.1607347	3.34	0.001	.2225927	.8526612
trtphase	11.13768	1.941483	5.74	0.000	7.332445	14.94292
_cons	349.0109	10.6413	32.80	0.000	328.1543	369.8674

---

4. See Singer and Willett (2003) for a detailed discussion of how to select the appropriate random effects.

To help with the interpretation of this model, I will show you a graph of the predicted means as a function of observation day (see figure 13.5). I will illustrate how to make this kind of graph later in this section.

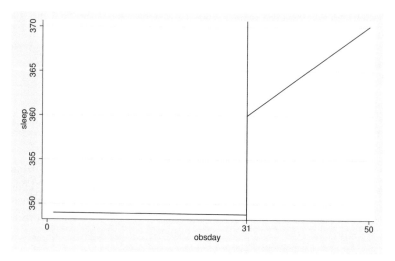

Figure 13.5: Predicted means of sleep across observation days using piecewise modeling of time

Let's now relate the graph in figure 13.5 to the results from the `mixed` command. The coefficient for `obsday1` represents the slope during the baseline phase. This value is -0.01 and is not significantly different from 0 $[z(N = 75) = -0.05, p = 0.961]$. We see this in figure 13.5 as a nearly flat line from day 1 to 31.

The coefficient for `obsday2` reflects the slope during the treatment phase. This slope is 0.538 and is significantly different from 0 $[z(N = 75) = 3.34, p = 0.001]$. During the treatment phase, each additional day of observation is associated with sleeping 0.538 additional minutes of sleep per night (or about 3.77 additional minutes per week).

We can estimate the change in the slope for the treatment phase versus the baseline phase using the `lincom` command, as shown below.

```
. lincom obsday2 - obsday1
 ( 1)   - [sleep]obsday1 + [sleep]obsday2 = 0
```

sleep	Coef.	Std. Err.	z	P>|z|	[95% Conf. Interval]	
(1)	.543422	.1663015	3.27	0.001	.217477	.8693669

This shows that the change in the slope from the baseline to treatment phase was 0.543 and that this change is significant [$z(N = 75) = 3.27$, $p = 0.001$]. The slope during the treatment phase is significantly greater than the slope during the baseline phase.

Now let's consider the coefficient for `trtphase`. This represents the jump in sleep at the beginning of the treatment phase (when `obsday` is 31). This coefficient is 11.1 and is statistically significant [$z(N = 75) = 5.74$, $p < 0.001$]. We can see this 11.1 minute jump in sleep that occurs when `obsday` is 31 in figure 13.5.

Let's now consider how to make the graph shown in figure 13.5. The difficulty in making this graph is because the x axis, which represents day of observation, does not correspond to a single variable. In fact, it is a reflection of the three predictors in our model, `obsday1`, `obsday2`, and `trtphase`. We need to express observation day in terms of `obsday1`, `obsday2`, and `trtphase` when `obsday` is 1, 31, and 50. This corresponds to the start of the study, the start of the treatment phase, and the approximate end of the study. The `showcoding` command (which I wrote for this book[5]) is used to show the values of `obsday1` and `obsday2` associated with `obsday`.

```
. showcoding obsday obsday1 obsday2 if inlist(obsday,1,31,50)
```

obsday	obsday1	obsday2
1	1	0
31	31	0
50	31	19

We can then use the `margins` command to compute the predicted means for these key days. When `obsday` equals 31, we estimate the predicted mean assuming `trtphase` is 0 and 1 to estimate the jump in the fitted values due to beginning the treatment phase. (Note the `noatlegend` option is included to save space.)

```
. margins, at(obsday1 = 1  obsday2 = 0  trtphase=0)
>          at(obsday1 = 31 obsday2 = 0  trtphase=0)
>          at(obsday1 = 31 obsday2 = 0  trtphase=1)
>          at(obsday1 = 50 obsday2 = 19 trtphase=1) noatlegend nopvalues
Adjusted predictions                            Number of obs    =        600
Expression   : Linear prediction, fixed portion, predict()
```

	Margin	Delta-method Std. Err.	[95% Conf. Interval]	
_at				
1	349.0051	10.67886	328.0749	369.9352
2	348.8312	12.30066	324.7224	372.9401
3	359.9689	12.15626	336.143	383.7947
4	370.0737	14.56641	341.5241	398.6233

---

5. See section 1.1.1 for information on how to download this command.

This estimates the four key values needed to create the graph shown in figure 13.5. Normally, we would use the `marginsplot` command to graph the means created by the `margins` command. But as you can see in figure 13.6, `marginsplot` creates a graph with the four estimated values side by side.

```
. marginsplot, xlabel(none) xtitle("")
  Variables that uniquely identify margins: _atopt
  Multiple at() options specified:
      _atoption=1: obsday1 = 1  obsday2 = 0  trtphase=0
      _atoption=2: obsday1 = 31 obsday2 = 0  trtphase=0
      _atoption=3: obsday1 = 31 obsday2 = 0  trtphase=1
      _atoption=4: obsday1 = 50 obsday2 = 19 trtphase=1
```

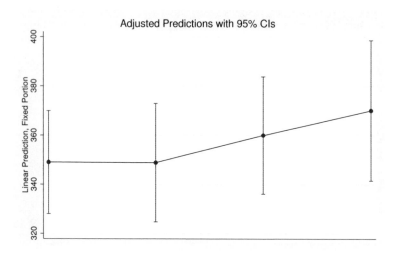

Figure 13.6: Predicted means of sleep across observation days using piecewise modeling of time

To create the kind of graph that we want (like the one shown in figure 13.5), we will manually input the means computed by the `margins` command into a dataset, as shown below. Then, we can use the `graph` command to graph the predicted means. The resulting graph is shown in figure 13.7.

```
. clear
. input obsday sleep

        obsday       sleep
  1.  1   349.00
  2.  31  348.83
  3.  31  359.97
  4.  50  370.07
  5.  end
. graph twoway line sleep obsday, xlabel(0 31 50) xline(31)
```

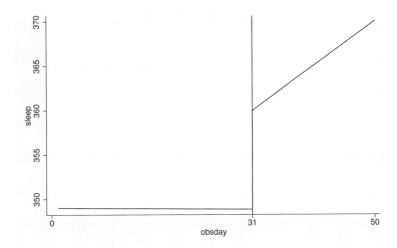

Figure 13.7: Predicted means of sleep across observation days using piecewise modeling of time

This example has illustrated the application of a piecewise model to a longitudinal design in which we had a baseline phase and a treatment phase. For this study, the treatment phase was composed of a combination of sleep medication and sleep education. In the next section, we consider an extension of this design that uses three groups (a control group, a medication group, and an education group).

13.5   Example 4: Piecewise effects of time by a categorical predictor

Let's consider an extension of the previous example that includes a baseline and treatment phase but where participants are randomly divided into different groups and receive one of three treatments during the treatment phase—medication, sleep education, or nothing (control). As in the previous example, sleep is measured approximately once a week for a total of eight measurements. The first 30 days of the study are a baseline period during which the sleep is observed but no treatment is administered to any of the groups. Starting on the 31st day, the medication group receives sleep medication; the sleep education group receives education about how to lengthen their sleep; and the control group receives nothing. The first phase of the study (days 1 to 30) is called the baseline phase, and the second phase (day 31 until the end of the study) is called the treatment phase.

In this study, time will be modeled in a piecewise fashion as we did in section 13.4. We will estimate a slope for the baseline phase and a slope for the treatment phase. We will also estimate any sudden increase or decrease in sleep at the start of the treatment phase. The key difference between this study and the previous study is that each of these terms will be estimated for each of the three treatment groups. We can then investigate the impact of the treatment group assignment (control, medication, education) on the slope in each phase as well as the jump or drop in sleep due to the start of the treatment phase.

The data for this example are stored in the dataset named sleep_cat3pw.dta. Let's begin by using this dataset, listing the first five observations, and showing summary statistics for the variables.

```
. use sleep_cat3pw, clear
. list in 1/5, sepby(id)
```

	id	group	obsday	sleep
1.	1	Control	1	353
2.	1	Control	9	345
3.	1	Control	15	340
4.	1	Control	21	337
5.	1	Control	26	324

```
. summarize
```

Variable	Obs	Mean	Std. Dev.	Min	Max
id	600	38	21.66677	1	75
group	600	2	.8171778	1	3
obsday	600	23.815	15.02877	1	54
sleep	600	362.4817	39.72333	207	531

The variable id identifies each person and ranges from 1 to 75. The variable group identifies the group assignment, coded as follows: 1 = control, 2 = medication, 3 = education. The variable obsday is the day of observation and ranges from 1 to 54. The variable sleep is the duration of sleep (in minutes) and ranges from 207 to 531.

As in the analysis in the previous section, we need to first make some variables before running the analysis. We use the mkspline command to create the variables obsday1 and obsday2, specifying 31 as the time that demarcates the baseline period from the treatment period.

```
. mkspline obsday1 31 obsday2 = obsday
```

To account for the jump in sleep at the start of the treatment phase, we use the `generate` and `replace` commands to create the variable `trtphase`, which is coded as follows: 0 = baseline, 1 = treatment.

```
. generate trtphase = 0 if obsday < 31
(222 missing values generated)
. replace trtphase  = 1 if obsday >= 31 & !missing(obsday)
(222 real changes made)
```

The command we use for this analysis is nearly the same as the one from the analysis in the previous section (section 13.4). The key difference is that we introduce `group` into the model as well as interactions of group with `obsday1`, `obsday2`, and `trtphase`. The `nolog` and `noretable` options are used to suppress the iteration log and random effects table to save space.

```
. mixed sleep i.group##c.obsday1 i.group##c.obsday2 i.group##i.trtphase,
> || id: trtphase obsday1 obsday2, cov(un) nolog noretable
```

Mixed-effects ML regression   Number of obs   =   600
Group variable: id   Number of groups =   75

                                               Obs per group:
                                                             min =   8
                                                             avg =   8.0
                                                             max =   8

                                               Wald chi2(11)   =   194.53
Log likelihood = -2427.8865   Prob > chi2   =   0.0000

sleep	Coef.	Std. Err.	z	P>|z|	[95% Conf. Interval]	
group						
Medication	-2.072872	3.411817	-0.61	0.543	-8.759911	4.614167
Education	16.31454	3.410371	4.78	0.000	9.630337	22.99874
obsday1	-.0415024	.2998388	-0.14	0.890	-.6291757	.5461709
group#						
c.obsday1						
Medication	.4016104	.4238517	0.95	0.343	-.4291237	1.232344
Education	-.1437658	.4240336	-0.34	0.735	-.9748565	.6873249
obsday2	-.1381301	.4038308	-0.34	0.732	-.9296238	.6533637
group#						
c.obsday2						
Medication	.1102917	.5733362	0.19	0.847	-1.013427	1.23401
Education	2.334265	.5706541	4.09	0.000	1.215804	3.452727
1.trtphase	-4.203853	2.793215	-1.51	0.132	-9.678453	1.270748
group#						
trtphase						
Medication #						
1	34.28463	3.997249	8.58	0.000	26.45016	42.11909
Education#1	1.647258	3.973667	0.41	0.678	-6.140986	9.435502
_cons	351.3643	2.414271	145.54	0.000	346.6324	356.0962

Note! More on the random effects

In terms of a multilevel model, the variables trtphase, obsday1, and obsday2 are level 1 variables. The variable group is a level 2 (person-level) variable. Cross-level interactions are formed by interacting group with trtphase, obsday1, and obsday2.

To help us interpret this model, I will show you a graph of the predicted means as a function of observation day and treatment group (see figure 13.8). I will illustrate how to make this kind of graph later in this section.

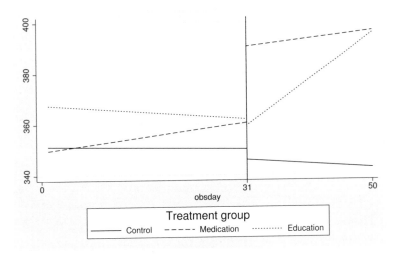

Figure 13.8: Predicted means of sleep by treatment group and day of observation using piecewise modeling of time

Looking at figure 13.8, we see that the sleep durations for the control group remain much the same across the entire study. For the medication group, their sleep durations grow slightly during the baseline phase, jump at the start of treatment, and then remain fairly steady during the treatment phase. The sleep durations for the education group are fairly steady during the baseline phase, show a small dip at the start of treatment, and a substantial increase in slope during the treatment phase. Let's use figure 13.8 as a reference as we interpret the results from this model, focusing first on the baseline slopes (section 13.5.1), then on the treatment slopes (section 13.5.2), and then on the jump at the introduction of the treatment (section 13.5.3).

13.5.1   Baseline slopes

To compute the baseline slope for each of the treatment groups, we can use the `margins` command combined with the `dydx(obsday1)` option, as shown below.

```
. margins group, dydx(obsday1)
```

Average marginal effects   Number of obs   =   600

Expression   : Linear prediction, fixed portion, predict()
dy/dx w.r.t. : obsday1

	dy/dx	Delta-method Std. Err.	z	P>|z|	[95% Conf. Interval]
obsday1					
group					
Control	-.0415024	.2998388	-0.14	0.890	-.6291757 .5461709
Medication	.360108	.299578	1.20	0.229	-.227054 .94727
Education	-.1852682	.2998353	-0.62	0.537	-.7729346 .4023982

The coefficients in this table represent the slopes associated with the three lines during the baseline condition. The slopes for the control, medication, and education groups are -0.04, 0.36, and -0.19 (respectively). None of these coefficients are significantly different from 0 ($ps \geq 0.229$).

We can further test the equality of all the baseline slopes using the contrast command below. This test is not significant [Wald $\chi^2(2, N = 75) = 1.78$, $p = 0.4108$].

```
. contrast group#c.obsday1
```

Contrasts of marginal linear predictions

Margins   : asbalanced

	df	chi2	P>chi2
sleep			
group#c.obsday1	2	1.78	0.4108

Let's now examine the slopes during the treatment phase.

13.5.2   Treatment slopes

Our interpretation and tests of the slopes during the treatment phase is much the same as the previous section regarding the slopes during the baseline phase. We first use the margins command to estimate the slopes for each group during the treatment phase.

```
. margins group, dydx(obsday2)

Average marginal effects                        Number of obs    =       600

Expression    : Linear prediction, fixed portion, predict()
dy/dx w.r.t. : obsday2
```

	dy/dx	Delta-method Std. Err.	z	P>|z|	[95% Conf. Interval]	
obsday2						
group						
Control	-.1381301	.4038308	-0.34	0.732	-.9296238	.6533637
Medication	-.0278384	.406983	-0.07	0.945	-.8255104	.7698336
Education	2.196135	.4031958	5.45	0.000	1.405886	2.986384

The slope during the treatment phase is not significant for the control group ($p = 0.732$) and not significant for the medication group ($p = 0.945$). The slope during the treatment phase for the education group is 2.20 (95% CI = $[1.41, 2.99]$) and is significant $[z(N = 75) = 5.45, p < 0.001]$.

We can test the equality of the treatment phase slope coefficients for all three groups using the contrast command. This test is significant [Wald $\chi^2(2, N = 75) = 21.27$, $p < 0.001$].

```
. contrast group#c.obsday2

Contrasts of marginal linear predictions

Margins      : asbalanced
```

	df	chi2	P>chi2
sleep			
group#c.obsday2	2	21.27	0.0000

Let's now make specific comparisons among the treatment phase slopes, comparing medication with control and comparing education with control. We can form these two comparisons by applying the r contrast coefficient to group.

```
. contrast r.group#c.obsday2
Contrasts of marginal linear predictions
Margins        : asbalanced
```

	df	chi2	P>chi2
sleep			
group#c.obsday2			
(Medication vs Control)	1	0.04	0.8475
(Education vs Control)	1	16.73	0.0000
Joint	2	21.27	0.0000

	Contrast	Std. Err.	[95% Conf. Interval]	
sleep				
group#c.obsday2				
(Medication vs Control)	.1102917	.5733362	-1.013427	1.23401
(Education vs Control)	2.334265	.5706541	1.215804	3.452727

The first contrast shows that the comparison of the treatment phase slope for the medication group is not significantly different from the control group [Wald $\chi^2(1, N = 75) = 0.04$, $p = 0.8475$]. The second comparison shows that the treatment phase slope is significantly greater for the education group than for the control group [Wald $\chi^2(1, N = 75) = 16.73$, $p < 0.001$]. The difference in the slopes (education versus control) is 2.33 (95% CI = $[1.22, 3.45]$).

13.5.3   Jump at treatment

Let's now test and interpret the jump (or drop) in sleep at the start of the treatment phase. We can estimate the size of the jump or drop in sleep durations at the start of the treatment phase for each group using the **contrast** command.

```
. contrast trtphase@group, cieffects
Contrasts of marginal linear predictions
Margins      : asbalanced
```

	df	chi2	P>chi2
sleep			
trtphase@group			
Control	1	2.27	0.1323
Medication	1	110.67	0.0000
Education	1	0.82	0.3657
Joint	3	113.76	0.0000

	Contrast	Std. Err.	[95% Conf. Interval]	
sleep				
trtphase@group				
(1 vs base) Control	-4.203853	2.793215	-9.678453	1.270748
(1 vs base) Medication	30.08077	2.859363	24.47653	35.68502
(1 vs base) Education	-2.556594	2.826302	-8.096044	2.982855

For the control and education groups, there is a drop in the sleep duration at the start of the treatment phase, but neither of these drops is significant ($ps \geq 0.1323$). For the medication group, sleep duration jumps by 30.1 minutes (95% CI $= [24.48, 35.69]$) at the start of the treatment phase, and that jump is significantly different from 0 [Wald $\chi^2(1, N = 75) = 110.67$, $p < 0.001$].

We can test the equality of the jump or drop for all three groups using the contrast command below. This test is significant [Wald $\chi^2(1, N = 75) = 92.58$, $p < 0.001$].

```
. contrast group#trtphase
Contrasts of marginal linear predictions
Margins      : asbalanced
```

	df	chi2	P>chi2
sleep			
group#trtphase	2	92.58	0.0000

We can further compare the jump or drop for the medication group with that of the control group and also compare the jump or drop for the education group with that of the control group using the contrast command.

```
. contrast r.group#trtphase, cieffects
Contrasts of marginal linear predictions
Margins        : asbalanced
```

	df	chi2	P>chi2
sleep			
group#trtphase			
(Medication vs Control) (joint)	1	73.57	0.0000
(Education vs Control) (joint)	1	0.17	0.6785
Joint	2	92.58	0.0000

	Contrast	Std. Err.	[95% Conf. Interval]	
sleep				
group#trtphase				
(Medication vs Control)				
(1 vs base)	34.28463	3.997249	26.45016	42.11909
(Education vs Control)				
(1 vs base)	1.647258	3.973667	-6.140986	9.435502

The first contrast compares the jump or drop in sleep duration at the start of the treatment phase for the medication group with that of the control group. This jump in sleep duration was 34.3 minutes more for the medication group than for the control group (95% CI = [26.45, 42.12]). This difference was significant [Wald $\chi^2(1, N = 75) = 73.57$, $p < 0.001$].

The second contrast makes the same comparison but compares the medication group with the control group. This contrast was not significant [Wald $\chi^2(1, N = 75) = 0.17$, $p = 0.6785$].

13.5.4 Comparisons among groups at particular days

We can make comparisons among the groups at particular days within the study. Let's focus on the day before the start of the treatment (when obsday is 30) as well as on the days that correspond to the obsday of 35, 40, and 45. The showcoding command is used to show the values of obsday1, obsday2, and trtphase for these days.

```
. showcoding obsday obsday1 obsday2 trtphase if inlist(obsday,30,35,40,45)
```

obsday	obsday1	obsday2	trtphase
30	30	0	0
35	31	4	1
40	31	9	1
45	31	14	1

Let's now use the `margins` command to compare each group with the control group
at each of these days. The `at()` option is used to specify the day of observation in terms
of `obsday1`, `obsday2`, and `trtphase` using the values we obtained from the `showcoding`
command above. Applying the `r.` contrast operator to `group` yields reference group
contrasts (at these specified days) and compares each group with the control group.
The `noatlegend` is used to save printed space.

```
. margins r.group, at(obsday1=30 obsday2=0  trtphase=0)
>                   at(obsday1=31 obsday2=4  trtphase=1)
>                   at(obsday1=31 obsday2=9  trtphase=1)
>                   at(obsday1=31 obsday2=14 trtphase=1)
>                   contrast(nowald pveffects) noatlegend

Contrasts of adjusted predictions

Expression    : Linear prediction, fixed portion, predict()
```

	Contrast	Delta-method Std. Err.	z	P>|z|
group@_at				
(Medication vs Control) 1	9.975439	12.37464	0.81	0.420
(Medication vs Control) 2	45.10284	12.87496	3.50	0.000
(Medication vs Control) 3	45.6543	13.50462	3.38	0.001
(Medication vs Control) 4	46.20576	14.67721	3.15	0.002
(Education vs Control) 1	12.00157	12.37866	0.97	0.332
(Education vs Control) 2	22.84212	12.86972	1.77	0.076
(Education vs Control) 3	34.51345	13.50242	2.56	0.011
(Education vs Control) 4	46.18477	14.67255	3.15	0.002

Focusing on the comparison of the medication group with the control group, we see
that the difference is not significant before the treatment phase, that is, when `obsday`
was 30 ($p = 0.420$). However, the difference is significant at each of the time points
tested during the treatment phase, that is, when `obsday` is 35, 40, and 45 ($p = 0.000$,
0.001, and 0.002, respectively).

Shifting our attention to the comparison of the education group with the control
group, we see that the comparison is not significant before the start of the treatment
phase, that is, when `obsday` equals 30 ($p = 0.332$). The difference remains nonsignificant
early in the treatment phase (when `obsday` equals 35, $p = 0.076$) but is significant later
in the treatment phase, when `obsday` equals 40 ($p = 0.011$) and 45 ($p = 0.002$).

Before we conclude this section, let's see how to graph the predicted means as a
function of observation day and group. As we did in the previous section (section 13.4),
we use the `showcoding` command to determine how to express observation day in terms
of `obsday1` and `obsday2`. Below we see the values of `obsday1` and `obsday2` for the 1st
day of the study (when `obsday` equals 1), the 1st day of the treatment phase (when
`obsday` equals 31), and the 50th day of the study (corresponding to the approximate
end of the study).

```
. showcoding obsday obsday1 obsday2 if inlist(obsday,1,31,50)
```

obsday	obsday1	obsday2
1	1	0
31	31	0
50	31	19

We can now use the `margins` command to compute the predicted mean of sleep for each group at four key points—for the start of the study, the 1st day of the treatment phase (with and without treatment), and the 50th day of the study. (The `noatlegend` is used to save space.)

```
. margins group, at(obsday1 = 1  obsday2 = 0  trtphase=0)
>               at(obsday1 = 31 obsday2 = 0  trtphase=0)
>               at(obsday1 = 31 obsday2 = 0  trtphase=1)
>               at(obsday1 = 49 obsday2 = 18 trtphase=1) noatlegend nopvalues

Adjusted predictions                              Number of obs    =        600
Expression    : Linear prediction, fixed portion, predict()
```

		Delta-method		
	Margin	Std. Err.	[95% Conf. Interval]	
_at#group				
1#Control	351.3228	2.362293	346.6928	355.9528
1#Medication	349.6516	2.359361	345.0273	354.2758
1#Education	367.4936	2.357094	362.8738	372.1134
2#Control	350.0778	9.041142	332.3574	367.7981
2#Medication	360.4548	9.037423	342.7418	378.1678
2#Education	361.9356	9.043397	344.2108	379.6603
3#Control	345.8739	9.052166	328.132	363.6158
3#Medication	390.5356	9.075721	372.7475	408.3237
3#Education	359.379	9.05883	341.624	377.1339
4#Control	342.6405	15.5975	312.07	373.2111
4#Medication	396.5164	15.59872	365.9435	427.0894
4#Education	395.5746	15.59009	365.0186	426.1306

As we did in section 13.4, we input the means shown by the `margins` command into a dataset. The difference here is that we input three variables that correspond to the means for the three treatment groups (control, medication, education). We then use the `graph` command to graph the predicted means to create figure 13.9.

```
. preserve
. clear
. input obsday sleep1 sleep2 sleep3
        obsday       sleep1      sleep2       sleep3
  1.   1 351.3 349.7 367.5
  2.  31 350.1 360.5 361.9
  3.  31 345.9 390.5 359.4
  4.  50 342.6 396.5 395.6
  5. end
. graph twoway line sleep1 sleep2 sleep3 obsday, xlabel(0 31 50) xline(31)
> legend(title(Treatment group)
> label(1 "Control") label(2 "Medication") label(3 "Education") rows(1))

. restore
```

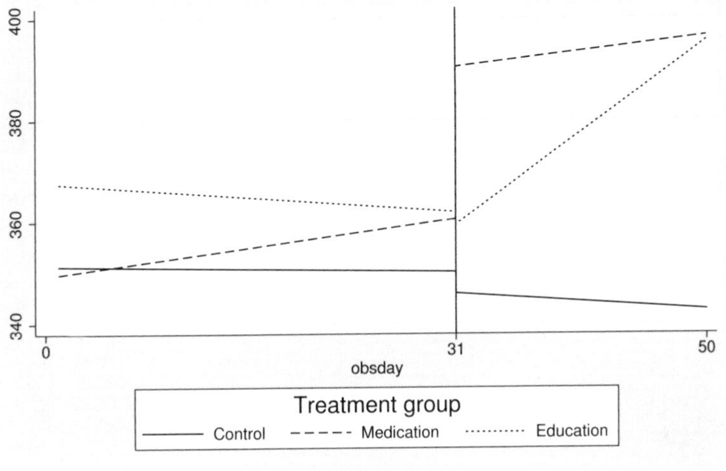

Figure 13.9: Predicted means of sleep by treatment group and day of observation using piecewise modeling of time

13.5.5 Summary of example 4

This example showed that medication and education both had an effect on sleep duration but exerted their effects in different ways.

Medication had an immediate effect on sleep duration at the start of the treatment phase by increasing the duration of sleep by 30.1 minutes more than for the control group. Beyond this initial jump, sleep durations did not further increase for the medication group during the treatment phase. Sleep durations were significantly greater in the medication group for the 35th, 40th, and 45th days of the study.

Education had a different kind of effect on sleep. It had no immediate impact on sleep duration at the start of the treatment phase. However, for every additional day in the treatment phase, sleep increased by 2.33 more minutes for the education group than for the control group. By the 40th day in the study (9 days into the treatment phase), sleep was significantly greater in the education group than in the control group.

Note: More on multilevel longitudinal models

For more information about modeling longitudinal data using multilevel models, see http://blog.stata.com/2013/02/18/multilevel-linear-models-in-stata-part-2-longitudinal-data/.

13.6   Closing thoughts

This chapter has illustrated four different examples of longitudinal models where time was treated as a continuous variable. Such models are useful when the time schedule of measurements differs among people and when you want to focus your attention on the role of time as it relates to the outcome. These four basic examples can be used and adapted to fit many commonly used longitudinal designs. For further information about fitting more complex models, as well as background information about using multilevel models for analyzing longitudinal data, I highly recommend Singer and Willett (2003) and Rabe-Hesketh and Skrondal (2012a).

Part IV

Regression models

This part of the book shows you how to use Stata commands to fit regression models, with an emphasis on the tools that I find useful as part of the day-to-day work performing regression analyses. This part emphasizes how to use these tools and defers questions of why and when you use such tools to your favorite regression book. My goal in this part of the book is to show how you can use the commands and features of Stata for performing common kinds of regression analyses and to illustrate especially useful and powerful features that Stata offers with respect to performing regression analyses.

The chapters in this part are ordered like a meal in which you decide to eat desert first. The sweet and delicious chapters are presented first; I defer nutritional topics like regression diagnostics and power analysis to the end.

This part begins with chapter 14, which shows you how to perform multiple regression using Stata by fitting a simple linear regression model and multiple regression models and how to test multiple coefficients within a multiple regression model.

Chapter 15 covers more details about using the `regress` command. It discusses options that you can use to customize the output of the `regress` command and the ability to redisplay results of the most recent regression command. It also shows how you can perform computations based on the sample of observations included in the most recent regression analysis like creating summary statistics for just the sample of observations included in a previous `regress` command. The chapter introduces the Stata concept of stored results following the `regress` command and shows you how to use these stored results. The chapter concludes with a discussion of how to store results from more than one model.

Chapter 16 focuses on tools that you can use within Stata to create formatted regression results present to others. It illustrates the use of the `estimates table` command for presenting a single regression model and for presenting the results from more than one regression model. This chapter also shows how you can use the user-written command `esttab` to create customized formatted output that can be used within a word processor like Word to include presentation-quality regression tables. The chapter concludes by discussing other user-written commands that can be used for creating formatted regression results.

Chapter 17 focuses on tools built into Stata that can be used for model building. It shows how you can fit multiple models using the same sample of observations and use the `nestreg` prefix command for fitting nested regression models and the `stepwise` prefix for fitting stepwise models.

Chapter 18 illustrates commands for performing regression diagnostics with Stata. The chapter demonstrates analytic and graphical methods that you can use for identifying outliers. The chapter also illustrates analytic and graphical methods that you can use for testing for nonlinearity. It also describes how you can detect multicollinearity, assess the homoskedasticity assumption, and evaluate the normality of the residuals.

This part of the book concludes with chapter 19, which illustrates how you can perform power analysis for regression models using Stata. It shows how you can compute power for a simple regression analysis and for a multiple regression analysis. This chapter also illustrates how to compute power for a nested multiple regression.

14 Simple and multiple regression

14.1   Chapter overview

This chapter shows you how to perform simple and multiple regression using Stata. Although the previous chapters have illustrated how to fit some complex models using the `anova`, `mixed`, and `regress` commands, I have yet to properly introduce the basics of simple and multiple regression models. In this chapter, I take a step back and introduce simple and multiple regression models. The chapter starts with a simple linear regression model and shows how to interpret the output from such a model and how to compute and graph predicted means (section 14.2). The next section presents a multiple regression model, showing how to compute and graph adjusted means and illustrating a variety of ways that you can describe the contribution of a predictor in explaining the outcome (section 14.3.4). The following section describes the Stata commands that you can use to perform tests on groups of coefficients (section 14.4). The chapter concludes with closing thoughts (section 14.5).

14.2   Simple linear regression

This section illustrates the use of a continuous predictor for predicting a continuous outcome using ordinary least-squares regression. This section also illustrates how to interpret and graph the results of such models. The examples in this section use gss2012_sbs.dta, which is read into memory with the use command.

```
. use gss2012_sbs
```

Let's use the self-reported happiness of the respondent as the outcome variable. Using the fre command, we can see the frequency distribution of this variable.[1] We can see that the values range from 1 (Completely unhappy) to 7 (Completely happy). A total of 690 responses were coded as missing (that is, .c, .i, or .n). About a third of the sample was not asked this question, in which case the response was coded as .i. This leaves 11 responses where the person could not choose (coded as .c) and 7 responses where the person gave no answer (coded as .n).

```
. fre happy7
```

happy7 — how happy R is (recoded)

			Freq.	Percent	Valid	Cum.
Valid	1	Completely unhappy	5	0.25	0.39	0.39
	2	Very unhappy	16	0.81	1.25	1.64
	3	Fairly unhappy	35	1.77	2.73	4.36
	4	Neither happy nor unhappy	77	3.90	6.00	10.36
	5	Fairly happy	440	22.29	34.27	44.63
	6	Very happy	563	28.52	43.85	88.47
	7	Completely happy	148	7.50	11.53	100.00
		Total	1284	65.05	100.00	
Missing	.c	Cannot choose	11	0.56		
	.i	Inapplicable	672	34.04		
	.n	No Answer	7	0.35		
		Total	690	34.95		
Total			1974	100.00		

Video tutorial: Simple linear regression

See a video demonstration of a simple linear regression at http://www.stata.com/sbs/simple-linear-regression.

The dataset includes a number of variables that can be used as predictors of the respondent's happiness. For the sake of creating a simple regression model, let's use

---

1. The fre command is a user-written alternative to the tabulate command. You can install it by typing ssc install fre, as described in section 1.1.1.

years of education as a predictor. This would examine whether years of education is linearly related to self-reported happiness. Before running the regression model predicting happiness from education, let's examine the distribution of the variable educ using the summarize command.

```
. summarize educ
    Variable |       Obs        Mean    Std. Dev.       Min        Max
-------------+--------------------------------------------------------
        educ |     1,972    13.52789     3.126576         0         20
```

We see that the years of education range from 0 to 20 with a mean of 13.5 and a standard deviation of 3.1. As we interpret this variable, let's assume that 12 years of education corresponds to graduating from high school, 16 years of education corresponds to graduating from college, and 20 years of education corresponds to completing a doctoral degree.

Let's now run a simple linear regression in which we predict self-reported happiness from the number of years of education. This is performed using the regress command below. The variable happy7 is the outcome variable, and educ is the predictor.

```
. regress happy7 educ
      Source |       SS           df       MS      Number of obs   =     1,284
-------------+----------------------------------   F(1, 1282)      =      6.10
       Model |  6.02653177         1  6.02653177   Prob > F        =    0.0137
    Residual |  1266.97035     1,282  .988276406   R-squared       =    0.0047
-------------+----------------------------------   Adj R-squared   =    0.0040
       Total |  1272.99688     1,283   .99220334   Root MSE        =    .99412

------------------------------------------------------------------------------
      happy7 |      Coef.   Std. Err.      t    P>|t|     [95% Conf. Interval]
-------------+----------------------------------------------------------------
        educ |   .0225038    .009113     2.47   0.014     .0046258    .0403818
       _cons |   5.195513   .1270012    40.91   0.000      4.94636    5.444666
------------------------------------------------------------------------------
```

The output of the regress command above shows us that the regression coefficient associated with educ is 0.0225. The constant is labeled as _cons and its coefficient is 5.20. We can write the regression equation corresponding to the output above as shown in (14.1).

$$\widehat{\text{happiness}} = 5.20 + .0225 \times \texttt{educ} \tag{14.1}$$

This equation tells us that the predicted mean of happiness given 0 years of education is 5.20 and that we expect happiness to increase by 0.0225 units for every additional year of education. The output from the regress command shows that the coefficient for educ is statistically significant. Let's consider the output in more detail below.

14.2.1   Decoding the output

Let's take a moment and decode the output from the previous `regress` command. The output can be divided into two parts, the upper part (which we can call the "header") and the lower part (which we can call the "table"). In fact, we can use the `noheader` option (as illustrated below) to omit display of the header so that we can focus just on the table.

```
. regress happy7 educ, noheader
```

happy7	Coef.	Std. Err.	t	P>|t|	[95% Conf. Interval]
educ	.0225038	.009113	2.47	0.014	.0046258 .0403818
_cons	5.195513	.1270012	40.91	0.000	4.94636 5.444666

The first column of the table shows us that the outcome variable is `happy7`, which is predicted by `educ`, and that the regression model also includes an intercept (constant), which is represented by `_cons`. The coefficient for `educ` is 0.0225, and the standard error for that estimate is 0.009. Dividing the coefficient for `educ` by its standard error yields the t-value of 2.47, which has a corresponding p-value of 0.014. Using a traditional alpha of 0.05, we could say that this coefficient is significantly different from 0. The following two columns show the 95% confidence interval (CI) for the coefficient. The 95% confidence interval for the coefficient for education ranges from 0.0046 to 0.0404. The estimates with respect to the intercept, `_cons`, are rarely interesting, so I will forgo discussing that part of the output.

Now, let's turn our attention to the content of the header portion of the output (shown below) by running the `regress` command with the `notable` option.

```
. regress happy7 educ, notable
```

Source	SS	df	MS			
				Number of obs	=	1,284
				F(1, 1282)	=	6.10
Model	6.02653177	1	6.02653177	Prob > F	=	0.0137
Residual	1266.97035	1,282	.988276406	R-squared	=	0.0047
				Adj R-squared	=	0.0040
Total	1272.99688	1,283	.99220334	Root MSE	=	.99412

The left portion of the output shows an analysis of variance table that decomposes the total variance of the outcome into two sources, that which can be explained by the predictors in the model (that is, `Model`) and that which is not explained by the predictors (that is, `Residual`). For this regression, the SS_{Total} is 1272.997, the SS_{Model} is 6.03, and the $SS_{Residual}$ is 1266.97. The next column shows the df_{Total} (which is $N-1$), the df_{Model} (which is 1, the number of predictors), and the $df_{Residual}$ (which is $df_{Total} - df_{Model}$, or $1283 - 1$). The MS column (mean squares) is the SS divided by df for each source (for example, $SS_{Residual}/df_{Residual}$ yields the $MS_{Residual}$). The MS_{Model} is 6.03 and the $MS_{Residual}$ is 0.988.

Moving our attention to the right column, we see the number of observations included in this analysis is $N = 1284$. Below the number of observations, we see an F statistic. This is computed by dividing the MS_{Model} by the $MS_{Residual}$. This F statistic tests the null hypothesis that all the regression coefficients in the model are equal to zero. We can reject that null hypothesis $[F(1, 1282) = 6.10,\ p = 0.0137]$. The R-squared value is computed by dividing the SS_{Model} by the SS_{Total}, which yields 0.0047. The adjusted R-squared is also displayed, which adjusts for the number of the predictors in the model. The last item in the right column in the header is labeled Root MSE, which is simply the square root of the mean square error (MS_{Error}). In other words, this is the square root of 0.9883.[2] This value is used when computing standard errors and CIs regarding predicted means of happiness.

14.2.2 Computing predicted means using the margins command

We could use the regression equation [that is, (14.1)] to compute the predicted mean of happiness for any given level of education. For example, let's compute the predicted mean for happiness given a high school education (that is, 12 years of education). We can predict the mean by substituting 12 into the regression equation. This substitution is performed below with the display command and shows that the predicted happiness given a high school education is 5.47.

```
. display 5.20 +  .0225*12
5.47
```

Rather than doing this computation by hand, we can use the margins command. margins computes the predicted mean of the outcome when educ equals 12 by specifying the at(educ=12) option.

```
. margins, at(educ=12) nopvalue
Adjusted predictions                           Number of obs     =      1,284
Model VCE    : OLS

Expression   : Linear prediction, predict()
at           : educ           =          12
```

	Margin	Delta-method Std. Err.	[95% Conf. Interval]	
_cons	5.465559	.0313401	5.404075	5.527042

As with our hand-computed value, margins shows that the predicted mean of happiness is 5.47 given 12 years of education. It also shows the standard error of the predicted mean (0.0313) as well as a 95% CI (5.40 to 5.53). We can summarize the output of margins by saying that the predicted mean of happiness, given 12 years of education, is 5.47 (95% CI $= [5.40, 5.53]$).

---

2. The regression table labels this as $MS_{Residual}$ which is just another name for MS_{Error}.

Note: The nopvalue option

You might have noticed that I included the `nopvalue` option on the `margins` command. This option suppresses a statistical test of whether the predicted mean significantly differs from zero. This is seldom interesting (and sometimes confusing), so I used the `nopvalue` option to suppress this part of the output. You will see this in other examples where I feel the *p*-values would not be meaningful.

Suppose we wanted to compute the predicted mean of happiness assuming a number of different values for education, say, 12, 16, or 20 years of education. Instead of running the `margins` command three separate times, we can run it once by specifying `at(educ=(12 16 20))`.

```
. margins, at(educ=(12 16 20)) nopvalue
Adjusted predictions                              Number of obs    =     1,284
Model VCE     : OLS

Expression    : Linear prediction, predict()
1._at         : educ              =          12
2._at         : educ              =          16
3._at         : educ              =          20
```

		Delta-method		
	Margin	Std. Err.	[95% Conf. Interval]	
_at				
1	5.465559	.0313401	5.404075	5.527042
2	5.555574	.0353293	5.486264	5.624883
3	5.645589	.064588	5.518879	5.772299

The `margins` output includes a legend indicating that the predicted means are computed for three _at values, when `educ` equals 12, 16, and 20. The table of results has a column labeled _at that links the output in the table to the _at values described in the legend. Thus the first line of output (labeled as 1 in the _at column) shows the predicted mean is 5.47 when `educ` equals 12. The second line shows the predicted mean is 5.56 when `educ` equals 16, and the third line of the table shows the predicted mean is 5.65 when `educ` equals 20.

   Sometimes, we might want to compute the predicted means given a range of values
for a predictor. For example, we might want to compute the predicted means when educ
equals 0, 4, 8, 12, 16, and 20. Rather than typing all of these values, we can specify
at(educ=(0(4)20)), which tells Stata that we want to specify the values of education
ranging from 0 to 20 in 4-unit increments. For more information on how to specify these
kinds of numbered lists, see help numlist. The margins command below computes
the predicted mean of happiness for the specified levels of education.[3]

```
. margins, at(educ=(0(4)20)) nopvalue vsquish
Adjusted predictions                          Number of obs     =        1,284
Model VCE    : OLS
Expression   : Linear prediction, predict()
1._at        : educ            =         0
2._at        : educ            =         4
3._at        : educ            =         8
4._at        : educ            =        12
5._at        : educ            =        16
6._at        : educ            =        20
```

	Margin	Delta-method Std. Err.	[95% Conf. Interval]	
_at				
1	5.195513	.1270012	4.94636	5.444666
2	5.285528	.0917757	5.105481	5.465575
3	5.375543	.0580839	5.261593	5.489493
4	5.465559	.0313401	5.404075	5.527042
5	5.555574	.0353293	5.486264	5.624883
6	5.645589	.064588	5.518879	5.772299

Note! Forgetting parentheses

A mistake I often make is typing the previous margins command as shown below.

```
. margins, at(educ=(0(4)20) nopvalue vsquish
```

As a result, Stata gives an error message.

```
) required
r(100);
```

It can be hard to see, but I forgot a close parenthesis on the at() option. I should
have specified at(educ=(0(4)20)) with two closing parentheses. If you see this
kind of message, double check your parentheses.

---

3. Note that I included the vsquish option, which vertically squishes the output, omitting extra blank
   lines. You do not need to type this when you run the command, but it saves space for this book.

Let's now consider how we can graph the predicted means computed by the `margins` command by using the `marginsplot` command.

14.2.3   Graphing predicted means using the marginsplot command

We can use the `marginsplot` command to create a graph showing the predicted means and CIs based on the most recent `margins` command. The last `margins` command used the `at(educ=(0(4)20))` option to compute the predicted mean given 0, 4, 8, 12, 16, or 20 years of education. The `marginsplot` command below graphs the predicted means of happiness as a function of the years of education, as shown in figure 14.1. Confidence intervals are shown for each value specified by the `at()` option (that is, 0, 4, 8, 16, and 20).

```
. marginsplot
  Variables that uniquely identify margins: educ
```

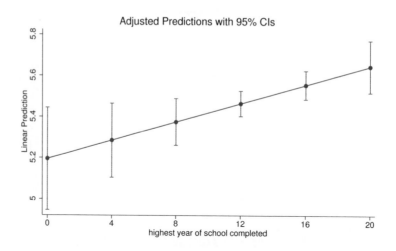

Figure 14.1: Predicted means (with CIs) for education ranging from 0 to 20 in 4-year increments

I like the graph shown in figure 14.1, but I wish I could see the CIs across the entire range of education values, not just at the specified values. To create such a graph, we first need to run the `margins` command, where we compute the predicted mean across the spectrum of education values. Thus we will rerun the `margins` command and specify `at(educ=(0(1)20))` to compute the predicted means at each level of `educ`. Let's run this `margins` command followed by the `marginsplot` command, which creates the graph shown in figure 14.2.

```
. margins, at(educ=(0(1)20)) vsquish
(output omitted)
. marginsplot
Variables that uniquely identify margins: educ
```

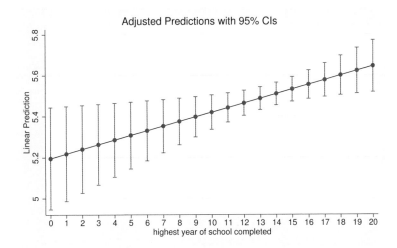

Figure 14.2: Predicted means (with confidence intervals) for education ranging from 0 to 20 in 1-year increments

Say that we wanted to modify the graph in figure 14.2 to display the regression line without markers and the confidence interval as a continuous shaded region. We can do this by adding options to the **marginsplot** command. The **recast()** option specifies that the fitted line should be displayed as a **line** graph (suppressing the markers). The **recastci()** option specifies that the confidence interval should be displayed as an **rarea** graph, with a shaded area for the confidence region. The resulting graph is shown in figure 14.3.

```
. marginsplot, recast(line) recastci(rarea)
  Variables that uniquely identify margins: educ
```

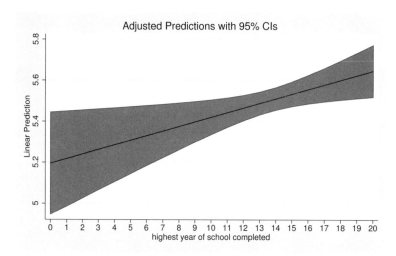

Figure 14.3: Predicted means and confidence interval shown as shaded region

Let's consider one final example of graphing the predicted means. Say that we wanted to graph the predicted means but wanted to omit the CIs. In this case, we can use the `marginsplot` command combined with the `noci` option to omit the display of the CIs. The resulting graph is shown in figure 14.4.

```
. marginsplot, noci
  Variables that uniquely identify margins: educ
```

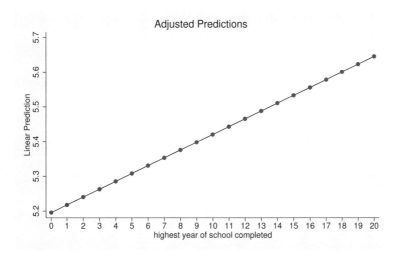

Figure 14.4: Predicted means by education omitting CIs

Note: More on marginsplot

For more information about how to customize graphs using the `marginsplot` command, see section 21.4.

14.3 Multiple regression

So far, all the examples have illustrated simple linear regression. Let's now consider a multiple regression model where self-reported happiness is still the outcome variable. The model will include the following predictor variables—age (`age`), education (`educ`), number of children (`children`), self-reported social class (`class`), weeks worked in the last year (`weekswrk`), and frequency of socializing with friends (`socfrend`).

With respect to the rationale for choosing these variables, we include the variables `educ`, `age`, `children`, and `class` as demographic variables. The variable `weekswrk` tells us about the contribution of how much one works, and `socfrend` tells us about the relationship between socializing and happiness.

Before running this model, let's look at descriptive statistics for these predictors to understand how they are coded.

14.3.1   Describing the predictors

The `summarize` command is used below to show summary statistics for the variables
`educ`, `age`, `children`, and `weekswrk`. Focusing on the minimum and maximum values
can help us quickly understand how these variables are coded. As we saw before, `educ`
is coded as the number of years of education and ranges from 0 to 20. The variable
`age` is coded as the age of the person (in years) and ranges from 18 to 89. The variable
`children` is coded as the number of children and ranges from 0 to 8. The variable
`weekswrk` is the number of weeks worked last year and ranges from 0 to 52.

```
. summarize educ age children weekswrk
    Variable |        Obs        Mean    Std. Dev.        Min        Max
-------------+----------------------------------------------------------
        educ |      1,972    13.52789     3.126576          0         20
         age |      1,969     48.1935     17.68711         18         89
    children |      1,971    1.891933      1.67453          0          8
    weekswrk |      1,963    30.73714     23.28468          0         52
```

For the remaining two variables, `class` and `socfrend`, I think we can understand
their meaning better by displaying the frequency distribution of these two variables.
Let's start by showing a frequency distribution of the variable `class` using the `fre`
command.

```
. fre class
class — subjective class identification
```

			Freq.	Percent	Valid	Cum.
Valid	1	lower class	200	10.13	10.22	10.22
	2	working class	853	43.21	43.59	53.81
	3	middle class	839	42.50	42.87	96.68
	4	upper class	65	3.29	3.32	100.00
		Total	1957	99.14	100.00	
Missing	.d dk		14	0.71		
	.n na		3	0.15		
		Total	17	0.86		
Total			1974	100.00		

The variable `class` is the subjective class identification of the person where 1 is
coded as lower class and 4 is coded as upper class.

Next, let's show the frequency distribution for socfrend.

```
. fre socfrend
socfrend — spend evening with friends
```

			Freq.	Percent	Valid	Cum.
Valid	1	never	141	7.14	10.85	10.85
	2	once a year	110	5.57	8.47	19.32
	3	sev times a year	219	11.09	16.86	36.18
	4	once a month	280	14.18	21.56	57.74
	5	sev times a mnth	255	12.92	19.63	77.37
	6	sev times a week	246	12.46	18.94	96.30
	7	Almost daily	48	2.43	3.70	100.00
		Total	1299	65.81	100.00	
Missing	.d	Don´t Know	2	0.10		
	.i	Inapplicable	672	34.04		
	.n	No Answer	1	0.05		
		Total	675	34.19		
Total			1974	100.00		

This variable measures how frequently one spends an evening socializing with friends. The scale ranges from a code of 1 (which represents never socializing with friends) to a code of 7 (which represents socializing with friends almost daily).

> **Note: Scales of measurement**
>
> Some people would rightly say that the variables class and socfrend are not interval variables. This is especially true for the variable socfrend, in which the gap between a code of 1 versus 2 (never versus once a year) is much greater than the gap between 6 and 7 (several times a week versus almost daily). This raises concerns regarding the linearity of the relationship between the predictor and outcome. Even if the relationship is linear, it raises concerns about the meaning of a one-unit change. For example, whether a one-unit change in socfrend from 1 to 2 is the same as a one-unit change from 6 to 7. I agree that these are concerns, and I ask that you temporarily overlook this issue regarding these predictors for the sake of illustration.

14.3.2 Running the multiple regression model

Having examined these predictors, let's fit a multiple regression model in which we predict happy7 from age, educ, children, class, weekswrk, and socfrend using the regress command.

```
. regress happy7 age educ children class weekswrk socfrend

      Source |       SS           df       MS            Number of obs   =     1,264
-------------+----------------------------------         F(6, 1257)      =     17.20
       Model | 93.7519138          6   15.625319         Prob > F        =    0.0000
    Residual | 1142.15236      1,257  .908633539         R-squared       =    0.0759
-------------+----------------------------------         Adj R-squared   =    0.0714
       Total | 1235.90427      1,263  .978546534         Root MSE        =    .95322

------------------------------------------------------------------------------
      happy7 |      Coef.   Std. Err.      t    P>|t|     [95% Conf. Interval]
-------------+----------------------------------------------------------------
         age |  -.0009607   .0017837    -0.54   0.590    -.0044601    .0025387
        educ |  -.0014373   .0097913    -0.15   0.883    -.0206463    .0177718
    children |   .0880061   .0179546     4.90   0.000     .0527818    .1232303
       class |   .2694915   .0397875     6.77   0.000     .1914344    .3475487
    weekswrk |   .0039221   .0012535     3.13   0.002     .0014629    .0063813
    socfrend |   .0812112    .017351     4.68   0.000     .0471711    .1152512
       _cons |   4.315559   .1764296    24.46   0.000      3.96943    4.661688
------------------------------------------------------------------------------
```

The multiple regression equation predicting self-reported happiness from the age, education, number of children, social class, weeks worked, and socializing with friends can be written as shown in (14.2).

$$\widehat{\text{happiness}} = 4.315559 + -0.001 \times \texttt{age} + -0.001 \times \texttt{educ} + 0.088 \times \texttt{children}$$
$$+ 0.269 \times \texttt{class} + 0.004 \times \texttt{weekswrk} + 0.081 \times \texttt{socfrend} \qquad (14.2)$$

The coefficients from this multiple regression model reflect the association between each predictor and the outcome after adjusting for all the other predictors. For example, the coefficient for `children` is 0.088, which means that for every additional child the respondent has, we would expect (holding all other variables constant) the happiness of the respondent to be 0.088 units higher.

14.3.3 Computing adjusted means using the margins command

As an aid to interpreting the coefficients from the multiple regression model, we can use the `margins` command to compute adjusted means of the outcome as a function of one or more predictors from the model. For example, to help interpret the coefficient for `children`, we can use the `margins` command to compute the adjusted mean of happiness given different values of `children`, adjusting for the other predictors in the model. The `margins` command below computes the adjusted mean of the happiness for the specified number of children (that is, 0, 1, 2, 3).

```
. margins, at(children=(0 1 2 3)) vsquish nopvalue
Predictive margins                            Number of obs    =      1,264
Model VCE    : OLS
Expression   : Linear prediction, predict()
1._at        : children       =            0
2._at        : children       =            1
3._at        : children       =            2
4._at        : children       =            3
```

		Delta-method		
	Margin	Std. Err.	[95% Conf. Interval]	
_at				
1	5.347033	.0425057	5.263643	5.430423
2	5.435039	.0307361	5.374739	5.495339
3	5.523045	.0269707	5.470133	5.575958
4	5.611051	.0339832	5.544381	5.677721

`margins` shows the adjusted means at four different levels of `children`. Among those with no children, the adjusted mean of happiness is 5.35 (after we adjust for all other predictors in the model). Among those with one child, the adjusted mean of happiness is 5.44 (after we adjust for all other predictors). As you can see below, if we take the adjusted mean given no children (5.347033) minus the adjusted mean given 1 child (5.435039), we obtain the coefficient associated with `children`.

```
. display 5.435039 - 5.347033
.088006
```

This affirms the meaning of the coefficient for `children`, showing it is the expected change in happiness for a one-unit increase in children. You would find the same result if you compared those with two children with those with one child, those with three children with those with two children, and so forth.

The `margins` command allows us to hold more than one variable constant at a time. The `margins` command below specifies the option `at(children=(0 1 2 3)` `weekswrk=0)`. This `margins` command computes the adjusted mean of the happiness for the specified number of children (that is, 0, 1, 2, 3) while also holding the weeks worked in the last year constant at 0.

```
. margins, at(children=(0 1 2 3) weekswrk=0) vsquish nopvalue
Predictive margins                              Number of obs     =      1,264
Model VCE    : OLS

Expression   : Linear prediction, predict()
1._at        : children      =           0
               weekswrk      =           0
2._at        : children      =           1
               weekswrk      =           0
3._at        : children      =           2
               weekswrk      =           0
4._at        : children      =           3
               weekswrk      =           0
```

		Delta-method		
	Margin	Std. Err.	[95% Conf. Interval]
_at			
1	5.224154	.0578387	5.110683 5.337624
2	5.31216	.0498528	5.214356 5.409964
3	5.400166	.0476449	5.306693 5.493638
4	5.488172	.0519567	5.38624 5.590103

Note how the legend informs us that the first _at value corresponds to `children` being zero and `weekswrk` being zero. The second, third, and fourth _at values correspond to `children` being 1, 2, and 3 (respectively) and `weekswrk` being 0. The other variables in the model are not mentioned, which indicates that the adjusted means are adjusted for those variables.

Compared with the results of the previous `margins` command, these results show that the adjusted means are lower when the number of weeks worked is held constant at zero. However, the effect of `children` remains exactly the same. In the previous `margins` command, we saw that the change in the adjusted means for a 1-child increase was 0.088006. Let's compare the adjusted means for one child with that for zero children for this most recent `margins` command.

```
. display 5.31216 - 5.224154
.088006
```

Even when weeks of work is held constant at 0, the increase in happiness due to a gain in 1 child remains the same, 0.088006.

14.3.4   Describing the contribution of a predictor

This section considers different ways to describe the relationship between the predictors and the outcome variable. The examples will be based on the multiple regression model fit from the previous section.

```
. regress happy7 age educ children class weekswrk socfrend

      Source |       SS           df       MS      Number of obs   =     1,264
-------------+----------------------------------   F(6, 1257)      =     17.20
       Model |  93.7519138         6   15.625319   Prob > F        =    0.0000
    Residual |  1142.15236     1,257  .908633539   R-squared       =    0.0759
-------------+----------------------------------   Adj R-squared   =    0.0714
       Total |  1235.90427     1,263  .978546534   Root MSE        =    .95322

------------------------------------------------------------------------------
      happy7 |      Coef.   Std. Err.      t    P>|t|     [95% Conf. Interval]
-------------+----------------------------------------------------------------
         age | -.0009607   .0017837    -0.54   0.590    -.0044601    .0025387
        educ | -.0014373   .0097913    -0.15   0.883    -.0206463    .0177718
    children |  .0880061   .0179546     4.90   0.000     .0527818    .1232303
       class |  .2694915   .0397875     6.77   0.000     .1914344    .3475487
    weekswrk |  .0039221   .0012535     3.13   0.002     .0014629    .0063813
    socfrend |  .0812112    .017351     4.68   0.000     .0471711    .1152512
       _cons |  4.315559   .1764296    24.46   0.000      3.96943    4.661688
------------------------------------------------------------------------------
```

One-unit change

The simplest way to describe the relationship between a predictor and the outcome is by using the regression coefficient presented in the regression table. For example, focusing on the variable `children`, we see that happiness is expected to increase by 0.088 units for every additional child. Those with 2 children are expected to be 0.088 units happier than those with one child (after we account for the contribution of all the other predictors in the model).

Multiple-unit change

Sometimes, it can be more meaningful to interpret a regression coefficient in terms of a change in more than one unit of the predictor. For example, consider the variable `children`. We might want to express the change in happiness that results from an increase in having two children. We can simply multiply the coefficient for children by two. We can do so manually, as shown below.

```
. display .088*2
.176
```

Or we can use the `lincom` command to compute this for us. In the example below, `lincom` multiplies the coefficient for `children` by two and displays that value, along with a standard error, significance test, and confidence interval.

```
. lincom children*2
( 1)  2*children = 0
```

happy7	Coef.	Std. Err.	t	P>|t|	[95% Conf. Interval]	
(1)	.1760121	.0359091	4.90	0.000	.1055637	.2464606

Or consider the predictor `weekswrk`. The coefficient for that variable is 0.0039221. This might make it sound as if this variable has a very small contribution because it has such a small coefficient. But we have to account for the scaling of the variable. It is the number of weeks worked in the last year, and the coefficient corresponds to a change in one additional week of work. Let's say that we consider a 26-unit change, which would represent an increase in work by half a year. We can compute this contribution by using the `lincom` command and multiplying the coefficient for `weekswrk` by 26.

```
. lincom weekswrk*26
( 1)  26*weekswrk = 0
```

happy7	Coef.	Std. Err.	t	P>|t|	[95% Conf. Interval]	
(1)	.1019751	.0325909	3.13	0.002	.0380365	.1659137

The output above shows us that when we increase the weeks worked by 26 weeks, happiness is expected to increase by 0.102 units.

At first, it seemed that the contribution of `children` was much more substantial than the contribution of `weekswrk` because the coefficient for children was 0.088 compared with 0.0039 for `weekswrk`. However, the predicted increase in happiness due to working 26 more weeks is 0.102, which is a bit larger than the contribution of having 1 additional child.

Milestone change in units

Sometimes, it can be useful to describe the adjusted mean of the outcome as a function of certain milestones. Referring to the previous example, regarding the number of weeks worked in the last year, we see the change in happiness resulting from working no weeks in the last year versus working the entire year (that is, working 0 weeks versus 52 weeks). We could compute the expected change in happiness due to an increase in 52 weeks of work using the `lincom` command.

```
. lincom weekswrk*52

( 1)   52*weekswrk = 0
```

happy7	Coef.	Std. Err.	t	P>|t|	[95% Conf. Interval]
(1)	.2039502	.0651819	3.13	0.002	.076073 .3318275

Another way that we could present this is by computing the adjusted mean specifying 0 weeks of work in the last year as well as computing the adjusted mean specifying 52 weeks of work in the last year. This is computed using the `margins` command combined with the `at(weekswrk=(0 52))` option.

```
. margins, at(weekswrk=(0 52))

Predictive margins                            Number of obs    =      1,264
Model VCE     : OLS

Expression    : Linear prediction, predict()
1._at         : weekswrk      =            0
2._at         : weekswrk      =           52
```

		Delta-method				
	Margin	Std. Err.	t	P>|t|	[95% Conf. Interval]
_at					
1	5.385823	.0475515	113.26	0.000	5.292534 5.479112
2	5.589773	.0372851	149.92	0.000	5.516625 5.662921

This shows that the adjusted mean of happiness given 0 weeks of work last year is 5.39 compared with 5.59 given 52 weeks of work last year. Let's use the `display` command to compute the difference in the adjusted mean given 52 weeks of work minus the adjusted mean given 0 weeks of work.

```
. display 5.589773 - 5.385823
.20395
```

Note how this difference corresponds to the value obtained from the `lincom` command, which computed the expected increase in happiness due to 52 additional weeks of work.

One SD change in predictor

A common way to compare the relative strength of different predictors is by computing standardized regression coefficients, often called Beta coefficients. These coefficients express the change in the outcome variable that is expected for a one standard-deviation change in the predictor. It is as though each of the predictors were scaled in standard units where a one-unit change corresponds to a one standard-deviation change. This common scaling facilitates the comparison of the strength of the different predictors by placing all the predictors on a common scale. By adding the `beta` option to the `regress` command, we can request the display of standardized regression coefficients. This is illustrated in the `regress` command below. The column labeled `Beta` contains the standardized regression coefficient for each variable. For example, for a one standard-deviation increase in weeks of work, happiness is expected to increase by 0.092 standard deviations.

```
. regress happy7 age educ children class weekswrk socfrend, beta
```

Source	SS	df	MS		
Model	93.7519138	6	15.625319	Number of obs =	1,264
Residual	1142.15236	1,257	.908633539	F(6, 1257) =	17.20
				Prob > F =	0.0000
				R-squared =	0.0759
Total	1235.90427	1,263	.978546534	Adj R-squared =	0.0714
				Root MSE =	.95322

happy7	Coef.	Std. Err.	t	P>|t|	Beta
age	-.0009607	.0017837	-0.54	0.590	-.017136
educ	-.0014373	.0097913	-0.15	0.883	-.0044033
children	.0880061	.0179546	4.90	0.000	.1497917
class	.2694915	.0397875	6.77	0.000	.1961329
weekswrk	.0039221	.0012535	3.13	0.002	.0915807
socfrend	.0812112	.017351	4.68	0.000	.1354949
_cons	4.315559	.1764296	24.46	0.000	.

Partial and semipartial correlation

Another way of expressing the size of the relationship between a predictor and an outcome is via the use of squared partial correlations, which are sometimes called "partial eta-squared". These represent the proportion of variance that is explained by a predictor after accounting for the contribution of the other predictors. We can use the `estat esize` command to compute the eta-squared (squared partial correlations) for the entire model as well as for each variable in the model, as shown below.

```
. estat esize
Effect sizes for linear models
```

Source	Eta-Squared	df	[95% Conf. Interval]	
Model	.0758569	6	.0467745	.1011903
age	.0002307	1	.	.00485
educ	.0000171	1	.	.0024588
children	.018755	1	.0067787	.0361324
class	.0352123	1	.0179287	.0573225
weekswrk	.0077284	1	.0010757	.0201797
socfrend	.0171295	1	.0058041	.033914

I like to express these values as percentages. Using that metric, we see that the output above shows that the overall model explains 7.59% of the variance in happiness. Note how this corresponds to the R-squared value from the original **regress** command. This output also shows a 95% confidence interval for this value, which ranges from 4.68% to 10.1%.

Let's now focus on a single predictor, the number of children. The number of children explains 1.88% of the variance in happiness after we account for all the other variables in the model. The 95% confidence interval for this estimate ranges from 0.68% to 3.61%.

Rather than computing squared partial correlations, we may prefer to compute adjusted-squared partial correlations, sometimes called "partial omega-squared". Like adjusted R-squared, the omega-squared value adjusts for the number of predictors in the model to give a more accurate representation of the population value. Consider the **estat esize** command combined with the **omega** option below. The output of this command shows the overall omega-squared for the entire model as well as partial omega-squared for each predictors.

```
. estat esize, omega
Effect sizes for linear models
```

Source	Omega-Squared	df	[95% Conf. Interval]	
Model	.0714458	6	.0422245	.0969001
age	0	1	.	.0040583
educ	0	1	.	.0016653
children	.0179744	1	.0059886	.0353656
class	.0344447	1	.0171474	.0565726
weekswrk	.006939	1	.000281	.0194002
socfrend	.0163476	1	.0050131	.0331454

Again, using percentages to describe these results, we see that the omega-squared for the entire model is 7.14%. Note how this corresponds to the adjusted R-squared value from the **regress** command. The confidence interval for this value ranges from 4.22% to 9.69%.

Focusing on the variable **children**, we see that this variable explains 1.80% of the variance in happiness (95% CI = [0.60%, 3.54%]).

Note: Semipartial correlations

If you wish to compute semipartial correlations between the predictors and the outcome variable, you can use the **pcorr** command. For this model, you can issue the **pcorr** command as shown below.

```
. pcorr happy7 age educ children class weekswrk socfrend
```

14.4   Testing multiple coefficients

The previous examples have focused on evaluating each coefficient on its own. But there are times when you are interested in testing the contribution of two or more coefficients. This section illustrates how you can use the `test`, `testparm`, and `lincom` commands to form tests among multiple coefficients.

14.4.1   Testing whether coefficients equal zero

Past research suggests that social connections play an important role in one's happiness. The previous model included the variable `socfrend`, which measured the frequency with which the respondent socializes with friends. Let's extend that model to include three additional variables measuring the frequency that the respondent socializes with relatives (`socrel`), with neighbors (`socommun`), or at a bar (`socbar`). The `regress` command with predictors representing four different kinds of socializing behaviors is shown below.

```
. regress happy7 age educ children class weekswrk
>         socrel socommun socfrend socbar

      Source |       SS           df       MS           Number of obs   =     1,259
-------------+----------------------------------        F(9, 1249)      =     12.50
       Model | 101.838831          9   11.3154256       Prob > F        =    0.0000
    Residual | 1130.78706      1,249   .905353934       R-squared       =    0.0826
-------------+----------------------------------        Adj R-squared   =    0.0760
       Total | 1232.62589      1,258   .979829804       Root MSE        =     .9515

      happy7 |      Coef.   Std. Err.      t    P>|t|     [95% Conf. Interval]
-------------+----------------------------------------------------------------
         age | -.0007588   .0018118    -0.42   0.675    -.0043133    .0027957
        educ |  .0001877    .009849     0.02   0.985    -.0191346     .01951
    children |  .0802695   .0181918     4.41   0.000     .0445797    .1159592
       class |  .2737261   .0397983     6.88   0.000     .1956473    .3518049
     weekswrk |  .0042701   .0012715     3.36   0.001     .0017756    .0067646
      socrel |  .0332493   .0169544     1.96   0.050     -.000013    .0665116
    socommun |  .0277576   .0138519     2.00   0.045      .000582    .0549332
    socfrend |  .0683486   .0186387     3.67   0.000      .031782    .1049152
      socbar | -.0217074   .0178683    -1.21   0.225    -.0567627    .0133478
       _cons |  4.138308   .1927201    21.47   0.000     3.760218    4.516399
```

Looking at the output of the previous regression, we could look at the coefficients associated with each of the individual socializing variables. But our interest is broader than that; we are interested in the overall joint contribution of these four variables. Such a joint test asks whether the four variables, when considered as a group, significantly improves our ability to predict happiness above and beyond the contribution of the other variables in the model (that is, `educ`, `age`, `children`, `class`, `weekswrk`, and `health`). Another way to frame this question is by simultaneously testing the following four null hypotheses.

$$H_0\#1 : \beta_{\texttt{socrel}} = 0$$
$$H_0\#2 : \beta_{\texttt{socommun}} = 0$$
$$H_0\#3 : \beta_{\texttt{socfrend}} = 0$$
$$H_0\#4 : \beta_{\texttt{socbar}} = 0$$

The `test` command below jointly tests all four of these hypotheses at once, assessing the joint contribution of all four of these variables in the regression model.

```
. test socrel socommun socfrend socbar
 ( 1)   socrel = 0
 ( 2)   socommun = 0
 ( 3)   socfrend = 0
 ( 4)   socbar = 0
       F(  4,  1249) =    7.87
            Prob > F =    0.0000
```

The first portion of the output shows the null hypothesis that is tested. The test of this hypothesis is significant [$F(4, 1249) = 7.87$, $p < 0.001$]. We can reject the null hypothesis that all the coefficients are equal to zero. In the section on nested regression models, we will see another way we can approach this kind of test (see section 17.3).

14.4.2   Testing the equality of coefficients

Let's consider a different question. Let's ask whether the coefficients associated with the four socialization variables are all equal to one another. That is, let's test the following null hypothesis.

$$H_0 : \beta_{\texttt{socrel}} = \beta_{\texttt{socommun}} = \beta_{\texttt{socfrend}} = \beta_{\texttt{socbar}}$$

Another way we can express this null hypothesis is shown below.

$$H_0\#1 : \beta_{\texttt{socrel}} = \beta_{\texttt{socommun}}$$
$$H_0\#2 : \beta_{\texttt{socrel}} = \beta_{\texttt{socfrend}}$$
$$H_0\#3 : \beta_{\texttt{socrel}} = \beta_{\texttt{socbar}}$$

This hypothesis can be tested using the `testparm` command. Note the inclusion of the `equal` option. This specifies a test of the equality of the named coefficients.

```
. testparm socrel socommun socfrend socbar, equal
 ( 1)   - socrel + socommun = 0
 ( 2)   - socrel + socfrend = 0
 ( 3)   - socrel + socbar = 0
       F(  3,  1249) =    3.63
            Prob > F =    0.0126
```

This test is significant [$F(3, 1249) = 3.63$, $p = 0.0126$]. We can reject the null hypothesis that the coefficients are all equal to one another. We cannot say more specifically which coefficients are different from one another. A visual inspection of these coefficients from the regress output suggests that socbar might be different from the remaining coefficients and that the remaining coefficients may be equal to one another. Let's consider the question of whether there are significant differences among the coefficients for socrel, socommun, and socfrend using the testparm command.

```
. testparm socrel socommun socfrend, equal
 ( 1)   - socrel + socommun = 0
 ( 2)   - socrel + socfrend = 0
        F(  2,  1249) =    1.37
           Prob > F =     0.2546
```

The test of the equality of these 3 coefficients is not significant [$F(2, 1249) = 1.37$, $p = 0.2546$]. Let's next explore how the size of these three coefficients compares with the fourth coefficient for socbar.

14.4.3   Testing linear combinations of coefficients

Having found that the coefficients for socrel, socommun, and socfrend are not significantly different from one another, let's consider the comparison of the average of these coefficients with the fourth coefficient, socbar. That is, let's test the null hypothesis shown below.

$$H_0 : \frac{(\beta_{\text{socrel}} + \beta_{\text{socommun}} + \beta_{\text{socfrend}})}{3} = \beta_{\text{socbar}}$$

Specifying the null hypothesis shows how we should formulate the test command, as shown below.

```
. test (socrel + socommun + socfrend)/3 = socbar
 ( 1)   .3333333*socrel + .3333333*socommun + .3333333*socfrend - socbar = 0
        F(  1,  1249) =    9.58
           Prob > F =     0.0020
```

This test is significant. The average of the coefficients for socrel, socommun, and socfrend is significantly different from the coefficient for socbar. To quantify the degree of this difference, we can use the lincom command.

```
. lincom  (socrel + socommun + socfrend)/3 - socbar
 ( 1)   .3333333*socrel + .3333333*socommun + .3333333*socfrend - socbar = 0
```

happy7	Coef.	Std. Err.	t	P>|t|	[95% Conf. Interval]	
(1)	.0648259	.020946	3.09	0.002	.0237326	.1059192

This `lincom` command shows the difference in the magnitude of the average of the coefficients for `socrel`, `socommun`, and `socfrend` as compared with `socbar`. The average of the first 3 coefficients is 0.061 units greater than the coefficient for `socbar`.

Note: More on lincom

Let's dissect the previous `lincom` command. It first computes the average of the three regression coefficients associated with `socrel`, `socommun`, and `socfrend`. We can manually compute this value as $(0.0332493 + 0.0277576 + 0.0683486)/3$, which yields 0.0431185. Then, we subtract the coefficient for `socbar` (taking $0.0431185 - -0.0217074$) to obtain 0.0648259. As you see, the estimate computed by `lincom` is the same as the one we would obtain by manually averaging the coefficients for `socrel`, `socommun`, and `socfrend` and subtracting the coefficient for `socbar`. However, doing so with `lincom` is simpler, and it also provides a significance test.

14.5   Closing thoughts

This chapter has illustrated how to use the `regress` command to fit simple and multiple regression models. We have seen how to compute adjusted means using the `margins` command and how to create graphs of the adjusted means using the `marginsplot` command. The chapter also illustrated how you can perform tests regarding multiple coefficients in the model.

For more details about how to customize the look of graphs created using the `marginsplot` command, refer to section 21.4. Also, for information about how to perform regression diagnostics using Stata, see chapter 18.

15 More details about the regress command

15.1 Chapter overview

This chapter covers more details about using the **regress** command. Specifically, it discusses options that you can use to customize the output of the **regress** command (section 15.2) and the ability to redisplay results of the most recent regression command (section 15.3). It also shows how you can perform computations based on the sample of observations included in the most recent regression analysis, such as creating summary statistics for just the sample of observations included in the previous **regress** command (section 15.4). The chapter next discusses the concept of stored results following the **regress** command and how you can use these stored results (section 15.5). The chapter concludes with a discussion of how to store results from more than one model (section 15.7), a key tool that will be used in the following chapter on presenting results of regression models.

15.2 Regression options

Let's explore more about the nature of the **regress** command itself, focusing on some of the options that can be used with this command. Let's start by using the gss2012_sbs dataset and running a regression model that predicts happiness from socializing with friends, social class, and health. The **use** command below uses the gss2012_sbs dataset, and the **regress** command fits a model predicting **happy7** from **socfrend**, **class**, and **health**.

```
. use gss2012_sbs

. regress happy7 socfrend class health
```

Source	SS	df	MS		
				Number of obs	= 1,269
				F(3, 1265)	= 59.27
Model	153.939466	3	51.3131552	Prob > F	= 0.0000
Residual	1095.2536	1,265	.865813122	R-squared	= 0.1232
				Adj R-squared	= 0.1212
Total	1249.19307	1,268	.985168033	Root MSE	= .93049

happy7	Coef.	Std. Err.	t	P>|t|	[95% Conf. Interval]
socfrend	.0409546	.0160549	2.55	0.011	.0094574 .0724518
class	.179613	.037124	4.84	0.000	.1067817 .2524443
health	.2503335	.0241089	10.38	0.000	.2030356 .2976314
_cons	4.05819	.1215313	33.39	0.000	3.819765 4.296615

Say we wanted to see the coefficients as standardized regression coefficients. We can specify the **beta** option with the **regress** command, and standardized regression coefficients are displayed in lieu of confidence intervals (CIs).

```
. regress happy7 socfrend class health, beta
```

Source	SS	df	MS		
				Number of obs	= 1,269
				F(3, 1265)	= 59.27
Model	153.939466	3	51.3131552	Prob > F	= 0.0000
Residual	1095.2536	1,265	.865813122	R-squared	= 0.1232
				Adj R-squared	= 0.1212
Total	1249.19307	1,268	.985168033	Root MSE	= .93049

happy7	Coef.	Std. Err.	t	P>|t|	Beta
socfrend	.0409546	.0160549	2.55	0.011	.0681293
class	.179613	.037124	4.84	0.000	.1299843
health	.2503335	.0241089	10.38	0.000	.2823907
_cons	4.05819	.1215313	33.39	0.000	.

We really do not need to see the header portion of the output when showing the standardized coefficients, so we can request that the header be suppressed with the **noheader** option. Using that option, shown below, displays just the regression table.

```
. regress happy7 socfrend class health, beta noheader
```

happy7	Coef.	Std. Err.	t	P>|t|	Beta
socfrend	.0409546	.0160549	2.55	0.011	.0681293
class	.179613	.037124	4.84	0.000	.1299843
health	.2503335	.0241089	10.38	0.000	.2823907
_cons	4.05819	.1215313	33.39	0.000	.

We could, if we like, display just the header portion of the regression table (omitting the regression table) by specifying the **notable** option.

```
. regress happy7 socfrend class health, notable
      Source |       SS           df       MS          Number of obs   =       1,269
-------------+----------------------------------        F(3, 1265)      =       59.27
       Model | 153.939466          3   51.3131552       Prob > F        =      0.0000
    Residual |  1095.2536      1,265   .865813122       R-squared       =      0.1232
-------------+----------------------------------        Adj R-squared   =      0.1212
       Total | 1249.19307      1,268   .985168033       Root MSE        =      .93049
```

Say we want to display the *p*-values with two decimal places. The `pformat()` option gives us control of the formatting of the *p*-values in the table. The example below uses the `pformat()` option to display the *p*-values using a fixed format with a total width of four digits with two decimal places.[1]

```
. regress happy7 socfrend class health, noheader pformat(%4.2f)
```

happy7	Coef.	Std. Err.	t	P>|t|	[95% Conf. Interval]	
socfrend	.0409546	.0160549	2.55	0.01	.0094574	.0724518
class	.179613	.037124	4.84	0.00	.1067817	.2524443
health	.2503335	.0241089	10.38	0.00	.2030356	.2976314
_cons	4.05819	.1215313	33.39	0.00	3.819765	4.296615

The `sformat()` option controls the display of the test statistic (in this case, the *t*-value). The example below displays the *t*-values using a fixed width of four digits with one decimal place.

```
. regress happy7 socfrend class health, noheader sformat(%4.1f)
```

happy7	Coef.	Std. Err.	t	P>|t|	[95% Conf. Interval]	
socfrend	.0409546	.0160549	2.6	0.011	.0094574	.0724518
class	.179613	.037124	4.8	0.000	.1067817	.2524443
health	.2503335	.0241089	10.4	0.000	.2030356	.2976314
_cons	4.05819	.1215313	33.4	0.000	3.819765	4.296615

The `cformat()` option controls the display of the coefficient, standard error, and confidence interval. This option is used to display these columns with a fixed width of four digits using two decimal places.

---

1. For the sake of saving space, I have included the `noheader` option on this example and subsequent ones.

```
. regress happy7 socfrend class health, noheader cformat(%4.2f)
```

happy7	Coef.	Std. Err.	t	P>|t|	[95% Conf. Interval]	
socfrend	0.04	0.02	2.55	0.011	0.01	0.07
class	0.18	0.04	4.84	0.000	0.11	0.25
health	0.25	0.02	10.38	0.000	0.20	0.30
_cons	4.06	0.12	33.39	0.000	3.82	4.30

These options are combined in the following example to control the display of all columns at once. The coefficients, standard errors, CIs, and p-values are displayed with a fixed width of four columns with two decimal places; the t-values are displayed with a fixed width of four columns and one decimal place.

```
. regress happy7 socfrend class health, noheader
>    cformat(%4.2f) pformat(%4.2f) sformat(%4.1f)
```

happy7	Coef.	Std. Err.	t	P>|t|	[95% Conf. Interval]	
socfrend	0.04	0.02	2.6	0.01	0.01	0.07
class	0.18	0.04	4.8	0.00	0.11	0.25
health	0.25	0.02	10.4	0.00	0.20	0.30
_cons	4.06	0.12	33.4	0.00	3.82	4.30

In each of these instances, the options were used to merely change the appearance of the output of the regress command. If our dataset contained millions of observations, it would seem wasteful to spend time refitting the model merely to change the formatting of the regression output. As we will see in the following section, we can use a shortcut to save such computation time.

15.3 Redisplaying results

After fitting a regression model with the regress command, we can redisplay the results of the last regression by simply typing the regress command, as shown below. We can do this because Stata holds the estimation results from the most recent estimation command in memory. When we type regress, Stata does not refit the model; it merely redisplays the estimation results using the results stored in memory.

```
. regress
      Source |       SS           df       MS          Number of obs   =     1,269
-------------+----------------------------------       F(3, 1265)      =     59.27
       Model |  153.939466         3  51.3131552       Prob > F        =    0.0000
    Residual |   1095.2536     1,265  .865813122       R-squared       =    0.1232
-------------+----------------------------------       Adj R-squared   =    0.1212
       Total |  1249.19307     1,268  .985168033       Root MSE        =    .93049

------------------------------------------------------------------------------
      happy7 |      Coef.   Std. Err.      t    P>|t|     [95% Conf. Interval]
-------------+----------------------------------------------------------------
    socfrend |   .0409546   .0160549     2.55   0.011     .0094574    .0724518
       class |    .179613    .037124     4.84   0.000     .1067817    .2524443
      health |   .2503335   .0241089    10.38   0.000     .2030356    .2976314
       _cons |    4.05819   .1215313    33.39   0.000     3.819765    4.296615
------------------------------------------------------------------------------
```

This feature works even after you type other commands, so long as they are not other estimation commands.[2] To demonstrate this, let's use the summarize command to compute summary statistics for the variables educ and children.

```
. summarize educ children
    Variable |        Obs        Mean    Std. Dev.       Min        Max
-------------+--------------------------------------------------------
        educ |      1,972    13.52789    3.126576          0         20
    children |      1,971    1.891933     1.67453          0          8
```

We can now type regress, and the results can still be redisplayed.

```
. regress
      Source |       SS           df       MS          Number of obs   =     1,269
-------------+----------------------------------       F(3, 1265)      =     59.27
       Model |  153.939466         3  51.3131552       Prob > F        =    0.0000
    Residual |   1095.2536     1,265  .865813122       R-squared       =    0.1232
-------------+----------------------------------       Adj R-squared   =    0.1212
       Total |  1249.19307     1,268  .985168033       Root MSE        =    .93049

------------------------------------------------------------------------------
      happy7 |      Coef.   Std. Err.      t    P>|t|     [95% Conf. Interval]
-------------+----------------------------------------------------------------
    socfrend |   .0409546   .0160549     2.55   0.011     .0094574    .0724518
       class |    .179613    .037124     4.84   0.000     .1067817    .2524443
      health |   .2503335   .0241089    10.38   0.000     .2030356    .2976314
       _cons |    4.05819   .1215313    33.39   0.000     3.819765    4.296615
------------------------------------------------------------------------------
```

This does not seem that useful because you can simply scroll back into the output to see the previous regression results. What makes this useful is when you include options with the regress command to change the way the results are displayed. For example, to display the latest regression results using standardized regression coefficients, we can

---

2. Examples of estimation commands include the anova, regress, logistic, or rreg command. You can see section 4.5 for more information about estimation commands and postestimation commands. You can also type help estimation commands for a list of all estimation commands.

specify the beta option after the regress command. Note that this does not refit the model.

```
. regress, beta

      Source |       SS           df       MS      Number of obs   =     1,269
-------------+----------------------------------   F(3, 1265)      =     59.27
       Model |  153.939466         3  51.3131552   Prob > F        =    0.0000
    Residual |   1095.2536     1,265  .865813122   R-squared       =    0.1232
-------------+----------------------------------   Adj R-squared   =    0.1212
       Total |  1249.19307     1,268  .985168033   Root MSE        =    .93049

------------------------------------------------------------------------------
      happy7 |      Coef.   Std. Err.      t    P>|t|                      Beta
-------------+----------------------------------------------------------------
    socfrend |   .0409546   .0160549     2.55   0.011                  .0681293
       class |    .179613    .037124     4.84   0.000                  .1299843
      health |   .2503335   .0241089    10.38   0.000                  .2823907
       _cons |    4.05819   .1215313    33.39   0.000                         .
------------------------------------------------------------------------------
```

Or we could redisplay the results using 99% CIs with the level(99) option.

```
. regress, level(99)

      Source |       SS           df       MS      Number of obs   =     1,269
-------------+----------------------------------   F(3, 1265)      =     59.27
       Model |  153.939466         3  51.3131552   Prob > F        =    0.0000
    Residual |   1095.2536     1,265  .865813122   R-squared       =    0.1232
-------------+----------------------------------   Adj R-squared   =    0.1212
       Total |  1249.19307     1,268  .985168033   Root MSE        =    .93049

------------------------------------------------------------------------------
      happy7 |      Coef.   Std. Err.      t    P>|t|     [99% Conf. Interval]
-------------+----------------------------------------------------------------
    socfrend |   .0409546   .0160549     2.55   0.011    -.0004626     .0823718
       class |    .179613    .037124     4.84   0.000     .0838436     .2753825
      health |   .2503335   .0241089    10.38   0.000     .1881391     .3125278
       _cons |    4.05819   .1215313    33.39   0.000     3.744673     4.371707
------------------------------------------------------------------------------
```

When we typed the regress command above, Stata accessed the saved regression estimation results and redisplayed the regression table using a 99% confidence interval. Saving the estimation results is a key design feature of Stata. We have taken advantage of this feature when we have issued postestimation commands like margins and contrast. These postestimation commands access the saved results from the most recent estimation command, allowing these programs to perform further computations based on the most recent estimation command. We can take advantage of this design feature ourselves; that is, we can access the estimation results from the most recent estimation command. Let's use this feature to perform computations in the context of the most recent estimation command, as illustrated in the following section.

15.4   Identifying the estimation sample

When you write up your results for a regression model, it is customary to provide descriptive statistics for the sample. Let's illustrate this based on the regression model we have been using in this chapter, repeated below.

```
. regress happy7 socfrend class health
```

Source	SS	df	MS			
				Number of obs	=	1,269
				F(3, 1265)	=	59.27
Model	153.939466	3	51.3131552	Prob > F	=	0.0000
Residual	1095.2536	1,265	.865813122	R-squared	=	0.1232
				Adj R-squared	=	0.1212
Total	1249.19307	1,268	.985168033	Root MSE	=	.93049

happy7	Coef.	Std. Err.	t	P>|t|	[95% Conf. Interval]	
socfrend	.0409546	.0160549	2.55	0.011	.0094574	.0724518
class	.179613	.037124	4.84	0.000	.1067817	.2524443
health	.2503335	.0241089	10.38	0.000	.2030356	.2976314
_cons	4.05819	.1215313	33.39	0.000	3.819765	4.296615

We can use the `estat summarize` command to provide summary statistics for the variables used in the regression model.

```
. estat summarize
```
Estimation sample regress   Number of obs =   1,269

Variable	Mean	Std. Dev.	Min	Max
happy7	5.506698	.9925563	1	7
socfrend	4.040189	1.651152	1	7
class	2.398739	.7183039	1	4
health	3.404255	1.119661	1	5

Note how the number of observations, 1,269, matches the number of observations from the regression model. Stata refers to these observations as the "estimation sample", the sample of people who were included in the latest estimation command (in this case, the `regress` command). Unfortunately, this command shows only the variables that were included in our model. To describe our sample, I would like to include summary statistics for `educ` and `children`. To that end, I use the `summarize` command to obtain summary statistics for these variables.

```
. summarize educ children
```

Variable	Obs	Mean	Std. Dev.	Min	Max
educ	1,972	13.52789	3.126576	0	20
children	1,971	1.891933	1.67453	0	8

The sample size for these summary statistics reflects Ns of 1,972 and 1,971. But our estimation sample size was only 1,269 because only part of the overall sample was asked the question about their happiness. I want to report summary statistics just for those included in our estimation sample. Fortunately, Stata has a way of helping us identify those who belong to our estimation sample. We can include the if e(sample) specification shown below, and the summarize command is performed only on the observations included in the estimation sample (that is, $N = 1269$).

```
. summarize educ children if e(sample)
    Variable │        Obs        Mean    Std. Dev.        Min        Max
─────────────┼─────────────────────────────────────────────────────────
        educ │      1,269    13.60993     3.016243          0         20
    children │      1,269    1.840032     1.687897          0          8
```

This allows us to report summary statistics for the observations that were in our estimation sample. We could report that in our analytic sample of $N = 1269$ observations, the average years of education was 13.6, and the average number of children was 1.84. In both cases, these values were computed based on the $N = 1269$ observations included in the estimation sample for the regression analysis.

Likewise, let's compute frequencies for gender for those who were in the estimation sample. We can use the tabulate command with the if e(sample) specification as shown below to obtain the frequency distribution of gender just for the $N = 1269$ observations included in the estimation sample.

```
. tabulate female if e(sample)
   Is R female │
      (yes=1   │
       no=0)?  │      Freq.     Percent        Cum.
──────────────┼───────────────────────────────────────
         Male │        577       45.47       45.47
       Female │        692       54.53      100.00
──────────────┼───────────────────────────────────────
        Total │      1,269      100.00
```

Among the $N = 1269$ observations in the estimation sample, 54.53% were female.

15.5 Stored results

Let me elaborate a bit more on this mysterious e(sample). After we run an estimation command (like regress), Stata holds on to pieces of information about the results from the estimation command. Stata calls these pieces of information "stored results". We can see the particular stored results for our most recent estimation command by typing the command ereturn list.

```
. ereturn list
scalars:
                    e(N) =  1269
                 e(df_m) =  3
                 e(df_r) =  1265
                    e(F) =  59.26585522812927
                   e(r2) =  .1232311239727175
                 e(rmse) =  .9304907965413636
                  e(mss) =  153.9394655088854
                  e(rss) =  1095.253599896949
                 e(r2_a) =  .1211518301955777
                   e(ll) =  -1707.207153847416
                 e(ll_0) =  -1790.651429429895
                 e(rank) =  4
macros:
              e(cmdline) : "regress happy7 socfrend class health"
                e(title) : "Linear regression"
             e(marginsok) : "XB default"
                  e(vce) : "ols"
               e(depvar) : "happy7"
                  e(cmd) : "regress"
           e(properties) : "b V"
              e(predict) : "regres_p"
                e(model) : "ols"
            e(estat_cmd) : "regress_estat"
matrices:
                   e(b) :  1 x 4
                   e(V) :  4 x 4
functions:
               e(sample)
```

You can surmise the meaning of a number of these return values. For example, e(N) holds the sample size, while e(r2) holds the value of R-squared. By referring back to the results of the regress command, you can indeed confirm that the sample size is 1,269 (corresponding to the value stored in e(N)) and that the R-squared is indeed 0.1232 (corresponding to the value stored in e(r2)).

Rather than playing a guessing game to figure out the meaning of these stored results, we can consult help regress, and the help file includes a section titled *Stored results*, which describes each of the stored results. The help file for every Stata estimation command has a section titled *Stored results*, which describes the stored results for that command. I have copied the *Stored results* section from the help file for the regress command and show that section on the following page.

<u>Stored results</u>

regress stores the following in e():

Scalars
  e(N)   number of observations
  e(mss)   model sum of squares
  e(df_m)   model degrees of freedom
  e(rss)   residual sum of squares
  e(df_r)   residual degrees of freedom
  e(r2)   R-squared
  e(r2_a)   adjusted R-squared
  e(F)   F statistic
  e(rmse)   root mean squared error
  e(ll)   log likelihood under additional assumption of
                     i.i.d. normal errors
  e(ll_0)   log likelihood, constant-only model
  e(N_clust)   number of clusters
  e(rank)   rank of e(V)

Macros
  e(cmd)   regress
  e(cmdline)   command as typed
  e(depvar)   name of dependent variable
  e(model)   ols or iv
  e(wtype)   weight type
  e(wexp)   weight expression
  e(title)   title in estimation output when vce() is not ols
  e(clustvar)   name of cluster variable
  e(vce)   vcetype specified in vce()
  e(vcetype)   title used to label Std. Err.
  e(properties)   b V
  e(estat_cmd)   program used to implement estat
  e(predict)   program used to implement predict
  e(marginsok)   predictions allowed by margins
  e(asbalanced)   factor variables fvset as asbalanced
  e(asobserved)   factor variables fvset as asobserved

Matrices
  e(b)   coefficient vector
  e(V)   variance-covariance matrix of the estimators
  e(V_modelbased)   model-based variance

Functions
  e(sample)   marks estimation sample

I frequently use the e(sample) stored result. In addition, I often use the stored results categorized as Scalars. We will revisit these stored values in chapter 16, which discusses the presentation of results from regression models. You will be able to refer to the names of these stored results to add their contents to a regression table (for example, referring to N for the sample size or r2 for the R-squared or r2_a for the adjusted R-squared).

15.6   Storing results

The stored results for a given estimation command linger until you run another estimation model. Once you run a new estimation model, the stored results from the previous model are discarded. Sometimes, we might want to store the results from a model that we like for future reference. This section demonstrates the `estimates store` command, which allows us to store (in memory) the estimation results from a command for access later in our Stata session.

Consider the model we fit at the start of this chapter.

```
. regress happy7 socfrend class health
```

Source	SS	df	MS		Number of obs	=	1,269
					F(3, 1265)	=	59.27
Model	153.939466	3	51.3131552		Prob > F	=	0.0000
Residual	1095.2536	1,265	.865813122		R-squared	=	0.1232
					Adj R-squared	=	0.1212
Total	1249.19307	1,268	.985168033		Root MSE	=	.93049

happy7	Coef.	Std. Err.	t	P>|t|	[95% Conf. Interval]	
socfrend	.0409546	.0160549	2.55	0.011	.0094574	.0724518
class	.179613	.037124	4.84	0.000	.1067817	.2524443
health	.2503335	.0241089	10.38	0.000	.2030356	.2976314
_cons	4.05819	.1215313	33.39	0.000	3.819765	4.296615

Say that we really like these results and want to hold on to them for use later in our Stata session. We can use the `estimates store` command below, and the results for this model are stored using the name `hap1`.

```
. estimates store hap1
```

Let's now run a model that adds the variables `age`, `educ`, and `children` to the above model.

```
. regress happy7 socfrend class health age educ children
```

Source	SS	df	MS		
Model	180.555684	6	30.092614	Number of obs	= 1,267
Residual	1068.13729	1,260	.847728009	F(6, 1260)	= 35.50
				Prob > F	= 0.0000
				R-squared	= 0.1446
Total	1248.69298	1,266	.986329365	Adj R-squared	= 0.1405
				Root MSE	= .92072

happy7	Coef.	Std. Err.	t	P>|t|	[95% Conf. Interval]
socfrend	.0655262	.0167668	3.91	0.000	.0326322 .0984202
class	.2001397	.0389712	5.14	0.000	.1236842 .2765952
health	.263699	.0245636	10.74	0.000	.2155089 .3118892
age	.0005277	.0016648	0.32	0.751	-.0027383 .0037937
educ	-.009905	.0095189	-1.04	0.298	-.0285797 .0087696
children	.0827224	.0173036	4.78	0.000	.0487753 .1166695
_cons	3.821898	.1752121	21.81	0.000	3.478159 4.165638

Say that we really like this model too. We can use the `estimates store` command to save the results for this model as well. Using that command below, we save the results using the name `hap2`.

```
. estimates store hap2
```

Using the `estimates dir` command, we can see a list of the stored results that we have stored in memory. This shows that we have two stored results in memory, one with the name `hap1` and the other named `hap2`. The stored results named `hap1` arose from the `regress` command in which the dependent variable was `happy7`; it had four parameters (coefficients on three predictors and the constant). The stored results named `hap2` also arose from the `regress` and also used `happy7` as the dependent variable. That model had seven parameters (coefficients on six predictors and the constant).

```
. estimates dir
```

name	command	depvar	npar	title
hap1	regress	happy7	4	
hap2	regress	happy7	7	

You might be rightly asking yourself why we would be interested in storing estimation results and wondering how we can access the stored results. As we will see in the next section, we can refer to these stored results to display the results from different models.

15.7   Displaying results with the estimates table command

Having stored the results of these two regression models, we can display the regression coefficients for either or both of these models using the `estimates table` command. If we type the `estimates table` command alone (as shown below), a table is shown with the regression coefficients for the currently active model. As illustrated below, the results are from the model named `hap2` and reflect the most recently fit (regression) model.

```
. estimates table
```

Variable	hap2
socfrend	.06552622
class	.20013971
health	.26369902
age	.00052769
educ	-.00990501
children	.08272239
_cons	3.8218982

The `estimates table` command below specifies that the results should be displayed from the model named `hap1`. Indeed, this command displays the regression coefficients from that model.

```
. estimates table hap1
```

Variable	hap1
socfrend	.04095457
class	.17961304
health	.25033347
_cons	4.0581901

Note, however, that this merely displays the results for the stored results of `hap1`. By just typing the command `estimate table`, we can see a table of the active results for the current model. The active results are for the model named `hap2`.

```
. estimates table
```

Variable	hap2
socfrend	.06552622
class	.20013971
health	.26369902
age	.00052769
educ	-.00990501
children	.08272239
_cons	3.8218982

This illustrates that the `estimates table` command can display the results for different stored results without altering the currently stored results.

We can display the regression coefficients for these two models together by using the `estimates table` command, shown below. I have included the `stats(r2)` option to display the R-squared value for each model.[3]

```
. estimates table hap1 hap2, stats(r2)
```

Variable	hap1	hap2
socfrend	.04095457	.06552622
class	.17961304	.20013971
health	.25033347	.26369902
age		.00052769
educ		-.00990501
children		.08272239
_cons	4.0581901	3.8218982
r2	.12323112	.14459574

This is just a small appetizer regarding the ways that we can display results from regression models using Stata. In the following chapter, we will explore the power of the `estimates table` command (as well as other commands) for flexibly displaying the results from regression models.

15.8   Closing thoughts

In this chapter, we have seen some of the options that we can use when running the `regress` command. Beyond that, we have learned about how every estimation command saves stored results. We can access these stored results to perform further computations, as we saw by creating summary statistics just for the observations that were included in the estimation sample. Furthermore, we can save the stored results and access the stored results using the `estimates table` command. In the following chapter, I will elaborate on the use of the `estimates table` command (and other commands) for creating nicely formatted regression tables.

---

3. I will elaborate on the use of the `stats()` option as well as many other options that you can use with the `estimates table` command in section 16.2.

16 Presenting regression results

16.1 Chapter overview

The ultimate goal of running a regression model is to present it to others. The format we use to present regression models is usually quite different from the standard output produced by Stata. The focus of this chapter is about showing tools that you can use within Stata to create formatted regression results for presentation to others. This chapter begins by illustrating the use of the `estimates table` command for presenting a single regression model (section 16.2). This is followed by examples of how you can use the `estimates table` command for presenting the results from more than one regression model (section 16.3). Section 16.4 presents a user-written command called `esttab` and shows how to use this command to export customized formatted output that can be used within a word processor like Word. Section 16.5 shows other user-written commands that can be used for creating formatted regression results. The chapter concludes with section 16.6, which presents closing thoughts about tools for creating formatted regression results using Stata. All the examples in this chapter will be based on the `gss2012_sbs.dta` dataset, with the variable `happy7` (self-rated happiness) as the outcome variable.

16.2   Presenting a single model

For this section, let's say that we want to report the results of the regression analysis predicting self-reported happiness from the following six predictor variables—frequency of socializing with friends, age, education, number of children, social class, and self-reported health. This regression model is fit using the `regress` command.

```
. use gss2012_sbs
. regress happy7 socfrend age educ children class health

      Source |       SS          df        MS       Number of obs   =    1,267
-------------+------------------------------       F(6, 1260)      =    35.50
       Model |  180.555684         6   30.092614   Prob > F        =    0.0000
    Residual |  1068.13729     1,260  .847728009   R-squared       =    0.1446
-------------+------------------------------       Adj R-squared   =    0.1405
       Total |  1248.69298     1,266  .986329365   Root MSE        =    .92072

      happy7 |      Coef.   Std. Err.      t     P>|t|     [95% Conf. Interval]
-------------+----------------------------------------------------------------
    socfrend |   .0655262   .0167668     3.91   0.000     .0326322    .0984202
         age |   .0005277   .0016648     0.32   0.751    -.0027383    .0037937
        educ |   -.009905   .0095189    -1.04   0.298    -.0285797    .0087696
    children |   .0827224   .0173036     4.78   0.000     .0487753    .1166695
       class |   .2001397   .0389712     5.14   0.000     .1236842    .2765952
      health |    .263699   .0245636    10.74   0.000     .2155089    .3118892
       _cons |   3.821898   .1752121    21.81   0.000     3.478159    4.165638
```

As we saw in section 15.7 of the previous chapter, we can use the `estimates table` command to create a table of the coefficients from the most recent estimation command, as shown below.

```
. estimates table

    Variable |     active
-------------+-----------
    socfrend |  .06552622
         age |  .00052769
        educ | -.00990501
    children |  .08272239
       class |  .20013971
      health |  .26369902
       _cons |  3.8218982
```

We can add numerous options to the `estimates table` command to customize the display of the regression results. In the following examples, I will illustrate how to apply options to the `estimates table` command to create a formatted table like the ones we would see in a journal article. Let's start this process by adding the `varlabel` option, which specifies that the variable label should be displayed in place of the variable name.[1] To accommodate these longer labels, I have also included the `varwidth()` option to increase the label width to 35 characters.

```
. estimates table, varlabel varwidth(35)
```

Variable	active
spend evening with friends	.06552622
age of respondent	.00052769
highest year of school completed	-.00990501
number of children	.08272239
subjective class identification	.20013971
rs Health in General (recoded)	.26369902
Constant	3.8218982

We are often uninterested in the constant (intercept) term of the regression model. We can use the `drop()` option to omit variables from the display of the table. The name Stata uses for the constant (intercept) term is _cons. So, if we want to drop the constant, we can specify the option `drop(_cons)`.

```
. estimates table, varlabel varwidth(35) drop(_cons)
```

Variable	active
spend evening with friends	.06552622
age of respondent	.00052769
highest year of school completed	-.00990501
number of children	.08272239
subjective class identification	.20013971
rs Health in General (recoded)	.26369902

The following example uses the `b()` option to control the formatting of the regression coefficients. By specifying `b(%6.3f)`, the regression coefficients are displayed using a fixed width of 6 columns with 3 decimal places.

```
. estimates table, varlabel varwidth(35) drop(_cons) b(%6.3f)
```

Variable	active
spend evening with friends	0.066
age of respondent	0.001
highest year of school completed	-0.010
number of children	0.083
subjective class identification	0.200
rs Health in General (recoded)	0.264

---

1. If you wish to change the labels displayed for the variables, you can use the `label variable` command, as described in section 22.4.

The example below adds the `se` option to add standard errors to the output. Note how the standard error is displayed below the regression coefficient and is displayed using the same formatting as the regression coefficient.[2]

```
. estimates table, varlabel varwidth(35) drop(_cons) b(%6.3f) se
```

Variable	active
spend evening with friends	0.066
	0.017
age of respondent	0.001
	0.002
highest year of school completed	-0.010
	0.010
number of children	0.083
	0.017
subjective class identification	0.200
	0.039
rs Health in General (recoded)	0.264
	0.025

legend: b/se

Rather than displaying standard errors, let's display stars next to the coefficients to indicate the level of significance. The `star()` option is used below to display 1 star for a result significant at 0.05, 2 stars for a result significant at 0.01, and 3 stars for a result significant at 0.001.

```
. estimates table, varlabel varwidth(35) drop(_cons) b(%6.3f)
> star(0.05 0.01 0.001)
```

Variable	active
spend evening with friends	0.066***
age of respondent	0.001
highest year of school completed	-0.010
number of children	0.083***
subjective class identification	0.200***
rs Health in General (recoded)	0.264***

legend: * p<.05; ** p<.01; *** p<.001

We can use the `stats()` option to specify additional statistics that we want displayed at the bottom of the regression table. Say that we wanted the table to include the sample size and R-squared. Back in section 15.5, we saw how you can see the names and values of the stored results via the `ereturn list` command. In that section, we saw that the stored result for the sample size is called `e(N)` and that the stored result for R-squared is called `e(r2)`. Focusing on the name inside the parentheses, we see that N stands for

---

2. If you wished the standard error to be formatted differently from the coefficient, you could specify the `se()` option, placing formatting information within the parentheses. For example, if you wanted the standard error to be formatted with a total width of 4 digits with 2 decimal places, you could specify the `se(%4.2f)` option.

the sample size and `r2` stands for the R-squared. We can specify the option `stats(N r2)`, as shown below, to display the sample size and R-squared at the bottom of the regression table.

```
. estimates table, varlabel varwidth(35) drop(_cons) b(%6.3f)
> star(0.05 0.01 0.001) stats(N r2)
```

Variable	active
spend evening with friends	0.066***
age of respondent	0.001
highest year of school completed	-0.010
number of children	0.083***
subjective class identification	0.200***
rs Health in General (recoded)	0.264***
N	1267
r2	0.145

legend: * p<.05; ** p<.01; *** p<.001

In all the examples in this section, we have focused on the display of a single regression model. Sometimes, when we report regression results, we want to report the results of more than one model in a single table. This is commonly used to display the relationship between a focal predictor and the outcome after adjusting for different sets of covariates. We consider such an example in the following section.

16.3   Presenting multiple models

Let's consider an example in which our interest is in predicting happiness from the frequency with which one socializes with friends. Say that we want to display three different regression models, one in which the only predictor is our focal predictor, frequency of socializing with friends. The second model includes socializing as well as demographic variables as a set of covariates (that is, age, education, number of children, and social class). The third model is the same as the second model, except that self-reported health is included as an additional covariate in the model. Let's start this process by running each of these three regression models, which I will call model 1, model 2, and model 3 (respectively).

We first run model 1, which predicts happiness from the frequency of socializing with friends. This model shows a positive relationship between how often one socializes with friends and one's level of happiness.

```
. * Model 1
. regress happy7 socfrend
```

Source	SS	df	MS		
Model	18.4387453	1	18.4387453		
Residual	1254.55814	1,282	.978594493		
Total	1272.99688	1,283	.99220334		

Number of obs = 1,284
F(1, 1282) = 18.84
Prob > F = 0.0000
R-squared = 0.0145
Adj R-squared = 0.0137
Root MSE = .98924

happy7	Coef.	Std. Err.	t	P>|t|	[95% Conf. Interval]
socfrend	.0723247	.0166618	4.34	0.000	.0396373 .1050121
_cons	5.210231	.0725706	71.80	0.000	5.067861 5.352601

Let's now fit model 2, which adds age, education, number of children, and social class to the previous model. We find that the variable socfrend remains significantly related to happy7, even after adjusting for age, education, number of children, and social class.

```
. * Model 2
. regress happy7 socfrend age educ children class
```

Source	SS	df	MS		
Model	82.3714574	5	16.4742915		
Residual	1177.60889	1,266	.930180797		
Total	1259.98035	1,271	.991329934		

Number of obs = 1,272
F(5, 1266) = 17.71
Prob > F = 0.0000
R-squared = 0.0654
Adj R-squared = 0.0617
Root MSE = .96446

happy7	Coef.	Std. Err.	t	P>|t|	[95% Conf. Interval]
socfrend	.0780266	.0174486	4.47	0.000	.0437952 .1122579
age	-.0026743	.0017126	-1.56	0.119	-.0060342 .0006856
educ	.0037173	.0098036	0.38	0.705	-.0155159 .0229505
children	.0868444	.0180997	4.80	0.000	.0513356 .1223532
class	.2723057	.0401906	6.78	0.000	.1934583 .3511532
_cons	4.453079	.1727716	25.77	0.000	4.114129 4.792029

Let's now run model 3, in which we add self-reported health to the predictors of model 2. Even after we adjust for all of these covariates, the amount of time one socializes with friends is positively (and significantly) related to happiness.

```
. * Model 3
. regress happy7 socfrend educ age children class health

      Source │       SS           df       MS            Number of obs   =     1,267
─────────────┼──────────────────────────────────        F(6, 1260)      =     35.50
       Model │  180.555684          6   30.092614        Prob > F        =    0.0000
    Residual │  1068.13729      1,260   .847728009        R-squared       =    0.1446
─────────────┼──────────────────────────────────        Adj R-squared   =    0.1405
       Total │  1248.69298      1,266   .986329365        Root MSE        =    .92072

──────────────┬───────────────────────────────────────────────────────────────────
       happy7 │      Coef.   Std. Err.      t    P>|t|     [95% Conf. Interval]
──────────────┼───────────────────────────────────────────────────────────────────
     socfrend │   .0655262   .0167668     3.91   0.000     .0326322    .0984202
         educ │   -.009905   .0095189    -1.04   0.298    -.0285797    .0087696
          age │   .0005277   .0016648     0.32   0.751    -.0027383    .0037937
     children │   .0827224   .0173036     4.78   0.000     .0487753    .1166695
        class │   .2001397   .0389712     5.14   0.000     .1236842    .2765952
       health │    .263699   .0245636    10.74   0.000     .2155089    .3118892
        _cons │   3.821898   .1752121    21.81   0.000     3.478159    4.165638
──────────────┴───────────────────────────────────────────────────────────────────
```

We are now so excited about these results that we would like to write them up for publication. For this publication, let's display model 1, model 2, and model 3 in a single table. We could manually create such a table, or we could use the `estimates table` command to do the work for us. With this in mind, let's begin our analysis again, repeating it but doing so in such a way that we can report the results of these three models side by side. We will run the three models we ran previously, but after running each model, we will use the `estimates store` command to store the results from each model. (For more on the `estimates store` command, see section 15.6.)

We first run regression model 1 (but with the output omitted to save space). On the following command, we use the `estimates store` command to store the results from this model. The stored results are named `mod1`.

```
. regress happy7 socfrend
  (output omitted)
. estimates store mod1
```

We then run model 2 and then use the `estimates store` command, naming the results of this model `mod2`.

```
. regress happy7 socfrend age educ children class
  (output omitted)
. estimates store mod2
```

Finally, we run model 3. We then use the `estimates store` command to store the results of this model with the name `mod3`.

```
. regress happy7 socfrend age educ children class health
(output omitted)
. estimates store mod3
```

The results of these three regression models are stored using the names mod1, mod2, and mod3. We can now use the estimates table command (shown below) to display the results of the three models side by side. This shows the regression coefficients for the variables estimated in mod1, mod2, and mod3.

```
. estimates table mod1 mod2 mod3
```

Variable	mod1	mod2	mod3
socfrend	.07232471	.07802658	.06552622
age		-.00267429	.00052769
educ		.0037173	-.00990501
children		.0868444	.08272239
class		.27230571	.20013971
health			.26369902
_cons	5.210231	4.4530791	3.8218982

Using this formatted table, we easily see the results of these three models side by side. For example, we can see whether the magnitude of the coefficient for our focal predictor, socfrend, substantially changes across these three models.

Let's now customize the format of this table by adding options to the estimates table command. With the options illustrated in the previous section, the results are displayed using 35 columns for the labels (displayed using the variable labels instead of variable names). The constant is dropped from the table, and the coefficients are formatted with 6 columns and 3 decimal places. Furthermore, stars are used to indicate the significance level ($* = 0.05$, $** = 0.01$, and $*** = 0.001$), and the bottom of the regression table displays the sample size and R-squared for each model.

```
. estimates table mod1 mod2 mod3, varlabel varwidth(35) drop(_cons)
> b(%6.3f) star(0.05 0.01 0.001) stats(N r2)
```

Variable	mod1	mod2	mod3
spend evening with friends	0.072***	0.078***	0.066***
age of respondent		-0.003	0.001
highest year of school completed		0.004	-0.010
number of children		0.087***	0.083***
subjective class identification		0.272***	0.200***
rs Health in General (recoded)			0.264***
N	1284	1272	1267
r2	0.014	0.065	0.145

legend: * p<.05; ** p<.01; *** p<.001

Note: Differing sample sizes

If you consider the above table more carefully, you will notice that the sample size for the three models are not the same. This could raise a concern that comparisons across these models could be tainted by the differing sample sizes. One possible method to address this, especially in a case like this where the sample sizes are so similar, is to refit the three models on the same sample. In this case, that would mean fitting all three models using the $N = 1267$ observations used in Model 3. Such a strategy is illustrated in section 17.2.

This table looks just about good enough to publish. The only fly in the ointment is that while the output looks very nice as it is displayed on the screen, it will require some effort to format it within a word processor to make the results ready for publication. It would be nice if this process of creating a publication-quality table could be automated. The next section shows how we can do this with the user-written `esttab` command.

16.4   Creating regression tables using esttab

As we have seen in the previous sections, the `estimates table` command is very useful for creating tables that have the overall look of a publication-quality table. As I analyze data, I will use the `estimates table` command to create tables on the screen for displaying and comparing regression models. But for the actual task of creating a publication-quality table, we can turn to one of many user-written programs that can do this task for us. In this section, I focus on illustrating the `esttab` command, written by Ben Jann (2005, 2007b), a Stata user from the University of Bern. I chose to illustrate the `esttab` command in this section because I feel its syntax and usage has the greatest resemblance to the `estimates table` command. This permits me to build upon what we have learned from using the `estimates table` command in the previous sections. In this section, I will focus on illustrating how you can use the `esttab` command to create customized publication-quality tables that you can directly integrate into your existing Word documents.[3]

Before we begin, we first need to install the `esttab` command. You can install it over the Internet using the `ssc` command.[4]

```
. ssc install estout, replace
```

---

3. I would encourage you to consider the other user-written commands for creating and formatting regression output as described in the following section (section 16.5). Each of these programs has its particular design philosophy and strengths, and depending on your goal, you might find one of those programs more suitable to your needs.

4. The `ssc` command below refers to `estout` because the main program is called `estout`, while we will be focusing on the user-friendly tool called `esttab`.

16.4.1   Presenting a single model with esttab

Let's begin by illustrating how to use the `esttab` command to display the results of a regression model that predicts happiness from socializing with friends, age, education, number of children, social class, and self-reported health. The `regress` command for fitting this model is shown below.

```
. regress happy7 socfrend age educ children class health
```

Source	SS	df	MS			
Model	180.555684	6	30.092614			
Residual	1068.13729	1,260	.847728009			
Total	1248.69298	1,266	.986329365			

```
                                      Number of obs =     1,267
                                      F(6, 1260)    =     35.50
                                      Prob > F      =    0.0000
                                      R-squared     =    0.1446
                                      Adj R-squared =    0.1405
                                      Root MSE      =    .92072
```

happy7	Coef.	Std. Err.	t	P>|t|	[95% Conf. Interval]	
socfrend	.0655262	.0167668	3.91	0.000	.0326322	.0984202
age	.0005277	.0016648	0.32	0.751	-.0027383	.0037937
educ	-.009905	.0095189	-1.04	0.298	-.0285797	.0087696
children	.0827224	.0173036	4.78	0.000	.0487753	.1166695
class	.2001397	.0389712	5.14	0.000	.1236842	.2765952
health	.263699	.0245636	10.74	0.000	.2155089	.3118892
_cons	3.821898	.1752121	21.81	0.000	3.478159	4.165638

As we did before, let's use the `estimates table` command to display the results from this regression to the screen.

```
. estimates table
```

Variable	active
socfrend	.06552622
age	.00052769
educ	-.00990501
children	.08272239
class	.20013971
health	.26369902
_cons	3.8218982

Instead of using the `estimates table` command, we can display this table using the `esttab` command. Note how, as with the `estimates table` command, we merely type `esttab`, and the results of the most recent estimation command are displayed as a formatted regression table.

```
. esttab
```

	(1) happy7
socfrend	0.0655***
	(3.91)
age	0.000528
	(0.32)
educ	-0.00991
	(-1.04)
children	0.0827***
	(4.78)
class	0.200***
	(5.14)
health	0.264***
	(10.74)
_cons	3.822***
	(21.81)
N	1267

```
t statistics in parentheses
* p<0.05, ** p<0.01, *** p<0.001
```

Also note how the esttab command produces a customized-style output, as compared with the estimates table command that produces (by default) a very bare output. The default output of esttab shows the coefficient with stars to indicate the significance level. Below each coefficient, the t statistic is shown in parentheses. Further, a legend is displayed at the bottom of the table explaining the interpretation of the stars and indicating that the t-values are included in parentheses.

This is not the central benefit of the esttab command. The huge benefit is that we can export these results in a format that is ready for publication or presentation. The example below adds using table2a.rtf to the esttab command, and the results are saved into a file named table2a.rtf, which can be read by many word processors (including Word).[5]

```
. esttab using table2a.rtf
(output written to table2a.rtf)
```

---

5. The .rtf extension stands for "Rich Text Format". If you are using Word, an .rtf file works very much like a .doc or .docx file. Once you open an .rtf file in Word, it looks just like a normal Word document, and you can save the file as a .doc or .docx file or copy and paste the contents into an existing .doc or .docx file.

Note: File already exists

If you run the above `esttab` command again, you will receive the following error.

```
esttab using table2a.rtf
file table2a.rtf already exists
r(602);
```

In that case, you can add the `replace` option if you want to overwrite the existing file, as shown below.

```
. esttab using table2a.rtf, replace
```

I opened this file in Word, and it is formatted as a nice-looking table. Figure 16.1 shows how this file looked while I was editing it in Word. While I am editing the table, the **Table Tools** group is shown, revealing the **Design** and **Layout** ribbon bars for editing and customizing tables. In this image, I have selected the **Layout** ribbon bar, showing some of the tools that are available for editing this table. For example, I clicked on the **View Gridlines** button, which shows the gridlines for the table, emphasizing that Word considers this to be a table and showing you its tabular structure.

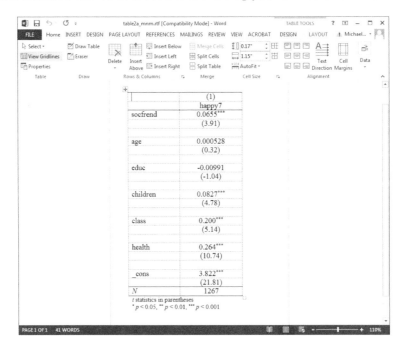

Figure 16.1: View of `table2a.rtf` within Word

It took me only a few minutes to customize this table to create a presentation-quality table, as shown in figure 16.2. In particular, I created customized labels for the predictors, but these labels were too wide for the first column. But because this is a table, I simply widened the first column to accommodate the width of the new labels. Again, because this is a table, it was easy to delete the blank rows between each predictor. I also merged the two cells above the coefficients to create a single cell with the title "Happiness Rating". As I noted before, this customized version of this table is shown in figure 16.2.

Table 1. Regression model predicting happiness

	Happiness Rating
Socializing with friends	0.0655***
	(3.91)
Age	0.000528
	(0.32)
Education	-0.00991
	(-1.04)
Number of children	0.0827***
	(4.78)
Social class	0.200***
	(5.14)
Self-rated health	0.264***
	(21.81)
N	1267

t statistics in parentheses
$^{*} p < 0.05$, $^{**} p < 0.01$, $^{***} p < 0.001$

Figure 16.2: Edited version of `table2a.rtf`

While it did not take me very long to make the customizations shown in figure 16.2, it would be more efficient to make such customizations as part of creating the table via the `esttab` command. Fortunately, the `esttab` command provides many options for customizing the display of the table, frequently allowing you to create a table that is ready for presentation (or one that needs only minimal changes before presentation). Consider the consequences of adding a new covariate to our model. If we manually customize the table, we would need to export the table anew and then manually customize the entire table all over again. If, instead, the customizations are applied as options on the `esttab` command, adding a new covariate does not mean additional work manually customizing the table: the customizations are automatically applied via the options on the `esttab` command. The remaining examples in this section will focus on showing how to use options with the `esttab` command to create a table that is ready for publication or that requires the fewest changes possible.

In the example below, the `nocons` option is used to omit the display of the constant. The `ar2` option specifies that the adjusted R-squared should be displayed at the bottom of the table. Further, the `wide` option displays the table in a wide format. This customized table is saved as `table2b.rtf`. The formatted table, as it would appear within Word, is shown in figure 16.3.

```
. esttab using table2b.rtf, nocons ar2 wide
(output written to table2b.rtf)
```

| | (1) | |
	happy7	
socfrend	0.0655***	(3.91)
age	0.000528	(0.32)
educ	-0.00991	(-1.04)
children	0.0827***	(4.78)
class	0.200***	(5.14)
health	0.264***	(10.74)
N	1267	
adj. R^2	0.141	

t statistics in parentheses
$^* p < 0.05$, $^{**} p < 0.01$, $^{***} p < 0.001$

Figure 16.3: View of `table2b.rtf` omitting constant, adding adj-*R*2, wide format

Let's further customize this table by presenting confidence intervals (CIs) instead of *t*-values. Simply including the `ci` option displays the 95% confidence intervals for the coefficient instead of the *t*-values. This customized table is saved as `table2c.rtf`, and the version as it would appear in Word is shown in figure 16.4. As you inspect this figure, note how the 95% confidence interval is displayed in the right column surrounded by brackets. Also note that the legend for the table has been updated to reflect that the confidence intervals are shown in brackets.

```
. esttab using table2c.rtf, nocons ar2 wide ci
(output written to table2c.rtf)
```

| | (1) | |
	happy7	
socfrend	0.0655***	[0.0326,0.0984]
age	0.000528	[-0.00274,0.00379]
educ	-0.00991	[-0.0286,0.00877]
children	0.0827***	[0.0488,0.117]
class	0.200***	[0.124,0.277]
health	0.264***	[0.216,0.312]
N	1267	
adj. R^2	0.141	

95% confidence intervals in brackets
$^* p < 0.05$, $^{**} p < 0.01$, $^{***} p < 0.001$

Figure 16.4: View of `table2c.rtf`, including CIs

Let's further customize this table by specifying the formatting of the coefficients via the `b()` option and the formatting of the CIs via the `ci()` option. In both cases, let's display the values with a total width of 6 columns with 3 digits after the decimal place.

This formatted table is saved as `table2d.rtf`. You can see the resulting table as it would appear in Word in figure 16.5.

```
. esttab using table2d.rtf, nocons ar2 wide b(%6.3f) ci(%6.3f)
(output written to table2d.rtf)
```

	(1) happy7	
socfrend	0.066***	[0.033,0.098]
age	0.001	[-0.003,0.004]
educ	-0.010	[-0.029,0.009]
children	0.083***	[0.049,0.117]
class	0.200***	[0.124,0.277]
health	0.264***	[0.216,0.312]
N	1267	
adj. R^2	0.141	

95% confidence intervals in brackets
* $p < 0.05$, ** $p < 0.01$, *** $p < 0.001$

Figure 16.5: View of `table2d.rtf` with formatted coefficients and CIs

While I like this table, I am not very satisfied with using the variable names for labeling the table. By specifying the `label` option (as shown below), I can display the variable labels instead of the variable names. This version of the table is saved as `table2e.rtf`. I opened this file in Word, and I widened the first column to accommodate the longer labels. The resulting table, after this minor customization, is shown in figure 16.6.

```
. esttab using table2e.rtf, nocons ar2 wide b(%6.3f) ci(%6.3f) label
(output written to table2e.rtf)
```

	(1) how happy R is (recoded)	
spend evening with friends	0.066***	[0.033,0.098]
age of respondent	0.001	[-0.003,0.004]
highest year of school completed	-0.010	[-0.029,0.009]
number of children	0.083***	[0.049,0.117]
subjective class identification	0.200***	[0.124,0.277]
rs Health in General (recoded)	0.264***	[0.216,0.312]
Observations	1267	
Adjusted R^2	0.141	

95% confidence intervals in brackets
* $p < 0.05$, ** $p < 0.01$, *** $p < 0.001$

Figure 16.6: View of `table2e.rtf` with variable labels

While the variable labels are more descriptive than the variable names, I don't feel they are suitable for publication. Let's use the `label variable` command to change the labels for these variables and apply labels that will be more appropriate for a publication-quality table.[6]

```
. label variable happy7 "Happiness"
. label variable socfrend "Socializing with friends"
. label variable age "Age"
. label variable educ "Education (years)"
. label variable children "Number of children"
. label variable class "Social class"
. label variable health "Self-rated health"
```

Now, when we use the `esttab` command with the `label` option, these improved variable labels are used. The `esttab` command below saves the table with the improved variable labels as `table2f.rtf`. The resulting table as displayed in Word (after we widen the first column) is shown in figure 16.7.

```
. esttab using table2f.rtf, nocons ar2 wide b(%6.3f) ci(%6.3f) label
(output written to table2f.rtf)
```

| | (1) | |
	Happiness	
Socializing with friends	0.066***	[0.033,0.098]
Age	0.001	[-0.003,0.004]
Education (years)	-0.010	[-0.029,0.009]
# of children	0.083***	[0.049,0.117]
Social class	0.200***	[0.124,0.277]
Self rated health	0.264***	[0.216,0.312]
Observations	1267	
Adjusted R^2	0.141	

95% confidence intervals in brackets
* $p < 0.05$, ** $p < 0.01$, *** $p < 0.001$

Figure 16.7: View of `table2f.rtf` with improved variable labels

I feel that the table stored in the file `table2f.rtf` and shown in figure 16.7 is very suitable for presentations and could be used for a publication. The only manual customization that I applied to this table was to widen the first column to accommodate the longer variable labels. If you were happy with this table, you could easily integrate this into your manuscript using standard methods of copying the table to the clipboard and then pasting it into your manuscript.

6. Note that any other Stata commands would use these new variable labels as well. You can see more about labeling data in Stata in section 22.4.

16.4.2   Presenting multiple models with esttab

Let's now consider the three regression models that we fit earlier in this chapter, model 1, model 2, and model 3. These regression models were stored using the `estimates store` command. As we did before, let's use the `estimates table` command to display the results of these three models side by side.

```
. estimates table mod1 mod2 mod3
```

Variable	mod1	mod2	mod3
socfrend	.07232471	.07802658	.06552622
age		-.00267429	.00052769
educ		.0037173	-.00990501
children		.0868444	.08272239
class		.27230571	.20013971
health			.26369902
_cons	5.210231	4.4530791	3.8218982

Say that we wanted to create a publication-quality version of the above table. We can use the `esttab` command to display these three models together in a single table. Let's start by just using the default settings from `esttab` and saving the resulting table in `table2g.rtf`. Figure 16.8 shows how this table looks when we edit it within Word.

```
. esttab mod1 mod2 mod3 using table2g.rtf
(output written to table2g.rtf)
```

	(1) happy7	(2) happy7	(3) happy7
socfrend	0.0723***	0.0780***	0.0655***
	(4.34)	(4.47)	(3.91)
age		-0.00267	0.000528
		(-1.56)	(0.32)
educ		0.00372	-0.00991
		(0.38)	(-1.04)
children		0.0868***	0.0827***
		(4.80)	(4.78)
class		0.272***	0.200***
		(6.78)	(5.14)
health			0.264***
			(10.74)
_cons	5.210***	4.453***	3.822***
	(71.80)	(25.77)	(21.81)
N	1284	1272	1267

t statistics in parentheses
* $p < 0.05$, ** $p < 0.01$, *** $p < 0.001$

Figure 16.8: Display of three regression models with default settings

Let's now customize the display of this table using many of the customizations we have previously used. Namely, let's omit the constant (via the `nocons` option) and display the adjusted R-squared (via the `ar2` option). Further, let's use the `b()` and `ci()` options to display and format the regression coefficients and CIs. Further, let's use the `label` option to specify that the variable labels should be used in lieu of variable names. Then, let's apply two new options we have not seen before. The `nogaps` option is used to omit the gaps (blank rows) between each of the coefficients. Finally, the `mtitle()` option is used to specify names for each of the three models. The resulting table, after we edit the column widths in Word, is displayed in figure 16.9.

```
. esttab mod1 mod2 mod3 using table2h.rtf, nocons ar2 b(%6.3f) ci(%6.3f)
> label nogaps mtitle("Model 1" "Model 2" "Model 3")
(output written to table2h.rtf)
```

	(1) Model 1	(2) Model 2	(3) Model 3
Socializing with friends	0.072***	0.078***	0.066***
	[0.040,0.105]	[0.044,0.112]	[0.033,0.098]
Age		-0.003	0.001
		[-0.006,0.001]	[-0.003,0.004]
Education (years)		0.004	-0.010
		[-0.016,0.023]	[-0.029,0.009]
Number of children		0.087***	0.083***
		[0.051,0.122]	[0.049,0.117]
Social class		0.272***	0.200***
		[0.193,0.351]	[0.124,0.277]
Self rated health			0.264***
			[0.216,0.312]
Observations	1284	1272	1267
Adjusted R^2	0.014	0.062	0.141

95% confidence intervals in brackets
* $p < 0.05$, ** $p < 0.01$, *** $p < 0.001$

Figure 16.9: Display of three regression models with customized settings

16.4.3   Exporting results to other file formats

In the examples I have illustrated so far, I have focused on showing how you can use the `.rtf` extension on the file name to create a file that can be read into a word processor like Word. I have emphasized that option because I feel it is a file format most readers would be likely to use. However, the `esttab` command supports a wide variety of other file formats as well. For example, you can export the results to a comma-separated file (which can be read into Excel) by specifying a `.csv` extension. You can create an HTML-formatted file (for example, for display within a webpage) by using the `.html` extension. Or if you wanted to integrate the regression output into a LaTeX document, you can specify a `.tex` extension.

The `esttab` command below is identical to the previous example, except for one change and one addition. The change is that the results are saved as a LaTeX table named `table2h.tex`. The addition is that I added the `title()` option to supply a title for the LaTeX table as well as a label for the table. The results of the command are shown in table 16.1. Note how the title command not only allowed me to add a title to the table but also permitted me to reference this table by specifying \ref{tab2h} because I included \label{tab2h} in the title for the table.

```
. esttab mod1 mod2 mod3 using table2h.tex, nocons ar2 b(%6.3f) ci(%6.3f)
> label nogaps mtitle("Model 1" "Model 2" "Model 3")
> title("Display of table created in table2h.tex"\label{tab2h})
(output written to table2h.tex)
```

Table 16.1: Display of table created in table2h.tex

	(1) Model 1	(2) Model 2	(3) Model 3
Socializing with friends	0.072***	0.078***	0.066***
	[0.040,0.105]	[0.044,0.112]	[0.033,0.098]
Age		-0.003	0.001
		[-0.006,0.001]	[-0.003,0.004]
Education (years)		0.004	-0.010
		[-0.016,0.023]	[-0.029,0.009]
Number of children		0.087***	0.083***
		[0.051,0.122]	[0.049,0.117]
Social class		0.272***	0.200***
		[0.193,0.351]	[0.124,0.277]
Self-rated health			0.264***
			[0.216,0.312]
Observations	1284	1272	1267
Adjusted R^2	0.014	0.062	0.141

95% confidence intervals in brackets

* $p < 0.05$, ** $p < 0.01$, *** $p < 0.001$

Tip: More information on esttab

You can find out more about the `esttab` command by typing `help esttab`. In addition, you can visit http://repec.org/bocode/e/estout/esttab.html, where you can find webpages describing the use of `esttab`. There you can see that `esttab` is actually a simplified version of an even more powerful and flexible program called `estout`. You can also see Jann (2007b) for more about this suite of commands (available via the *Stata Journal* website at http://www.stata-journal.com/sjpdf.html?articlenum=st0085_1).

In this section, I have emphasized using the `esttab` command, but this is not the only user-written program that you can use for formatting and exporting regression tables. I emphasized this program because it most mirrors the built-in `estimates` `table` command in terms of syntax and function. Before closing this chapter, I would like to let you know about other programs you could use to format and export regression results.

16.5   More commands for presenting regression results

16.5.1   outreg

One of the earliest, if not *the* earliest, commands to address the problem of presenting regression results was written by John Luke Gallup of Portland State University. The command, called `outreg`, must have been downloaded and used at least a zillion times. The original version of `outreg` has been rewritten and replaced by a new version, also called `outreg`. You can download the latest and greatest version of this program using the `ssc` command shown below.

    . ssc install outreg, replace

After downloading this program, you can type `help outreg` for more information. (You can also type `help outreg_update` for a little bit of history about the new version versus the old version.) The `outreg` command is actually part of a broad and comprehensive system for formatting statistical output. For more information, you can type `help frmttable` (assuming you have installed `outreg`). You can also see the *Stata Journal* articles about this system for formatting output (see Gallup [2012b] and Gallup [2012a]).

16.5.2   outreg2

As implied by the name of the command, `outreg2` is a variant of the `outreg` command. Written by Roy Wada (2005), this command is not just about creating publication-quality tables—it has tools that emphasize the ability to quickly and easily view the formatted regression results. You can download `outreg2` using the `ssc` command as shown below.

    . ssc install outreg2, replace

You can type `help outreg2` for more information about using this command.

16.5.3   xml_tab

If your aim is to save output into a spreadsheet, then you will want to consider the `xml_tab` command. This command saves Stata output directly into an XML file that can be opened with Excel (or OpenOffice Calc). The `xml_tab` command allows you to format the elements of the output to create publication-quality tables that you can print from a spreadsheet or that you can link into your word-processing program. You can download this program using the `ssc` command shown below.

```
. ssc install xml_tab, replace
```

After downloading the program, you can type `help xml_tab` for more information about using this command. You can also see Lokshin and Sajaia (2008) for more information (or see the online version at the *Stata Journal* website at http://www.stata-journal.com/sjpdf.html?articlenum=dm0037).

16.5.4   coefplot

Rather than presenting a tabular display of your regression results, perhaps you would like to create a graphical display instead. The `coefplot` command, written by Jann (2014), the author of `esttab`, creates a graphical presentation of your regression results. You can download this program using the `ssc` command shown below.

```
. ssc install coefplot, replace
```

Let me show you just one example using the `coefplot` command. The `coefplot` command below accesses the results stored as `mod1`, `mod2`, and `mod3` to create the graph shown in figure 16.10. I added a number of options to illustrate the customizations that are possible. Namely, I used the `drop()` option to omit the constant, the `plotlabels()` and `legend()` options to customize the legend, the `xline(0)` option to add a reference line at 0, and the `title()` option to add a title to the graph.

```
. coefplot mod1 mod2 mod3, drop(_cons)
> plotlabels("Model 1" "Model 2" "Model 3") legend(rows(1))
> xline(0) title("Predicting happiness using 3 models")
```

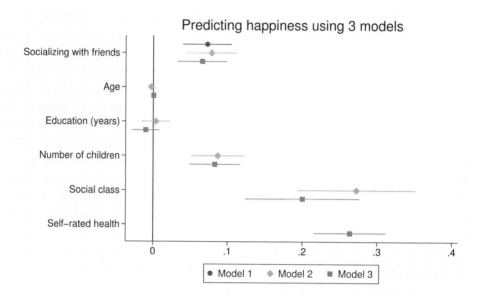

Figure 16.10: Plot of coefficients for mod1, mod2, and mod3

After downloading this command, you can learn more about using it by typing **help coefplot**, by visiting http://ideas.repec.org/p/bss/wpaper/1.html, or by seeing Jann (2014).

Video tutorial: Exporting results directly to Excel

See how you can export results directly to Excel from within Stata at http://www.stata.com/sbs/export-to-excel.

16.6   Closing thoughts

This chapter has presented a number of methods that you can use for saving formatted regression results. The heart of the chapter has focused on the use of the **estimates table** command (built into Stata) and the user-written **esttab** command. The **estimates table** command has the benefit of being built into Stata, and it can be used as an analytic tool for comparing and summarizing regression results in an easy-to-read format. By contrast, the **esttab** command has the advantage of being able to directly create formatted tables in a variety of formats that are ready for, or require very few modifications before, publication.

This is a difficult topic to cover in a book, because past history has shown considerable fluidity with respect to features available for displaying formatted regression output. A variety of user-written programs have been developed for presenting regression results, and the `outreg` command has been enhanced and radically expanded as `frmttable` to tackle the general problem of writing formatted tables of statistical results. Stata itself has shown considerable changes too. Recent advances have included the ability to directly save Stata datasets as Excel workbooks with the `export excel` command and the ability to write contents to the cells of a spreadsheet using the `putexcel` command. Furthermore, the Stata `mata` programming language has added features for saving Excel workbooks and Office Open XML (`.docx`) files (see `help mf_docx` and `help mf_xl`). While you might not use these capabilities, I can easily imagine them being exploited in the future by Stata users and StataCorp to provide new commands with even more features and ease of use for formatting and saving regression results. Even so, the examples shown in this chapter will still continue to work even if they are superseded by different or more powerful tools. And I hope that the examples shown in this chapter still convey useful and key concepts to help you save time and effort as you create formatted regression tables using Stata.

17 Tools for model building

17.1 Chapter overview

This chapter focuses on tools built into Stata for model building. The first section shows how you can fit multiple models using the same sample of observations (section 17.2). The next section illustrates the use of the `nestreg` prefix command for fitting nested regression models (section 17.3). Section 17.4 illustrates the use of the `stepwise` prefix for fitting stepwise models. The examples are based on using the General Social Survey dataset, which is read into memory with the `use` command.

```
. use gss2012_sbs
```

17.2 Fitting multiple models on the same sample

Let's consider one of the issues that arise when we fit regression models with different sets of predictors. When we fit and compare models with different sets of predictors, the sample size can change because of missing values for the predictors. This can make it difficult to compare coefficients across models because of the change in sample size. Let's illustrate this using the three regression models that were illustrated in section 16.3 from the previous chapter. The first of these three models predicts happiness from how often one socializes with friends. The `regress` command below runs the regression model, and then the `estimates store` command stores the results from this model with the name `mod1`.

```
. * Model 1
. regress happy7 socfrend
```

Source	SS	df	MS		
Model	18.4387453	1	18.4387453		
Residual	1254.55814	1,282	.978594493		
Total	1272.99688	1,283	.99220334		

Number of obs = 1,284
F(1, 1282)   = 18.84
Prob > F   = 0.0000
R-squared   = 0.0145
Adj R-squared = 0.0137
Root MSE   = .98924

happy7	Coef.	Std. Err.	t	P>|t|	[95% Conf. Interval]
socfrend	.0723247	.0166618	4.34	0.000	.0396373 .1050121
_cons	5.210231	.0725706	71.80	0.000	5.067861 5.352601

```
. estimates store mod1
```

Let's now fit the second model that adds the variables `age`, `educ`, `children`, and `class`. The results are stored using the name `mod2`.

```
. * Model 2
. regress happy7 socfrend age educ children class
```

Source	SS	df	MS		
Model	82.3714574	5	16.4742915		
Residual	1177.60889	1,266	.930180797		
Total	1259.98035	1,271	.991329934		

Number of obs = 1,272
F(5, 1266)   = 17.71
Prob > F   = 0.0000
R-squared   = 0.0654
Adj R-squared = 0.0617
Root MSE   = .96446

happy7	Coef.	Std. Err.	t	P>|t|	[95% Conf. Interval]
socfrend	.0780266	.0174486	4.47	0.000	.0437952 .1122579
age	-.0026743	.0017126	-1.56	0.119	-.0060342 .0006856
educ	.0037173	.0098036	0.38	0.705	-.0155159 .0229505
children	.0868444	.0180997	4.80	0.000	.0513356 .1223532
class	.2723057	.0401906	6.78	0.000	.1934583 .3511532
_cons	4.453079	.1727716	25.77	0.000	4.114129 4.792029

```
. estimates store mod2
```

Model 3 adds the variable `health` to the model. The `estimates store` saves the results using the name `mod3`.

```
. * Model 3
. regress happy7 socfrend educ age children class health
      Source |       SS           df       MS      Number of obs   =     1,267
-------------+----------------------------------   F(6, 1260)      =     35.50
       Model |  180.555684         6   30.092614   Prob > F        =    0.0000
    Residual |  1068.13729     1,260  .847728009   R-squared       =    0.1446
-------------+----------------------------------   Adj R-squared   =    0.1405
       Total |  1248.69298     1,266  .986329365   Root MSE        =    .92072

------------------------------------------------------------------------------
      happy7 |      Coef.   Std. Err.      t    P>|t|     [95% Conf. Interval]
-------------+----------------------------------------------------------------
    socfrend |   .0655262   .0167668     3.91   0.000     .0326322    .0984202
        educ |   -.009905   .0095189    -1.04   0.298    -.0285797    .0087696
         age |   .0005277   .0016648     0.32   0.751    -.0027383    .0037937
    children |   .0827224   .0173036     4.78   0.000     .0487753    .1166695
       class |   .2001397   .0389712     5.14   0.000     .1236842    .2765952
      health |    .263699   .0245636    10.74   0.000     .2155089    .3118892
       _cons |   3.821898   .1752121    21.81   0.000     3.478159    4.165638
------------------------------------------------------------------------------

. estimates store mod3
```

In each of these models, the focus of interest is the coefficient for `socfrend`. As we did at the end of section 16.3, let's present the results of these three models using the `estimates table` command. When we add the `stats(N r2)` option, the sample size and R-squared are shown for each model. Let's focus on the sample size for each model, noting how the sample size differs for each. This is due to missing values on the predictors and leads to the omission of additional observations in model 2 and in model 3.

```
. estimates table mod1 mod2 mod3,
> b(%5.3f) star(0.05 0.01 0.001) drop(_cons) stats(N r2)

----------------------------------------------
    Variable |   mod1        mod2        mod3
-------------+--------------------------------
    socfrend | 0.072***    0.078***    0.066***
         age |             -0.003       0.001
        educ |              0.004      -0.010
    children |              0.087***    0.083***
       class |              0.272***    0.200***
      health |                          0.264***
-------------+--------------------------------
           N |   1284        1272        1267
          r2 |  0.014       0.065       0.145
----------------------------------------------
        legend: * p<.05; ** p<.01; *** p<.001
```

The comparison of model 1 (with $N = 1284$) with model 2 (with $N = 1272$) is not strictly an apples to apples comparison. Model 2 lost 12 observations compared with model 1. If we were comparing the estimate of `socfrend` between model 1 and model 2, the differences could be attributable to the loss of the 12 observations in model 2 instead of due to the additional predictors added in model 2. A reviewer might ask us to rerun the models using the same sample size for all three models. We can do that by running the models again, but this time we will run them in reverse order.

We first run model 3. We run this model first because we are going to run the subsequent models only for the observations for which complete data exist on model 3.

```
. * Model 3
. regress happy7 socfrend age educ children class health
```

Source	SS	df	MS
Model	180.555684	6	30.092614
Residual	1068.13729	1,260	.847728009
Total	1248.69298	1,266	.986329365

Number of obs = 1,267
F(6, 1260)   = 35.50
Prob > F   = 0.0000
R-squared   = 0.1446
Adj R-squared = 0.1405
Root MSE   = .92072

happy7	Coef.	Std. Err.	t	P>|t|	[95% Conf. Interval]	
socfrend	.0655262	.0167668	3.91	0.000	.0326322	.0984202
age	.0005277	.0016648	0.32	0.751	-.0027383	.0037937
educ	-.009905	.0095189	-1.04	0.298	-.0285797	.0087696
children	.0827224	.0173036	4.78	0.000	.0487753	.1166695
class	.2001397	.0389712	5.14	0.000	.1236842	.2765952
health	.263699	.0245636	10.74	0.000	.2155089	.3118892
_cons	3.821898	.1752121	21.81	0.000	3.478159	4.165638

Let's now create a variable that specifies whether an observation was included in model 3. As we saw in section 15.4, we can use e(sample) to identify whether an observation was contained within the estimation sample. In the generate command below, the variable frommod3 is assigned the value of e(sample), which means that a 1 is assigned to frommod3 if the observation was used in the estimation of model 3, and 0 if it was omitted from estimating model 3.

```
. generate frommod3 = e(sample)
```

Using the tabulate command, we can see that 1,267 observations have a 1, consistent with the number of observations that were included in model 3.

```
. tabulate frommod3
```

frommod3	Freq.	Percent	Cum.
0	707	35.82	35.82
1	1,267	64.18	100.00
Total	1,974	100.00	

Now, let's run model 2 again, but this time only including the observations that were present in model 3, by adding if frommod3==1 to the regress command. We can see that this model is fit with $N = 1267$ observations, like model 3. The `estimates store` command is used to save the results from this model; the results are called `mod2a`.

```
. * Model 2a
. regress happy7 socfrend age educ children class if frommod3==1

      Source |       SS           df       MS      Number of obs   =     1,267
-------------+----------------------------------   F(5, 1261)      =     17.92
       Model |  82.8571867         5  16.5714373   Prob > F        =    0.0000
    Residual |  1165.83579     1,261  .924532743   R-squared       =    0.0664
-------------+----------------------------------   Adj R-squared   =    0.0627
       Total |  1248.69298     1,266  .986329365   Root MSE        =    .96153

      happy7 |      Coef.   Std. Err.      t    P>|t|     [95% Conf. Interval]
-------------+----------------------------------------------------------------
    socfrend |   .0804612   .0174495     4.61   0.000     .0462279    .1146946
         age |  -.0027418   .0017092    -1.60   0.109     -.006095    .0006114
        educ |   .0056579   .0098248     0.58   0.565    -.0136169    .0249326
    children |   .0882432   .0180625     4.89   0.000     .0528074    .1236791
       class |   .2676707   .0401646     6.66   0.000     .1888738    .3464675
       _cons |   4.430778   .1731252    25.59   0.000     4.091133    4.770423

. estimates store mod2a
```

We then repeat this procedure for model 1 by rerunning the `regress` command for model 1, and adding `if frommod3==1`. This model is also fit with $N = 1267$ observations. The `estimates store` command is then used to save the results from this model, which are called `mod1a`.

```
. * Model 1a
. regress happy7 socfrend if frommod3==1

      Source |       SS           df       MS      Number of obs   =     1,267
-------------+----------------------------------   F(1, 1265)      =     19.30
       Model |  18.7654906         1  18.7654906   Prob > F        =    0.0000
    Residual |  1229.92748     1,265  .972274692   R-squared       =    0.0150
-------------+----------------------------------   Adj R-squared   =    0.0142
       Total |  1248.69298     1,266  .986329365   Root MSE        =    .98604

      happy7 |      Coef.   Std. Err.      t    P>|t|     [95% Conf. Interval]
-------------+----------------------------------------------------------------
    socfrend |   .0736871   .0167728     4.39   0.000     .0407815    .1065927
       _cons |   5.209052   .0731975    71.16   0.000     5.065451    5.352654

. estimates store mod1a
```

Let's now show the use of the `estimates table` command to show the results of model 1a, model 2a, and model 3 side by side. The `estimates table` command below displays the models named `mod1a`, `mod2a`, and `mod3`. Note how the sample size for all 3 models is now the same, $N = 1267$.

```
. estimates table mod1a mod2a mod3,
> b(%5.3f) star(0.05 0.01 0.001) drop(_cons) stats(N r2)
```

Variable	mod1a	mod2a	mod3
socfrend	0.074***	0.080***	0.066***
age		-0.003	0.001
educ		0.006	-0.010
children		0.088***	0.083***
class		0.268***	0.200***
health			0.264***
N	1267	1267	1267
r2	0.015	0.066	0.145

```
legend: * p<.05; ** p<.01; *** p<.001
```

We can create a table that shows the original models (mod1, mod2, and mod3) as well as the two updated models (mod1a and mod2a). The estimates table command below shows all five of these models side by side.

```
. estimates table mod1 mod2 mod3 mod1a mod2a,
> b(%5.3f) drop(_cons) stats(N r2)
```

Variable	mod1	mod2	mod3	mod1a	mod2a
socfrend	0.072	0.078	0.066	0.074	0.080
age		-0.003	0.001		-0.003
educ		0.004	-0.010		0.006
children		0.087	0.083		0.088
class		0.272	0.200		0.268
health			0.264		
N	1284	1272	1267	1267	1267
r2	0.014	0.065	0.145	0.015	0.066

We can see that the coefficient for model 1 is similar to the coefficient in model 1a. Likewise, we can see that the coefficients shown in model 2 are very similar to the coefficients shown in model 2a.

17.3   Nested models

This section describes nested regression models. These are models where variables are grouped into blocks, and the blocks of variables are entered sequentially. Comparison tests are reported that assess the contribution of each new block after accounting for the contribution of the previous blocks of variables. This modeling strategy is sometimes called a "nested regression" model (not to be confused with "nested factors") and sometimes called a "hierarchical" regression model (not to be confused with a "hierarchical linear" model).

Let's start by considering a very simple example of a nested regression model predicting happiness from three blocks of variables. After showing this simple example, I will then present a more realistic example (in section 17.3.2).

17.3.1 Example 1: A simple example

Let's start by considering a very simple example of a nested regression model. The outcome variable for this model is happy7. Say that we want to enter the variables educ and age as the first block of variables. For the second block, we want to enter health, and for the third block of variables, we want to enter two variables regarding socializing behaviors, socfrend and socrel (frequency of socializing with friends and frequency of socializing with relatives). This will allow us to assess the contribution of each block of variables after accounting for the contribution of the variables in the previous blocks.

The command below uses the nestreg prefix command to indicate that we want to use a nested modeling approach in combination with the regress command.[1] This is followed by the name of the dependent variable, happy7, followed by the list of predictor variables. Parentheses are used to group variables into blocks. The parentheses in the command below indicate that educ and age form the first block, health forms the second block, and socfrend and socrel form the third block.

---

1. You could use other estimation commands following the nestreg: prefix, such as qreg or logistic.

```
. nestreg: regress happy7 (educ age) (health) (socfrend socrel)
Block  1: educ age
```

Source	SS	df	MS
Model	6.84089303	2	3.42044652
Residual	1251.87597	1,272	.984179221
Total	1258.71686	1,274	.988003817

Number of obs =	1,275
F(2, 1272) =	3.48
Prob > F =	0.0312
R-squared =	0.0054
Adj R-squared =	0.0039
Root MSE =	.99206

happy7	Coef.	Std. Err.	t	P>|t|	[95% Conf. Interval]	
educ	.0238534	.009142	2.61	0.009	.0059183	.0417884
age	-.0005247	.0015711	-0.33	0.738	-.003607	.0025576
_cons	5.205589	.1487995	34.98	0.000	4.913669	5.497508

```
Block  2: health
```

Source	SS	df	MS
Model	132.534789	3	44.178263
Residual	1126.18207	1,271	.886059853
Total	1258.71686	1,274	.988003817

Number of obs =	1,275
F(3, 1271) =	49.86
Prob > F =	0.0000
R-squared =	0.1053
Adj R-squared =	0.1032
Root MSE =	.94131

happy7	Coef.	Std. Err.	t	P>|t|	[95% Conf. Interval]	
educ	.0008771	.0088862	0.10	0.921	-.0165562	.0183104
age	.0029197	.0015185	1.92	0.055	-.0000594	.0058988
health	.2923428	.0245452	11.91	0.000	.2441892	.3404964
_cons	4.359108	.1580662	27.58	0.000	4.049009	4.669207

```
Block  3: socfrend socrel
```

Source	SS	df	MS
Model	149.371371	5	29.8742743
Residual	1109.34549	1,269	.874188724
Total	1258.71686	1,274	.988003817

Number of obs =	1,275
F(5, 1269) =	34.17
Prob > F =	0.0000
R-squared =	0.1187
Adj R-squared =	0.1152
Root MSE =	.93498

happy7	Coef.	Std. Err.	t	P>|t|	[95% Conf. Interval]	
educ	-.0019573	.0089624	-0.22	0.827	-.0195401	.0156255
age	.0047216	.0015698	3.01	0.003	.001642	.0078012
health	.2836341	.02447	11.59	0.000	.2356281	.3316401
socfrend	.0463803	.0171846	2.70	0.007	.0126669	.0800937
socrel	.0449526	.0161108	2.79	0.005	.0133459	.0765593
_cons	3.949846	.1833816	21.54	0.000	3.590082	4.309611

Block	F	Block df	Residual df	Pr > F	R2	Change in R2
1	3.48	2	1272	0.0312	0.0054	
2	141.86	1	1271	0.0000	0.1053	0.0999
3	9.63	2	1269	0.0001	0.1187	0.0134

We can divide this output into two general sections, the first section containing the regression tables for block 1, block 2, and block 3 and the second portion that summarizes the three blocks and assesses the contribution of each block of variables.

Let's first consider the top portion of the output. The section titled `Block 1` shows the estimation of the regression model predicting the outcome using the variables from the first block, namely, `educ` and `age`. The section titled `Block 2` shows the regression model when the second block of variables is added, which yields a model predicting the outcome from the first block of variables with `health` added to the model. The section titled `Block 3` shows the regression model after the third block of variables is added, namely, `socfrend` and `socrel`.

Let's now turn our attention to the final portion of the output, which is the key part of the output. This shows the contribution of each block of variables after accounting for the previous block of variables.

Looking at the first row of the output, we see that the contribution of the first block of variables (`educ` and `age`) is significant [$F(2, 1272) = 3.48$, $p = 0.0054$]. This block of variables explains 0.54% of the variation in the dependent variable (`R-squared` = 0.0054).

The contribution of the variable in the second block, `health`, is significant after accounting for the first block of variables [$F(1, 1271) = 141.86$, $p < 0.001$]. The variable `health` explains an additional 9.99% of the variance in happiness after accounting for block 1 (`Change in R2` = 0.0999). At this stage, all the variables in the model (`educ`, `age`, and `health`) explain 10.53% of the variance in the dependent variable (`R-squared` = 0.1053).

The contribution of the third block of variables, `socfrend` and `socrel`, is significant after accounting for the first and second block of variables [$F(2, 1269) = 9.63$, $p = 0.0001$]. The variables in block 3 explain an additional 1.34% of the variance in happiness after accounting for blocks 1 and 2 (`Change in R2` = 0.0134). All the variables in the model at this stage (`educ`, `age`, `health`, `socfrend`, and `socrel`) explain 11.87% of the variance in the outcome (`R-squared` = 0.1187).

Let's consider a second example that is a bit more realistic. As part of this second example, I will illustrate how we can create stored results associated with each step of the nested regression model and how we can use the `estimates table` command to display the regression coefficients for each block, side by side, in one compact table.

17.3.2   Example 2: A more realistic example

Let's consider a model very similar to the one we considered in the previous section. Like before, the outcome variable is happiness, and the predictors are composed of demographic variables, health, and measures of socializing. Instead of having two measures of socializing, this model includes four measures of socializing.

Our interest is in assessing the amount of variance that socializing behaviors can explain in happiness after we account for demographics and health. Specifically, the goal is to first determine the percentage of variance explained by demographic factors and then the percentage of variance explained by health (after we account for demographic factors) and then finally to add socializing behaviors to the model and determine the amount of variance the socializing behaviors explain after we account for demographic and health factors.

As we did in the example before, we will use the `nestreg` prefix before the `regress` command. This is followed by the outcome variable, and then the blocks of variables are given using parentheses to identify the blocks. The first block is composed of the demographic variables `educ`, `age`, `children`, and `class`. The second block contains one variable, `health`. The third block contains four measures of socializing, `socfrend`, `socrel`, `socommun`, and `socbar`. The output from the command below is very lengthy, so it is omitted to save space.

```
. nestreg : regress happy7 (educ age children class)
> (health) (socfrend socrel socommun socbar)
  (output omitted)
```

Let's repeat the command above but include a couple of additional options. Let's first include the `store(nesthap)` option with the `nestreg` command. This will store the results from block 1 with the name `nesthap1`, store the results from block 2 with the name `nesthap2`, and store the results from block 3 with the name `nesthap3`.[2] Further, the `quietly` option is used to suppress the display of the regression tables from each block. Below you can see that the output shows the variables added at each block and the table summarizing the contribution of each block of variables.

```
. nestreg, store(nesthap) quietly: regress happy7 (age educ children class)
>      (health)
>      (socfrend socrel socommun socbar), notable
Block  1: age educ children class
Block  2: health
Block  3: socfrend socrel socommun socbar
```

| | | Block | Residual | | | Change |
Block	F	df	df	Pr > F	R2	in R2
1	16.67	4	1257	0.0000	0.0504	
2	123.40	1	1256	0.0000	0.1353	0.0850
3	6.37	4	1252	0.0000	0.1526	0.0172

Let's start by interpreting the summary of the results of predicting happiness from each block of variables. We see that the first block of variables (demographics) explained 5.04% of the variance in happiness and that this was significant $[F(4, 1257) = 16.67, p < 0.001]$. The second block, `health`, explained an additional 8.50% of the variance in happiness, and that increment in explained variance was significant $[F(1, 1256) = 123.40,$

---

2. You can refer back to section 15.6 for more about storing results.

$p < 0.001$]. The third block (composed of the four socializing variables) explained an additional 1.72% of the variance in happiness, and that increment in explained variance was also significant [$F(4, 1252) = 6.37$, $p < 0.001$].

Of course, this analysis is not complete without displaying and inspecting the regression tables for each of the three blocks of variables. Using the store() option, we have created a set of stored results named nesthap1 (for the first model), nesthap2 (for the second model), and nesthap3 (for the third model). As we saw in section 16.3, we can use the estimates table command to create a compact and formatted display showing the results of multiple models. The estimates table command is used below to show the coefficients for the models nesthap1, nesthap2, and nesthap3. Further, options are included to format the coefficients, add stars to signify the significance level, add the R-squared and sample size for each model, and omit the display of the constant.

```
. estimates table nesthap1 nesthap2 nesthap3, b(%4.3f)
> star(0.05 0.01 0.001) stats(r2 N) drop(_cons)
```

Variable	nesthap1	nesthap2	nesthap3
age	-0.004**	-0.001	0.001
educ	0.011	-0.007	-0.009
children	0.077***	0.073***	0.075***
class	0.273***	0.201***	0.203***
health		0.274***	0.266***
socfrend			0.050**
socrel			0.033*
socommun			0.030*
socbar			-0.013
r2	0.050	0.135	0.153
N	1262	1262	1262

```
legend: * p<.05; ** p<.01; *** p<.001
```

Using this method for running a nested regression model creates less output while also creating an output that is formatted to facilitate comparisons among the coefficients in the models.

Note: Displaying nested models with esttab

Another advantage of using the store() option is that we can then use the esttab command to create a formatted regression table using these stored results (as we saw in section 16.4.2). The esttab command below creates a formatted version of the regression tables displayed in the estimates table command above. The resulting table is saved as nestreg.rtf.

```
. esttab nesthap1 nesthap2 nesthap3 using nestreg.rtf,
> nocons ar2 b(%6.3f) ci(%6.3f) label nogaps
> mtitle("Model 1" "Model 2" "Model 3")
```

17.4   Stepwise models

Stata offers the ability to fit stepwise regression models. This is done by using the stepwise prefix. Imagine that we want to find the variables that are the best predictors of happiness (happy7). Our candidate variables are age, educ, children, class, health, socrel, socommun, socfrend, and socbar. In the example below, we use a forward selection procedure with 0.05 as the significance level for entering variables into the model. Typing only pe() specifies a forward selection method.

```
. stepwise, pe(.05): regress happy7 age educ children class
> health socfrend socrel socommun socbar
              begin with empty model
p = 0.0000 <  0.0500  adding  health
p = 0.0000 <  0.0500  adding  class
p = 0.0000 <  0.0500  adding  children
p = 0.0002 <  0.0500  adding  socfrend
p = 0.0130 <  0.0500  adding  socommun
p = 0.0430 <  0.0500  adding  socrel
```

Source	SS	df	MS		Number of obs	=	1,262
					F(6, 1255)	=	37.31
Model	188.507779	6	31.4179632		Prob > F	=	0.0000
Residual	1056.91298	1,255	.842161738		R-squared	=	0.1514
					Adj R-squared	=	0.1473
Total	1245.42076	1,261	.98764533		Root MSE	=	.91769

happy7	Coef.	Std. Err.	t	P>|t|	[95% Conf.	Interval]
health	.2592685	.0239306	10.83	0.000	.2123201	.3062169
class	.1939354	.0368361	5.26	0.000	.1216684	.2662025
children	.0826923	.0160225	5.16	0.000	.0512585	.1141261
socfrend	.0441151	.0172725	2.55	0.011	.0102289	.0780013
socommun	.0289651	.0132581	2.18	0.029	.0029545	.0549757
socrel	.0326699	.0161264	2.03	0.043	.0010322	.0643076
_cons	3.585914	.1441992	24.87	0.000	3.303016	3.868812

We could approach this by using a backward selection procedure. The example below uses backward selecting with 0.10 as the significance level for removal from the model. Typing only the pr() option specifies a backward selection procedure.

```
. stepwise, pr(.10): regress happy7 age educ children class
> health  socfrend socrel socommun socbar
                    begin with full model
p = 0.6950 >= 0.1000   removing age
p = 0.4011 >= 0.1000   removing socbar
p = 0.3372 >= 0.1000   removing educ
```

Source	SS	df	MS	Number of obs	=	1,262
				F(6, 1255)	=	37.31
Model	188.507779	6	31.4179632	Prob > F	=	0.0000
Residual	1056.91298	1,255	.842161738	R-squared	=	0.1514
				Adj R-squared	=	0.1473
Total	1245.42076	1,261	.98764533	Root MSE	=	.91769

happy7	Coef.	Std. Err.	t	P>|t|	[95% Conf. Interval]	
socommun	.0289651	.0132581	2.18	0.029	.0029545	.0549757
socrel	.0326699	.0161264	2.03	0.043	.0010322	.0643076
children	.0826923	.0160225	5.16	0.000	.0512585	.1141261
class	.1939354	.0368361	5.26	0.000	.1216684	.2662025
health	.2592685	.0239306	10.83	0.000	.2123201	.3062169
socfrend	.0441151	.0172725	2.55	0.011	.0102289	.0780013
_cons	3.585914	.1441992	24.87	0.000	3.303016	3.868812

We can specify the pr(), pe(), and forward options to perform a forward stepwise regression. In the example below, a forward stepwise regression is performed where the significance level for removal is 0.10 and the significance level for entry is 0.05.

```
. stepwise, pr(.10) pe(.05) forward: regress happy7 age educ children class
>    health socfrend socrel socommun socbar
                    begin with empty model
p = 0.0000 < 0.0500   adding    health
p = 0.0000 < 0.0500   adding    class
p = 0.0000 < 0.0500   adding    children
p = 0.0002 < 0.0500   adding    socfrend
p = 0.0130 < 0.0500   adding    socommun
p = 0.0430 < 0.0500   adding    socrel
```

Source	SS	df	MS	Number of obs	=	1,262
				F(6, 1255)	=	37.31
Model	188.507779	6	31.4179632	Prob > F	=	0.0000
Residual	1056.91298	1,255	.842161738	R-squared	=	0.1514
				Adj R-squared	=	0.1473
Total	1245.42076	1,261	.98764533	Root MSE	=	.91769

happy7	Coef.	Std. Err.	t	P>|t|	[95% Conf. Interval]	
health	.2592685	.0239306	10.83	0.000	.2123201	.3062169
class	.1939354	.0368361	5.26	0.000	.1216684	.2662025
children	.0826923	.0160225	5.16	0.000	.0512585	.1141261
socfrend	.0441151	.0172725	2.55	0.011	.0102289	.0780013
socommun	.0289651	.0132581	2.18	0.029	.0029545	.0549757
socrel	.0326699	.0161264	2.03	0.043	.0010322	.0643076
_cons	3.585914	.1441992	24.87	0.000	3.303016	3.868812

Although I am illustrating the use of stepwise models, I would urge you to consider the critiques of these methods. In fact, Stata has an excellent FAQ page describing some of the reasons to be skeptical about the usage of, and results from, stepwise models; see http://www.stata.com/support/faqs/statistics/stepwise-regression-problems/. One of the main concerns that I have about such models is the extent to which we can believe that the results would be replicated in another sample. I would highly discourage trusting models built using stepwise methods that have not been replicated on a sample independent of the one used for building the original stepwise model.

Note: Prefix commands

This chapter has illustrated the use of the `nestreg` prefix command for fitting nested regression models and the `stepwise` prefix command for fitting stepwise regression models. There are additional prefix commands beyond these commands. You can learn more about `prefix` commands by typing `help prefix`.

17.5 Closing thoughts

In my mind, this chapter really has two purposes. The obvious purpose was to illustrate how to fit multiple models using the same sample, how to fit nested regression models, and how to fit models using stepwise regression. But a deeper purpose was to illustrate how you can stitch together tools we have seen in previous chapters to perform tasks that are not so easily performed in isolation. For example, the section on fitting multiple models using the same sample illustrated how the `e(sample)` stored result can be assigned to a variable and then used to control the sample that is used for fitting other models. Furthermore, the section on fitting nested regression models illustrated how we can store the results of each block and then access those stored results to create a compact table of results using the `estimates table` command. Or you can go further and access these stored results to create a presentation-quality table using `esttab`. I hope these examples give you a flavor of what you can accomplish when drawing upon stored results (as described in section 15.5), storing results (as described in section 15.6), and presenting results (as described in chapter 16).

18 Regression diagnostics

18.1 Chapter overview

This chapter illustrates commands for performing regression diagnostics using Stata. The chapter begins with section 18.2, which illustrates tools that you can use with Stata for identifying outliers. The next section shows different methods that you can use for testing for nonlinearity (section 18.3). The following section describes how you can detect multicollinearity (section 18.4). This is followed by an illustration of how you can evaluate the homoskedasticity assumption (section 18.5). Then, I briefly consider how to evaluate the normality of the residuals (section 18.6). The chapter concludes with closing thoughts in section 18.7.

Note: Regression diagnostics

This chapter shows how to use tools within Stata for performing regression diagnostics. If you would like more information about the meaning of such diagnostics, guidelines for performing diagnostics, and remedies for dealing with misbehaved data, I would recommend consulting Fox (1991).

18.2   Outliers

The coefficients fit in a regression model can be influenced by a small number of observations or even a single observation. This section will illustrate commands that can be used within Stata for identifying outliers, especially those that can influence the regression coefficients.

For this section, let's use an example where we want to predict the level of education of the respondent as a function of the education of the respondent's father, the education of the respondent's mother, and the age of the respondent. For this study, we will focus only on males who are at least 25 years of age. The data for this analysis are contained in the `gss2012_sbs` dataset, which is read into memory with the `use` command below. Then, the `keep if` command is used to keep just the males who are at least 25 years old.

```
. use gss2012_sbs.dta, clear
. keep if age >=25 & female==0
(1,169 observations deleted)
```

Let's examine the distribution of the outcome and the predictors by using the `summarize` command. The `summarize` command below computes summary statistics for the respondent's education, the education of the respondent's father, the education of the respondent's mother, and the age of the respondent.

```
. summarize educ paeduc maeduc age
```

Variable	Obs	Mean	Std. Dev.	Min	Max
educ	803	13.58406	3.251525	0	20
paeduc	604	11.5298	4.333996	0	20
maeduc	699	11.55222	3.886064	0	20
age	805	50.39752	16.28343	25	89

We note that the values for `educ`, `maeduc`, and `paeduc` all range from 0 to 20. Examining these summary statistics is our first step in the data-screening process. We are looking for observations that seem out of range or miscoded. Noting the value of 0 for these variables, imagine that we consult with the data collection team and data entry team to inquire whether a value of 0 is a legitimate value. The team members let us know that a person who reports having never attended school is assigned a value of 0; thus 0 is a legitimate response code for years of education. The values for `age` range from 25 to 89, raising no red flags with respect to the values for `age`.

To further inspect the data before analysis, we can create a scatterplot matrix that shows all the scatterplots between each of these variables. A scatterplot matrix is like a visual correlation matrix. We create such a graph using the `graph matrix` command below. The resulting graph is shown in figure 18.1.

```
. graph matrix educ paeduc maeduc age
```

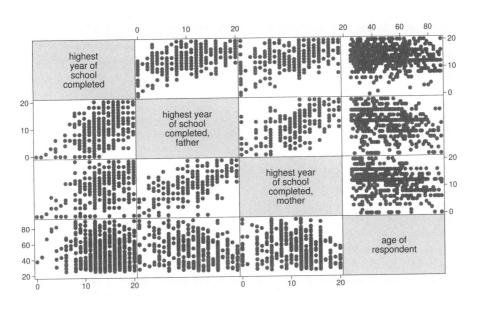

Figure 18.1: Scatterplot matrix of outcome and predictors

Looking at the scatterplots in figure 18.1, I focus on the plots in the top row, where the y axis is the outcome variable (`educ`) and the x axes are the predictor variables. This is structured this way because of how I ordered the variables on the `graph matrix` command, placing the outcome first, followed by the predictors. In looking at the scatterplots at the top, I see some observations that are outside of the pack but nothing that seems seriously concerning.

Let's now run the regression in which we predict `educ` from the father's education, the mother's education, and age. Remember that this is the subsetted dataset that contains just males who are at least 25 years old.

```
. regress educ paeduc maeduc age

      Source |       SS           df       MS            Number of obs   =       561
-------------+------------------------------              F(3, 557)       =     71.78
       Model |  1679.13343          3   559.711143        Prob > F        =    0.0000
    Residual |   4343.4655        557   7.7979632         R-squared       =    0.2788
-------------+------------------------------              Adj R-squared   =    0.2749
       Total |  6022.59893        560   10.7546409        Root MSE        =    2.7925

------------------------------------------------------------------------------
        educ |      Coef.   Std. Err.      t    P>|t|     [95% Conf. Interval]
-------------+----------------------------------------------------------------
      paeduc |   .3158505    .038816     8.14   0.000     .2396068    .3920941
      maeduc |    .14583    .0419552     3.48   0.001     .0634203    .2282398
         age |   .023109    .0078294     2.95   0.003     .0077302    .0384877
       _cons |   7.478382   .6230264    12.00   0.000     6.254613     8.70215
------------------------------------------------------------------------------
```

The results of this regression show that all three predictors are significant. Note that the sample size for this model is $N = 561$. This reflects the number of observations that had complete data on all variables and were included in the analysis.

Rather than focusing on the interpretation of the predictors, we will begin investigating for potentially outlying and influential observations.

18.2.1 Standardized residuals

Let's start by computing the standardized residuals for every observation included in this regression analysis. We can use the `predict` command with the `rstandard` option to create standardized residuals for every observation. The `predict` command below creates a new variable named `standres` that contains the standardized residual for every observation that was included in the analysis.[1] The standardized residuals are computed only for the observations included in the analysis because of the `if e(sample)` specification. (See section 15.4 for more information about `e(sample)`.)

```
. predict standres if e(sample), rstandard
(244 missing values generated)
```

---

1. Note that we could have used any valid variable name in place of `standres`. We could have typed `predict zres if e(sample), rstandard`, and the `predict` command would have created a new variable named `zres` that would have contained the standardized residual.

> **Note: Options on predict**
>
> In the `predict` command above, the option `rstandard` specifies that we want to compute standardized residuals. You can see a complete list of all the options that can be provided with the `predict` command by typing
>
>   . help regress postestimation
>
> and then clicking on the `predict` hotlink.

Let's examine the summary statistics for `standres` using the `summarize` command.

```
. summarize standres
    Variable |       Obs        Mean    Std. Dev.       Min        Max
-------------+--------------------------------------------------------
    standres |       561   -.0000349    1.001861  -3.028854   2.953894
```

Note how there are 561 observations corresponding to the number of observations in the regression analysis. Also note that the mean is approximately 0 and the standard deviation is approximately 1. This is why these are called standardized residuals. We see that the minimum and maximum values reflect some extreme values, -3.03 and 2.95. Let's use the user-written command called `extremes`[2] to inspect the five highest and five lowest values of `standres`.

```
. extremes standres

  obs:      standres

  208.     -3.028854
  488.     -3.006348
  364.      -2.78242
  695.     -2.713853
  692.     -2.704131


  642.      2.560448
  242.      2.608321
  509.       2.64558
  461.       2.90216
   21.      2.953894
```

This shows the five lowest values followed by the five highest values. I note that the highest value is not particularly higher than the second highest value, and so forth. Likewise, the lowest value is not substantially lower than the second lowest value, and so forth.

---

2. You can download this command by typing `ssc install extremes, replace`, as described in section 1.1.1.

Note: Cutoff values

Some regression books provide guidelines for cutoffs that identify values that merit further observation, such as a cutoff that means a standardized residual is so high that you should investigate it further. While such cutoffs can have their benefits, this chapter will focus on the extremity of the observations in the context of the other observations, a method I personally use in my work. For more perspective on this, I recommend the discussion by Fox (1991).

Let's add additional variables to the `extremes` command to show us the `id` associated with these extreme observations as well as the `educ`, `maeduc`, `paeduc`, and `age` variables associated with these observations.

```
. extremes standres id educ maeduc paeduc age
```

obs:	standres	id	educ	maeduc	paeduc	age
208.	-3.028854	512	0	0	0	38
488.	-3.006348	1202	8	20	16	39
364.	-2.78242	879	3	3	3	80
695.	-2.713853	1691	1	0	0	44
692.	-2.704131	1683	2	1	1	67

642.	2.560448	1562	18	0	6	67
242.	2.608321	585	20	0	14	42
509.	2.64558	1254	20	8	8	63
461.	2.90216	1123	20	6	9	32
21.	2.953894	62	19	5	3	71

This reveals something very interesting. Consider the response pattern for `id` 512 having the lowest standardized residual. The respondent reported 0 years of education as well as 0 years of education for the mother and father. This raises an interesting point. We earlier established that 0 is a legitimate response, but here we see a pattern that seems curious and might merit further investigation. That is, we might examine the entire response patterns for the respondent with the `id` value of 512 to determine whether we can further understand why all 3 of these measures of education are coded as 0. As we proceed forward, let's assume that we investigated this and found that we were satisfied that this response pattern seemed plausible.[3] Let's now consider some other statistics we can use to assess observations that might be outlying and influential.

---

3. It is often possible to use the internal consistency of responses in a survey to evaluate the plausibility of extreme responses or extreme response patterns. For example, I recently performed the data analysis for a survey in which we found a respondent who reported that she was 102 years old. I cross compared this value against the last time the respondent reported having her last menstrual period. She reported her last menstrual period was 71 years ago, a value quite consistent with being 102 years old.

18.2.2   Studentized residuals, leverage, Cook's D

Let's now consider three additional statistics we can use for identifying outlying observations, starting with the studentized residual. This is much like the standardized residual, except that the studentized residual is computed after omitting the observation in question. Thus, we compute the studentized residual for observation 1 with all the observations except observation 1. This has the benefit of identifying how extreme this observation is, omitting that observation from the computation of the regression coefficients. We can compute the studentized residual using the `predict` command with the `rstudent` option, as shown below. This `predict` command creates a new variable named `studres` that contains the studentized residual.

```
. predict studres if e(sample), rstudent
(244 missing values generated)
```

Let's use the `summarize` command to compute summary statistics for `studres`. I focus my attention on the minimum and maximum values. The `summarize` command shows that the studentized residuals range from −3.05 to 2.97. Let's keep these minimum and maximum values in mind as we later evaluate the studentized residuals.

```
. summarize studres
```

Variable	Obs	Mean	Std. Dev.	Min	Max
studres	561	−.0000556	1.003955	−3.051367	2.974632

Another measure of interest is the leverage of the observation. This refers to the degree to which the observation deviates from other observations with respect to the predictors. Observations that have high leverage have a greater potential to influence the regression coefficients because of their inclusion (versus exclusion) in the analysis. We can compute the leverage for each observation via the `predict` command combined with the `leverage` option. The `predict` command below creates a new variable called `lev` that contains the leverage for each observation included in the regression analysis.

```
. predict lev if e(sample), leverage
(244 missing values generated)
```

Let's then use the `summarize` command to create summary statistics for the variable `lev`. I note that the highest value is about 0.042.

```
. summarize lev
```

Variable	Obs	Mean	Std. Dev.	Min	Max
lev	561	.0071301	.0053642	.001798	.0419891

Let's compute one more measure that is useful for identifying influential observations, called Cook's *D*. This is an overall measure of the influence of an observation on the regression coefficients. Observations with greater values of Cook's D exert more influence on the regression coefficients. We can compute Cook's D for each observation via the `predict` command combined with the `cooksd` option. The `predict` command below creates a new variable named `d` that contains the value of Cook's D for each observation included in the regression analysis.

```
. predict d if e(sample), cooksd
(244 missing values generated)
```

Let's now use the `summarize` command to compute summary statistics for `d`, focusing on the maximum value. The highest value for Cook's D for this analysis is 0.075.

```
. summarize d
    Variable |        Obs        Mean    Std. Dev.         Min         Max
-------------+--------------------------------------------------------------
           d |        561     .0023007    .0060692    1.05e-08    .0745467
```

Note: Other options with the predict command

You can use other options with the `predict` command to compute other regression diagnostic statistics. Consider the three `predict` commands shown below.

```
. predict covrat, covratio
. predict mydf, dfits
. predict w, welsch
```

The first `predict` command creates the variable `covrat`, which contains the COV-RATIO statistic. The second `predict` command creates the variable named `mydf`, which contains the DFITS statistic. The third `predict` command creates the variable named `w`, which contains the Welsch distance statistic. You can type `help regress postestimation` and then click on the `predict` hotlink for more information.

Now, let's recap what we have done so far. We have used the `predict` command to create four new variables named `standres`, `studres`, `lev`, and `d`. Let's use the `describe` command for these variables, which shows the labels that describe the contents of each variable.

```
. describe standres studres lev d
              storage   display    value
variable name   type     format    label      variable label
-------------------------------------------------------------------------
standres        float    %9.0g                 Standardized residuals
studres         float    %9.0g                 Studentized residuals
lev             float    %9.0g                 Leverage
d               float    %9.0g                 Cook's D
```

Let's now show summary statistics for these four variables using the `summarize` command. This will help us to refer to the minimum and maximum values as we consider how extreme these values are in the following analyses.

```
. summarize standres studres lev d

    Variable |        Obs        Mean    Std. Dev.        Min        Max
-------------+--------------------------------------------------------------
    standres |        561   -.0000349    1.001861   -3.028854   2.953894
     studres |        561   -.0000556    1.003955   -3.051367   2.974632
         lev |        561    .0071301    .0053642    .001798   .0419891
           d |        561    .0023007    .0060692   1.05e-08   .0745467
```

Using the `extremes` command, we will look at the five observations that have the highest Cook's D value. At the same time, let's show the `id` of the observation, the standardized residual, studentized residual, and leverage.

```
. extremes d id standres studres lev, high n(5)

  obs:          d      id    standres      studres         lev
-------------------------------------------------------------------
  740.   .0314061    1805    -1.95731    -1.962312    .0317499
  642.   .0334173    1562    2.560448     2.573338    .0199818
  695.   .0411102    1691   -2.713853    -2.729521    .0218397
  208.   .0560533     512   -3.028854    -3.051367    .0238571
  242.   .0745467     585    2.608321     2.622041    .0419891
```

We can see that the observations that have the highest Cook's D correspond to observations that have among the highest (or *the* highest) residuals and leverage. Let's inspect these five observations more carefully by looking at their values for the outcome (that is, `educ`) and the predictors (that is, `paeduc`, `maeduc`, and `age`).

```
. extremes d id educ paeduc maeduc age, high n(5)

  obs:          d      id    educ    paeduc    maeduc    age
-------------------------------------------------------------------
  740.   .0314061    1805      8         5        16     86
  642.   .0334173    1562     18         6         0     67
  695.   .0411102    1691      1         0         0     44
  208.   .0560533     512      0         0         0     38
  242.   .0745467     585     20        14         0     42
```

These five observations fall into two patterns. The first pattern I see corresponds to the observations with `id` values of 512 and 1691. The observation for `id` = 512 appears as the observation with the second highest Cook's D, having 0 education for all 3 measures of education. Similarly, `id` = 1691 has 1 year of education for the respondent and 0 years of education for the mother and father.

The second pattern is observations that have values that have opposite patterns of extremes. The observation identified with `id = 585` has 20 years of education for the respondent but 0 years of education for the mother. Similarly, when `id = 1562`, the respondent has 18 years of education, and the mother has 0 years of education. Finally, the respondent identified with `id = 1805` is 86 years old with only 8 years of education, but the mother has 16 years of education.

18.2.3   Graphs of residuals, leverage, and Cook's D

Let's now explore some graphical techniques we can use with Stata to help us visually identify outliers. One commonly used graph is a leverage versus residual squared plot. Stata has a built-in command to create such a graph called `lvr2plot`, as shown below. The option `mlabel(id)` is added to label the observations with the variable (`id`). The graph created by this command is shown in figure 18.2.

```
. lvr2plot, mlabel(id)
```

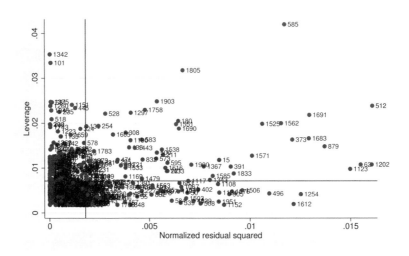

Figure 18.2: Leverage versus residual squared plot

In this kind of graph, influential observations are typified by having high leverage (on the y axis) combined with a high residual squared (on the x axis). Such observations are located in the upper right quadrant of the graph. Note how the upper right portion includes the observations that we identified in the previous `extremes` command; these have values of `id` of 585, 512, 1691, 1562, and 1805.

One of my favorite graphs to create is very similar to this but instead shows the studentized residual on the y axis, the leverage on the x axis, the markers sized proportional to Cook's D, and the observations identified using the `id` variable. (Such a graph is illustrated and described in Fox [1991].) This graph is created using the `twoway` command below[4] and is shown in figure 18.3.

```
. twoway (scatter studres lev [pw=d], msymbol(Oh))
>        (scatter studres lev       , msymbol(i) mlabel(id) mlabpos(center))
```

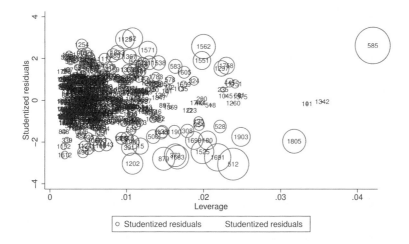

Figure 18.3: Leverage versus residual squared plot

I like this graph because it contains so much information. The most influential observations are designated with large circles (having greater values of Cook's D) and are located toward the top right corner (with a very positive residual and high leverage) or toward the bottom right corner (with a very negative residual and high leverage). As before, this graph calls attention to the same observations that we identified in the previous `extremes` command; these have values of `id` of 585, 512, 1691, 1562, and 1805.

18.2.4   DFBETAs and avplots

The statistics that we have considered so far, namely, the residuals, leverage, and Cook's D, do not specifically tell us about how an observation alters the coefficient of a particular predictor. By contrast, one of my favorite diagnostic statistics, DFBETA,

---

4. The first `scatter` command uses the `[pw=d]` specification to weight the size of the markers in proportion to the variable `d` and is combined with the `msymbol(Oh)` option to make the markers hollow circles. The second `scatter` command creates invisible markers (via the `msymbol(Oh)` option) and places the value of the variable `id` in the center via the combination of the `mlabel(id)` and `mlabpos(center)` options.

assesses the degree to which an observation changes the value of the regression coefficient because of the inclusion versus exclusion of the observation. We can compute a DFBETA value for each predictor in the model using the **dfbeta** command, shown below. The output shows that three variables are created—_dfbeta_1, which contains the DFBETA values for the predictor **paeduc**; _dfbeta_2, which contains the DFBETA values for **maeduc**; and _dfbeta_3, which contains the DFBETA values for **age**.

```
. dfbeta
(244 missing values generated)
                      _dfbeta_1: dfbeta(paeduc)
(244 missing values generated)
                      _dfbeta_2: dfbeta(maeduc)
(244 missing values generated)
                      _dfbeta_3: dfbeta(age)
```

Let's use the **summarize** command to compute summary statistics for _dfbeta_1, _dfbeta_2, and _dfbeta_3. In particular, let's note the minimum and maximum values for these DFBETA values. This will help us evaluate what constitutes an extreme value with respect to the minimum and maximum.

```
. summarize _dfbeta_1 _dfbeta_2 _dfbeta_3
```

Variable	Obs	Mean	Std. Dev.	Min	Max
_dfbeta_1	561	.0000437	.0468165	-.2186964	.3951509
_dfbeta_2	561	-.000035	.0537569	-.5319461	.2481079
_dfbeta_3	561	1.86e-06	.0456062	-.2832019	.2111628

Let's use a graphical technique (which is called an index plot) to graph the values of _dfbeta_1, _dfbeta_2, and _dfbeta_3 as a function of the variable **id**. The **mlabel(id id id)** option is used to label each of the three variables using **id**. The graph created by this command is shown in figure 18.4.

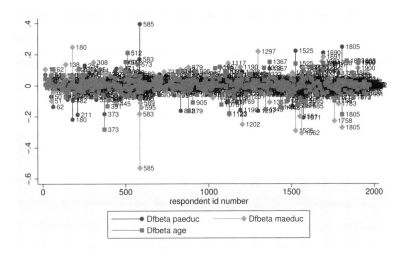

Figure 18.4: Index plot for DFBETAs

The legend at the bottom of figure 18.4 tells us that round markers are associated with the DFBETA for `paeduc`, diamonds are associated with the DFBETA for `maeduc`, and squares are associated with the DFBETA for `age`. We can see that for the observation labeled 585, there is an extremely high DFBETA value for `paeduc` (shown with a round marker near 0.4 at the top of the graph). This indicates that the inclusion of this observation increases the size of the coefficient for `paeduc`. For the observation labeled 585, we can also see an extremely low DFBETA value for `maeduc` (shown with a diamond marker near −0.6 at the bottom of the graph). This indicates that the inclusion of this observation decreases the size of the coefficient for `maeduc`.

I like to examine the index plot shown in figure 18.4 in combination with a graph of the added variable plots. The added variable plots show the relationship between each predictor and the outcome after accounting for all other predictors in the model. Using the `avplots` command below creates the graph shown in figure 18.5.

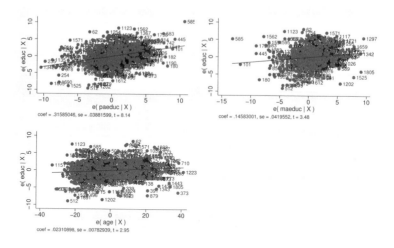

Figure 18.5: Added variable plot

Let's cross compare what we see in figures 18.4 and 18.5, focusing on the observation labeled as 585. Let's compare what we see for this observation with respect to the DFBETA for `paeduc` (the round observation in figure 18.4) with what we see in the top left panel of figure 18.5. Note how the DFBETA for this value is positive in figure 18.4, which indicates that it increases the coefficient for `paeduc`. Compare that with the position of this observation in the added variable plot in figure 18.5. In the added variable plot, the observation labeled 585 is in the top right corner, acting as a lever to increase the tilt of the regression line. Both of these plots show us how the observation labeled 585 is acting to increase the coefficient associated with `paeduc`.

Let's now perform the same kind of cross comparison for the observation labeled 585 but with respect to `maeduc`. This means comparing the very negative value shown as a diamond and labeled 585 in figure 18.4 with the observation labeled 585 in the top right panel of figure 18.5. The DFBETA for `maeduc` for this observation is negative (see figure 18.4). In the top right panel of figure 18.5, we can see that the observation labeled 585 is acting like a lever to diminish the tilt of the regression line, reducing the size of the coefficient for `maeduc`. In both of these cases, we can see that the observation labeled 585 is reducing the size of the coefficient for `maeduc`.

Although we have seen this observation before, pretend that we are not familiar with the values associated with the observation labeled 585. We can use the `list` command to show the exact values associated with this observation. Using the `list` command (shown below), we can identify the exact values for the outcomes and predictor for this observation and double check these values. It is possible that this observation might be influential because of an error (for example, a data entry error). In this case, assume we performed such checks and the data are correct.

```
. list id educ paeduc maeduc age if id==585
```

	id	educ	paeduc	maeduc	age
242.	585	20	14	0	42

I hope this illustrates why I like using these two graphs in combination. I find that the graph in figure 18.4 is very useful for identifying observations that are influencing a particular coefficient, but I cannot quite visualize how (or why) this observation is influencing the coefficient. So I can switch to figure 18.5, where I can more concretely understand how the particular observation is influencing the regression coefficient. I then often use the `list` command to inspect the values associated with the outlying observation of interest.

18.2.5 Running a regression with and without observations

One of the ways that I like to ultimately assess the influence of specific outlying observations is to run a regression analysis that includes all observations and compare that result with a regression analysis that omits one or more suspicious outlying observations. This sensitivity analysis tells us how robust (or sensitive) the results are with respect to the inclusion versus exclusion of the specific outlying observations.

Our investigation has suggested that the observation with the `id` value of 585 is a potentially influential observation. In addition, let's say the observations with the `id` values of 512 and 1691 are also potentially influential observations. These are the observations in which either all education values were zero or two of the education values were zero and the other education value was one.

Let's start this process by rerunning the original regression (with the output omitted to save space) and then storing the results of the regression with the name `original`.

```
. regress educ paeduc maeduc age
  (output omitted)
. estimates store original
```

Now, let's run this model again but omit the observations where the value of `id` is 585, 512, or 1691. The output of this is omitted to save space. The results of this model are stored using the name `deleted3` to indicate that it has deleted 3 observations.

```
. regress educ paeduc maeduc age if !inlist(id,585,512,1691)
  (output omitted)
. estimates store deleted3
```

Now, using the `estimates table` command, we can compare the results with and without these three observations. (See section 16.2 for more information about the `estimates table` command.)

```
. estimates table original deleted3, drop(_cons) b(%6.3f)
> star(0.05 0.01 0.001) stats(N r2)
```

Variable	original	deleted3
paeduc	0.316***	0.290***
maeduc	0.146***	0.151***
age	0.023**	0.021**
N	561	558
r2	0.279	0.259

```
legend: * p<.05; ** p<.01; *** p<.001
```

Even though these three observations seemed very problematic, we can see that the results of the original model are extremely similar to the results of the model in which these three observations are deleted. The size of the regression coefficients are very similar between the two models, and the R-squared is also very similar between the two models. I would conclude that the results of the original model are robust with respect to the inclusion (versus exclusion) of these three observations. Further sensitivity analyses could be performed using this same process to further examine the robustness of the original regression results.

Let's now turn our attention to the issue of examining the linearity assumption.

18.3   Nonlinearity

18.3.1   Checking for nonlinearity graphically

This section illustrates graphical approaches for checking for nonlinearity in the relationship between a predictor and an outcome variable. These approaches include 1) examining scatterplots of the predictor and outcome; 2) examining residual versus fitted plots; 3) creating plots based on locally weighted smoothers; and 4) plotting the mean of the outcome for each level of the predictor. Each of these approaches is considered in turn, starting with the use of scatterplots.

18.3.2   Using scatterplots to check for nonlinearity

Let's use a subset of the auto dataset called `autosubset`. It is the same as the `auto.dta` dataset that comes with Stata, but I have omitted three cars to make the nonlinear relationships a bit more clear.

```
. use autosubset
(1978 Automobile Data)
```

Let's look at a scatterplot of the size of the engine (`displacement`) by length of the car (`length`) with a line showing the linear fit, as shown in figure 18.6.

```
. graph twoway (scatter displacement length) (lfit displacement length),
> ytitle("Engine displacement (cu in.)") legend(off)
```

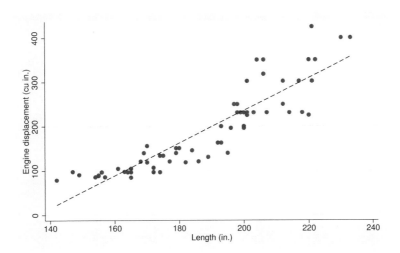

Figure 18.6: Scatterplot of engine displacement by length of car with linear fit line

The relationship between these two variables looks fairly linear, but the addition of the linear fit line helps us to see the nonlinearity. Note how for short cars (when length is below 160) that the fit line underpredicts and how for longer cars (when length is above 210) that the fit line also underpredicts.

Using a scatterplot like this can be a simple means of looking at the linearity of the simple relationship between a predictor and an outcome variable. However, this does not account for other predictors that you might want to include in a model. To this end, let's next look at how we can use the residuals for checking linearity.

18.3.3   Checking for nonlinearity using residuals

We can check for nonlinearity by looking at the relationship between the residuals and predicted values after accounting for other variables in the model. For example, let's run a regression predicting `displacement` from `length`, `trunk`, and `weight`, as shown below.

```
. regress displacement length trunk weight
```

Source	SS	df	MS		Number of obs	=	71
					F(3, 67)	=	189.26
Model	535697.677	3	178565.892		Prob > F	=	0.0000
Residual	63213.9848	67	943.49231		R-squared	=	0.8945
					Adj R-squared	=	0.8897
Total	598911.662	70	8555.88089		Root MSE	=	30.716

displacement	Coef.	Std. Err.	t	P>|t|	[95% Conf.	Interval]
length	-.7790159	.5796436	-1.34	0.183	-1.935989	.3779577
trunk	1.539695	1.230105	1.25	0.215	-.9156062	3.994995
weight	.1276394	.0155864	8.19	0.000	.0965288	.15875
_cons	-67.78624	61.93516	-1.09	0.278	-191.4093	55.83686

We can then look at the residuals versus the fitted values, as shown in figure 18.7. Note the U-shaped pattern of the residuals. This kind of pattern suggests that the relationship between the predictors and outcome is not linear.

```
. rvfplot
```

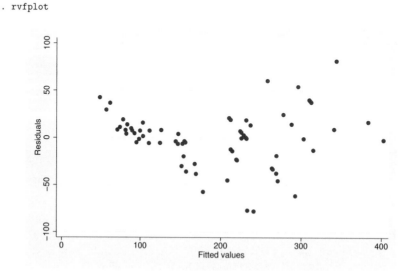

Figure 18.7: Residuals versus fitted values

For more information about the `rvfplot` command, see `help rvfplot`.

18.3.4   Checking for nonlinearity using a locally weighted smoother

With larger datasets, it can be harder to visualize nonlinearity using scatterplots or residual versus fitted plots. To show this, we will use an example from the gss2012_sbs dataset, which has a much larger number of observations.

```
. use gss2012_sbs
```

Suppose we want to determine the nature of the relationship between age (age) and education level (educ). The scatter command can be used (as shown below) to create a scatterplot of these two variables. The resulting scatterplot is shown in figure 18.8.

```
. scatter educ age, msymbol(oh)
```

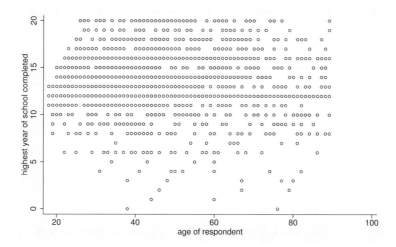

Figure 18.8: Scatterplot of education by age

It is very hard to discern the nature of the relationship between age and educ using this scatterplot. With such a large number of observations, the scatterplot is saturated with data points creating one big blotch that tells us very little about the shape of the relationship between the predictor and outcome.

Let's try a different strategy where we create a graph that shows the relationship between educ and age using a locally weighted regression (also called a locally weighted smoother, or lowess). The lowess command below creates a graph showing the locally weighted regression of educ on age, as shown in figure 18.9. The msymbol(p) option displays the markers using tiny points.

. lowess educ age, msymbol(p)

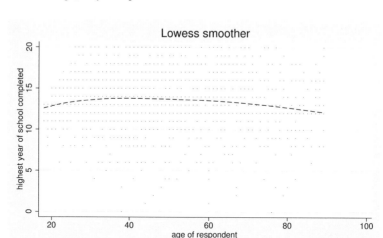

Figure 18.9: Locally weighted regression of education on age

The lowess graph suggests that there is nonlinearity in the relationship between `age` and `educ`. Education increases with age until about age 45, at which point the smoothed education values level out and then actually begin to decline. The graph produced by the `lowess` command is much more informative than the scatterplot alone.

18.3.5   Graphing an outcome mean at each level of predictor

Another way to visualize the relationship between the predictor and outcome is to create a graph showing the mean of the outcome at each level of the predictor. Using the example predicting `educ` from `age`, we would create a graph of the average of `educ` at each level of `age`. Although there are a large number of values the variable `age` can assume (from 18 to 89), the variable is composed of discrete integers with a reasonably large number of observations (usually more than 110) for each value. In such a case, we can explore the nature of the relationship between the predictor and outcome by creating a graph of the mean of the outcome variable (that is, education level) for each level of the predictor (that is, age). This kind of graph imposes no structure on the shape of the relationship between `age` and `educ` and allows us to observe the nature of the relationship between the predictor and the outcome.[5]

---

5. This technique requires that the predictor have whole number (integer) values. If the predictor were income measured to the penny, you could create income rounded to the nearest dollar. However, to have a large number of observations per income category, you might be even better off using income rounded to the nearest thousand dollars or income rounded to the nearest ten thousand dollars.

One simple way to create such a graph is to fit a regression model predicting the outcome and treating the predictor variable as a categorical (factor) variable. Following that, we use the `margins` command to obtain the predicted mean of the outcome for each level of the predictor. In this case, the predicted mean reflects the mean education level at each level of age.

```
. regress educ i.age
  (output omitted)
. margins age
  (output omitted)
```

Now, we can use the `marginsplot` command (shown below) to graph the average level of education as a function of age. The resulting graph is shown in figure 18.10.

```
. marginsplot
  Variables that uniquely identify margins: age
```

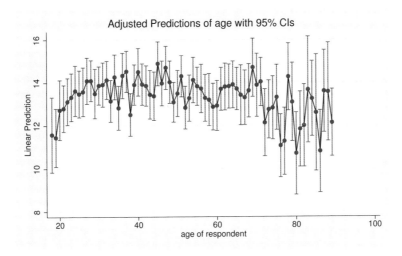

Figure 18.10: Mean education by age

This graph shows that education level generally increases with age until about age 45, where the level of education plateaus and then begins to decline. This is consistent with the graph created by the `lowess` command and shown in figure 18.9.

The predicted means vary erratically as age approaches 80 and beyond. This is due to the small number of observations in each of these age groups. For these ages, the confidence intervals are much wider compared with those of previous ages, which reflects greater uncertainty of the estimates because of the smaller number of observations.

If all the years had such a small number of observations, then the entire graph might be dominated by wild swings in the means and show little about the nature of the relationship between the predictor and outcome. For such a case, we could try grouping the predictor into larger bins such as by creating a variable that contains the decade of age (for example, 30 for 30–39, 40 for 40–49) based on the age. The gss2012_sbs dataset has a variable called `agedec`. The tabulation of this variable is shown below.[6]

```
. tabulate agedec

    Age (as
  decade) of
  respondent |      Freq.     Percent        Cum.
-------------+-----------------------------------
      18-20s |        331       16.81       16.81
         30s |        389       19.76       36.57
         40s |        357       18.13       54.70
         50s |        337       17.12       71.81
         60s |        285       14.47       86.29
         70s |        172        8.74       95.02
         80s |         98        4.98      100.00
-------------+-----------------------------------
       Total |      1,969      100.00
```

Let's graph the mean of `educ` by `agedec` by first using the `regress` command to predict `educ` from `i.agedec`. Then, we will use the `margins` command to compute the predicted mean of education as a function of `agedec` and then the `marginsplot` command to show the predicted mean of `educ` by `agedec`, as shown in figure 18.11.

---

6. I included people who are 18 and 19 years old in the bin with people in their 20s.

```
. regress educ i.agedec
  (output omitted)
. margins agedec
  (output omitted)
. marginsplot
  Variables that uniquely identify margins: agedec
```

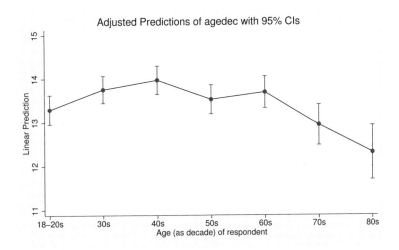

Figure 18.11: Mean education by decade of age

For each decade of age, the estimate of the mean of education is much more stable because of the higher sample size in each decade (compared with each year). However, if there were important changes in trend within a decade, this graph would conceal that information.

18.3.6 Summary

The previous sections have illustrated a number of ways to check for nonlinearity for a continuous predictor using visual approaches—namely, scatterplots, examination of residuals, and locally weighted smoothers—and examining the predicted means for each level of the predictor by treating the predictor as a categorical variable.

18.3.7 Checking for nonlinearity analytically

This section shows you how to check for nonlinearity using analytic approaches, namely, by adding power terms and by using factor variables.

Adding power terms

Another way to check for nonlinearity of a continuous variable is to add power (for example, quadratic, cubic) terms. Let's continue using the example looking at education as a function of the age of the respondent. Let's start by showing the linear model predicting educ from age.

```
. use gss2012_sbs

. regress educ age
      Source |       SS           df       MS       Number of obs   =     1,967
-------------+----------------------------------   F(1, 1965)      =      5.51
       Model |  53.7375843         1  53.7375843   Prob > F        =    0.0190
    Residual |   19166.664     1,965   9.7540275   R-squared       =    0.0028
-------------+----------------------------------   Adj R-squared   =    0.0023
       Total |  19220.4016     1,966  9.77639961   Root MSE        =    3.1231

-------------+----------------------------------------------------------------
        educ |      Coef.   Std. Err.      t    P>|t|     [95% Conf. Interval]
-------------+----------------------------------------------------------------
         age |  -.0093527   .0039846    -2.35   0.019    -.0171672   -.0015381
       _cons |   13.97678   .2044816    68.35   0.000     13.57576     14.3778
------------------------------------------------------------------------------
```

Clearly, this model has a strong linear component, which indicates that educ tends to increase as age increases. However, there may be nonlinearity in this relationship as well. Let's introduce a quadratic term (in addition to the linear term) by adding c.age#c.age to the model. This introduces a quadratic effect that would account for a single bend in the line relating age to educ. We would expect the quadratic term to be significant based on the graphs we saw in figures 18.9 and 18.10, which showed that there was a bend in the relationship between age and educ. The model with the linear and quadratic terms is shown below.[7]

```
. regress educ age c.age#c.age
      Source |       SS           df       MS       Number of obs   =     1,967
-------------+----------------------------------   F(2, 1964)      =     16.26
       Model |  313.072572         2  156.536286   Prob > F        =    0.0000
    Residual |  18907.3291     1,964  9.62694962   R-squared       =    0.0163
-------------+----------------------------------   Adj R-squared   =    0.0153
       Total |  19220.4016     1,966  9.77639961   Root MSE        =    3.1027

-------------+----------------------------------------------------------------
        educ |      Coef.   Std. Err.      t    P>|t|     [95% Conf. Interval]
-------------+----------------------------------------------------------------
         age |   .1023029   .0218739     4.68   0.000     .0594045    .1452013
             |
 c.age#c.age |  -.0010995   .0002118    -5.19   0.000     -.001515   -.0006841
             |
       _cons |   11.49294   .5198925    22.11   0.000     10.47334    12.51254
------------------------------------------------------------------------------
```

---

7. This model can also be specified as regress educ c.age##c.age.

Indeed, the quadratic term is significant in this model, which confirms what we graphically saw in figures 18.9 and 18.10. This provides analytic support for including a quadratic term for `age` when predicting `educ`.

A cubic term would imply that the line fitting `age` and `educ` has a tendency to have two bends in it. We can test for a cubic trend by specifying the model as shown below. Including `c.age##c.age##c.age` as a predictor is a shorthand for including the linear, quadratic, and cubic terms for `age`.

```
. regress educ c.age##c.age##c.age
```

Source	SS	df	MS		
Model	348.608072	3	116.202691		
Residual	18871.7936	1,963	9.61375117		
Total	19220.4016	1,966	9.77639961		

	Number of obs	=	1,967
	F(3, 1963)	=	12.09
	Prob > F	=	0.0000
	R-squared	=	0.0181
	Adj R-squared	=	0.0166
	Root MSE	=	3.1006

educ	Coef.	Std. Err.	t	P>|t|	[95% Conf. Interval]	
age	.2659484	.0878795	3.03	0.003	.0936014	.4382954
c.age#c.age	-.0045108	.0017869	-2.52	0.012	-.0080153	-.0010064
c.age#c.age#c.age	.0000218	.0000113	1.92	0.055	-4.37e-07	.0000439
_cons	9.142665	1.328279	6.88	0.000	6.53768	11.74765

The cubic term is not statistically significant ($p = 0.055$). Even if this were significant (say, $p = 0.049$), I would note that this dataset has a very large number of observations, so the model has the statistical power to detect very small effects. In fact, note how the R-squared for the cubic model is 0.0181 compared with 0.0163 for the quadratic model. This is a very trivial increase, suggesting that the cubic trend is not really an important term to include in this model.

We could continue this exploration process by adding power terms, searching for additional nonlinear components, but for this example, it seems that the major nonlinear component is strictly quadratic.

Using factor variables

The strategy of using factor variables to check for nonlinearity only makes sense when you have a relatively limited number of levels of the predictor that are coded as whole numbers. We have such an example in the predictor variable named `agedec`, which is the age of the respondent expressed as decade of life (for example, 30s, 40s, 50s, etc.). Let's explore the relationship between the decade of life and the years of education. We previously saw nonlinearities in the relationship between age and education, and in this section, we explore this by using the variable `agedec` as the age of the person (as a decade).

Below the `use` command reads in the `gss2012_sbs` dataset. Then, the `tabulate` command is used to show the distribution of `agedec`.

```
. use gss2012_sbs

. tabulate agedec
```

Age (as decade) of respondent	Freq.	Percent	Cum.
18-20s	331	16.81	16.81
30s	389	19.76	36.57
40s	357	18.13	54.70
50s	337	17.12	71.81
60s	285	14.47	86.29
70s	172	8.74	95.02
80s	98	4.98	100.00
Total	1,969	100.00	

Let's start by graphing the average years of education by each decade of life (using the strategy that was illustrated in section 18.3.5). The graph of the relationship between the mean of `educ` at each level of `agedec` is shown in figure 18.12.

```
. regress educ i.agedec
(output omitted)

. margins agedec
(output omitted)

. marginsplot
Variables that uniquely identify margins: agedec
```

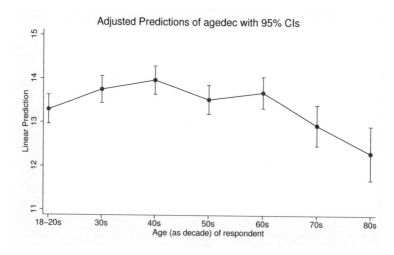

Figure 18.12: Average education by age (as a decade)

This graph shows nonlinearity in this relationship, but let's perform an analysis to detect any kind of nonlinearity in the relationship between decade of age and years of education. One approach is to formulate a regression model (as shown below) that includes both `c.agedec` and `i.agedec` as predictors in the same model.

```
. regress educ c.agedec i.agedec
note: 7.agedec omitted because of collinearity
```

Source	SS	df	MS
Model	318.006614	6	53.0011024
Residual	18902.395	1,960	9.64407909
Total	19220.4016	1,966	9.77639961

```
Number of obs =     1,967
F(6, 1960)    =      5.50
Prob > F      =    0.0000
R-squared     =    0.0165
Adj R-squared =    0.0135
Root MSE      =    3.1055
```

educ	Coef.	Std. Err.	t	P>|t|	[95% Conf. Interval]	
agedec	-.1644657	.0597587	-2.75	0.006	-.281663	-.0472684
agedec						
30s	.6241772	.2186029	2.86	0.004	.1954588	1.052896
40s	1.00196	.2260234	4.43	0.000	.5586885	1.445231
50s	.7255145	.2464901	2.94	0.003	.2421043	1.208925
60s	1.053018	.2850692	3.69	0.000	.4939477	1.612089
70s	.4681164	.3548572	1.32	0.187	-.2278207	1.164054
80s	0	(omitted)				
_cons	13.46054	.2059596	65.36	0.000	13.05662	13.86446

This unconventional-looking model divides up the relationship between `agedec` and the outcome into two pieces: the linear relationship, which is accounted for by `c.agedec`, and any remaining nonlinear components, which are explained by the indicator variables specified by `i.agedec`. Because of the inclusion of `c.agedec`, one of those indicators was no longer needed, and Stata excluded it for us. Let's now use the `testparm` command to perform a test of the indicator variables and to give us an overall test of the nonlinearity in the relationship between `agedec` and `educ`.

```
. testparm i.agedec

 ( 1)   2.agedec = 0
 ( 2)   3.agedec = 0
 ( 3)   4.agedec = 0
 ( 4)   5.agedec = 0
 ( 5)   6.agedec = 0

       F(  5,  1960) =    5.28
           Prob > F =    0.0001
```

This test is significant [$F(5, 1960) = 5.28$, $p = 0.001$], which indicates that overall, there is a significant contribution of nonlinear terms in the relationship between `agedec` and `educ`. This general strategy tells us that there is nonlinearity between `agedec` and `educ` but does not pinpoint the exact nature of the nonlinearity. Let's try another strategy that will help us pinpoint the nature of the nonlinearity. We start by running a model predicting `educ` from `i.agedec`.

```
. regress educ i.agedec
  (output omitted)
```

We now can use the `contrast` command with the `p.` contrast operator to obtain a detailed breakdown of all the different possible nonlinear trends in the relationship between `agedec` and `educ`.

```
. contrast p.agedec
Contrasts of marginal linear predictions
Margins        : asbalanced
```

	df	F	P>F
agedec			
(linear)	1	15.36	0.0001
(quadratic)	1	19.48	0.0000
(cubic)	1	0.04	0.8400
(quartic)	1	0.25	0.6194
(quintic)	1	0.26	0.6084
(sextic)	1	3.38	0.0662
Joint	6	5.50	0.0000
Denominator	1960		

	Contrast	Std. Err.	[95% Conf.	Interval]
agedec				
(linear)	-.3475788	.0886812	-.5214982	-.1736593
(quadratic)	-.3739165	.0847187	-.5400648	-.2077682
(cubic)	.0162022	.0802305	-.1411437	.1735482
(quartic)	-.0377058	.0759018	-.1865624	.1111509
(quintic)	.0362714	.0707777	-.1025361	.175079
(sextic)	.1213647	.0660246	-.008121	.2508505

These results show that in the relationship between `agedec` and `educ`, there are significant linear and quadratic effects. None of the other nonlinear trends are significant ($p > 0.05$). If we wished, we could limit our examination to just the nonlinear effects, as shown below.

```
. contrast p(2/6).agedec
Contrasts of marginal linear predictions
Margins        : asbalanced
```

	df	F	P>F
agedec			
(quadratic)	1	19.48	0.0000
(cubic)	1	0.04	0.8400
(quartic)	1	0.25	0.6194
(quintic)	1	0.26	0.6084
(sextic)	1	3.38	0.0662
Joint	5	5.28	0.0001
Denominator	1960		

	Contrast	Std. Err.	[95% Conf. Interval]	
agedec				
(quadratic)	-.3739165	.0847187	-.5400648	-.2077682
(cubic)	.0162022	.0802305	-.1411437	.1735482
(quartic)	-.0377058	.0759018	-.1865624	.1111509
(quintic)	.0362714	.0707777	-.1025361	.175079
(sextic)	.1213647	.0660246	-.008121	.2508505

Note how the test labeled `Joint` $[F(5, 1960) = 5.28, p = 0.001]$ matches the test we performed earlier regarding the contribution of the indicator variables (`i.agedec`) in the presence of the continuous version (`c.agedec`). Both of these tests ask the same question regarding the nonlinear contributions.

These tests suggest that linear and quadratic terms may be appropriate to use when you model the relationship predicting `educ` from `agedec`. You can see section 5.8 for more about the use of the `contrast` command with the `p.` contrast operator for performing polynomial contrasts.

18.4   Multicollinearity

Suppose we want to predict how many minutes one uses the web per week as a function of one's age (in decades, for example, 20s, 30s, 40s), education, and social class. We fit such a model as shown below.

```
. use gss2012_sbs

. regress webhh agedec educ class
```

Source	SS	df	MS		
Model	10389.0033	3	3463.00108		
Residual	175331.155	1,029	170.389849		
Total	185720.158	1,032	179.961394		

	Number of obs	=	1,033
F(3, 1029)	=	20.32
Prob > F	=	0.0000
R-squared	=	0.0559
Adj R-squared	=	0.0532
Root MSE	=	13.053

webhh	Coef.	Std. Err.	t	P>|t|	[95% Conf. Interval]	
agedec	-1.852324	.2612818	-7.09	0.000	-2.36503	-1.339618
educ	.6085497	.1585857	3.84	0.000	.2973613	.919738
class	-.1040551	.6213602	-0.17	0.867	-1.323333	1.115223
_cons	7.303095	2.29471	3.18	0.002	2.800249	11.80594

After fitting this model, we decide that how much one uses the web might depend on the year one was born, because it might relate to the kind of technology that you grew up with. This new model is fit below.

```
. regress webhh agedec educ class yrborn
```

Source	SS	df	MS
Model	11104.8438	4	2776.21095
Residual	174615.314	1,028	169.859255
Total	185720.158	1,032	179.961394

	Number of obs	=	1,033
F(4, 1028)	=	16.34
Prob > F	=	0.0000
R-squared	=	0.0598
Adj R-squared	=	0.0561
Root MSE	=	13.033

webhh	Coef.	Std. Err.	t	P>|t|	[95% Conf. Interval]	
agedec	.9866019	1.40729	0.70	0.483	-1.774887	3.748091
educ	.6266683	.1585844	3.95	0.000	.3154822	.9378545
class	-.1308061	.6205289	-0.21	0.833	-1.348454	1.086842
yrborn	.2833949	.1380475	2.05	0.040	.0125079	.5542819
_cons	-559.0759	275.9044	-2.03	0.043	-1100.476	-17.67585

My friend, who was looking over my shoulder while I did this, suggests that I try using the estat vif command to check to see whether I have an issue with multi-collinearity. This command with its output is shown below.

```
. estat vif
```

Variable	VIF	1/VIF
yrborn	30.11	0.033207
agedec	30.10	0.033221
class	1.15	0.867114
educ	1.15	0.873276
Mean VIF	15.63	

The output shows that the variables `agedec` and `yrborn` have very high variance inflation factor values (values over 10 or 20 deserve further consideration). This points out that these two variables are really the same thing. My experience is that genuine multicollinearity rarely arises from the inclusion of separate variables that are highly correlated. Instead, I find that it arises when you unintentionally include two variables that are literally measuring the same thing.

18.5   Homoskedasticity

Before we consider homoskedasticity, let's first consider its related assumption from the tradition of analysis of variance (ANOVA), the homogeneity of variance assumption.[8] The homogeneity of variance assumption states the variance of the dependent variable is equal across the levels of the independent variable. The homoskedasticity assumption is a generalization of this assumption to regression models.[9] The homoskedasticity assumption states that the variance of the residuals around the regression line is equal across all values of the predictors. One way to visually assess this assumption is to examine the distribution of the residuals as a function of the predicted values in a residuals versus fitted plot (Fox 1991).

Let's consider an example where our outcome variable is the number of minutes one spends on the world wide web per day, which we are predicting as a function of age, education, and social class. We first fit a model that predicts the number of hours one spends on the web from age, education, and social class using the `regress` command below.

```
. use gss2012_sbs

. regress webhh age educ class
```

Source	SS	df	MS		Number of obs	=	1,033
					F(3, 1029)	=	21.64
Model	11021.3593	3	3673.78643		Prob > F	=	0.0000
Residual	174698.799	1,029	169.775315		R-squared	=	0.0593
					Adj R-squared	=	0.0566
Total	185720.158	1,032	179.961394		Root MSE	=	13.03

webhh	Coef.	Std. Err.	t	P>|t|	[95% Conf. Interval]	
age	-.188292	.0255841	-7.36	0.000	-.2384949	-.1380891
educ	.6222584	.1584205	3.93	0.000	.3113943	.9331224
class	-.1115604	.6197681	-0.18	0.857	-1.327714	1.104593
_cons	9.888098	2.373387	4.17	0.000	5.230866	14.54533

Let's inspect the distribution of the residuals as a function of the fitted values to check for heteroskedasticity. The `rvfplot` command creates such a graph (as shown in

---

8. You can refer back to section 10.3.2 for more information about the homogeneity of variance assumption in ANOVA.

9. Or, more correctly, the homogeneity of variance assumption in ANOVA is a special case of the homoskedasticity assumption.

figure 18.13) in which the residuals are shown on the y axis and the fitted values are shown on the x axis.

```
. rvfplot, yline(0)
```

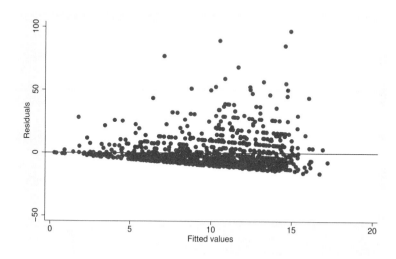

Figure 18.13: Residual versus fitted plot

The plot in figure 18.13 clearly shows that there is an issue of heteroskedasticity. The variance of the residuals is very small when the fitted values are small (for example, between 0 and 5) and is substantially greater when the fitted values are large (for example, 10 and above).

Note: Tests for heteroskedasticity

I am not a big fan of statistical tests regarding the distribution of variables (for example, tests of normality, or heteroskedasticity). Nevertheless, Stata offers two different commands for the test of heteroskedasticity, the `estat hettest` command and the `estat imtest` command. See `help hettest` and `help imtest` for more information.

Just as violation of the homogeneity of variance assumption can lead to untrustworthy p-values (Wilcox 1987), violation of the homoskedasticity assumption can likewise lead to untrustworthy p-values in regression models (Lumley et al. 2002). One way to address this issue is to include the vce(robust) option to obtain robust estimates of standard errors, which leads to trustworthy p-values.[10] Let's fit this model again using robust standard errors by adding the vce(robust) option, as shown below.

```
. regress webhh age educ class, vce(robust)
Linear regression                               Number of obs   =       1,033
                                                F(3, 1029)      =       24.64
                                                Prob > F        =      0.0000
                                                R-squared       =      0.0593
                                                Root MSE        =       13.03
```

webhh	Coef.	Robust Std. Err.	t	P>|t|	[95% Conf. Interval]	
age	-.188292	.0231973	-8.12	0.000	-.2338114	-.1427726
educ	.6222584	.1599957	3.89	0.000	.3083033	.9362135
class	-.1115604	.6864	-0.16	0.871	-1.458464	1.235343
_cons	9.888098	2.354389	4.20	0.000	5.268147	14.50805

Comparing the results using the vce(robust) option with the previous results (omitting that option), we see that the coefficients for the two models are identical but that the standard errors using the vce(robust) option are different. Despite the differences in the standard errors, the pattern of significance remains the same.

Note: More information on robust standard errors

You can find more information about how Stata estimates robust standard errors by typing help vce_option and then looking for the section about the vce(robust) option. Within that section, you can click on the link *Obtaining robust variance estimates*. That will take you to a section of the manual that describes how the vce(robust) option uses methods independently developed by Huber and White (see Huber [1967] and White [1980]), and this is often called a "sandwich estimator". The key is that the methods used by applying the vce(robust) option yield appropriate p-values even in the presence of heteroskedasticity.

18.6 Normality of residuals

This assumption states that the residuals are normally distributed. There are many different thoughts expressed about this assumption when fitting ordinary least-squares

---

10. Other remedies include checking the dataset for outliers or considering transformations of the outcome or predictors (see Fox [1991]).

regression models. Rather than describe the differing thoughts, I would recommend seeing Fox (1991), Weisberg (2014), and Lumley et al. (2002).

To illustrate how to evaluate the normality of residuals assumption, I will continue to use the example from the previous section in which the number of hours spent using the Internet is predicted as a function of age, education, and social class. The dataset for this example is read into memory with the `use` command below, and the `regress` command is used to fit this model (but the output is omitted to save space).

```
. use gss2012_sbs
. regress webhh age educ class
  (output omitted)
```

We can compute the standardized residuals for this model using the `predict` command with the `rstand` option.

```
. predict standres, rstand
(941 missing values generated)
```

Let's examine the distribution of these standardized residuals using the `kdensity` command with the `normal` option to show a normal curve overlay. The `kdensity` command creates the graph shown in figure 18.14.

```
. kdensity standres, normal
```

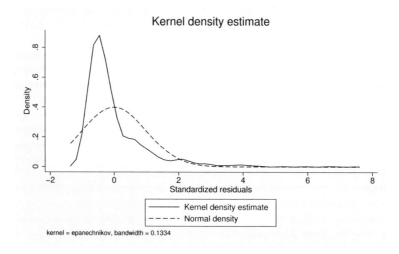

Figure 18.14: Distribution of residuals with normal overlay

It is clear to the eye that the distribution of the residuals from this regression analysis is substantially nonnormal.

> **Note: Tests for normality**
>
> If you wish to perform a statistical test regarding the normality of the residuals, you can use the `sktest` command or the `swilk` command. See `help sktest` and `help swilk` for more information.

18.7   Closing thoughts

This chapter has introduced commonly used Stata commands for performing regression diagnostics. This has included commands for checking for outliers, checking for nonlinearity, multicollinearity, homoskedasticity, and normality. For more details about the assumptions of regression analysis, I would recommend consulting your favorite regression book, Fox (1991) or Weisberg (2014). For more information about tools for regression diagnostics in Stata, see `help regress postestimation` and `help regress postestimation plots`.

19 Power analysis for regression

19.1 Chapter overview

This chapter illustrates how to compute power for a variety of regression models. The chapter begins by illustrating how to compute power for a simple regression analysis (section 19.2) as well as how to compute the power for a multiple regression analysis (section 19.3). This chapter also illustrates how to compute power for a nested multiple regression (section 19.4). The power analyses shown in this chapter use two commands that are not part of Stata and that I wrote for inclusion in this book. The `power multreg` command is used for the power analyses involving simple regression and multiple regression. The `power nestreg` command is used for computing power for a nested regression model. Although I wrote these commands, they inherit much of the functionality of the other built-in `power` commands (see `help power`). The `power` command was specifically designed so that user-written programs could be created to add custom methods to extend its functionality.

19.2 Power for simple regression

Let's start with an example of a power analysis for a simple regression. Say that we have one outcome and one predictor. Assume we believe that the predictor will explain 5% of the variance in the outcome and that we plan on having a sample size of 100. Let's then compute the power given these specifications. We can do so using the `power multreg` command below. The `r2(0.05)` option specifies that the predictor explains 5% of the variance in the outcome, and the `n(100)` option specifies that the sample size is 100. The results show that for an alpha of 0.05, with a sample size of 100 and 1 predictor that has an R-squared of 0.05, we would expect the power to be 0.6224.

```
. power multreg, r2(0.05) n(100)
```

Estimated power
Two-sided test

alpha	power	N	R-sq	K(# preds)
.05	.6224	100	.05	1

Note: Specifying alpha

Unless you specify otherwise, the `power multreg` and `power nestreg` commands assume an alpha of 0.05. You can specify your own alpha via the `alpha()` option. For example, the `power multreg` command below would use an alpha of 0.01.

```
. power multreg, r2(0.05) n(100) alpha(0.01)
```

Finding that this is not an acceptable power, let's compute the power associated with a range of sample sizes. By specifying n(100(10)200) in the `power multreg` command below, we obtain power estimates for sample sizes ranging from 100 to 200 in increments of 10. The output shows that the power for a sample size of 150 is 0.7972 and that the power for a sample size of 160 is 0.8223.

```
. power multreg, r2(0.05) n(100(10)200)
```

Estimated power
Two-sided test

alpha	power	N	R-sq	K(# preds)
.05	.6224	100	.05	1
.05	.6645	110	.05	1
.05	.7029	120	.05	1
.05	.7376	130	.05	1
.05	.769	140	.05	1
.05	.7972	150	.05	1
.05	.8223	160	.05	1
.05	.8447	170	.05	1
.05	.8646	180	.05	1
.05	.8822	190	.05	1
.05	.8977	200	.05	1

We can narrow down the smallest sample size that would give us at least 80% power by computing the power for a sample size of 150 to 160 in 1-unit increments by specifying the option n(150(1)160). The resulting output shows that a sample size of 152 is the smallest sample size we can use to achieve at least 80% power. With a sample size of 152 and an *R*-squared value of 5%, the power is 0.8024.

```
. power multreg, r2(0.05) n(150(1)160)
```

Estimated power
Two-sided test

alpha	power	N	R-sq	K(# preds)
.05	.7972	150	.05	1
.05	.7998	151	.05	1
.05	.8024	152	.05	1
.05	.805	153	.05	1
.05	.8076	154	.05	1
.05	.8101	155	.05	1
.05	.8126	156	.05	1
.05	.8151	157	.05	1
.05	.8175	158	.05	1
.05	.8199	159	.05	1
.05	.8223	160	.05	1

Say that our sample size was fixed at 100 and that we wanted to investigate the power that would be achieved as the R-squared value increases from 5% to 10% in half percent increments. Expressed as proportions, this is specified as r2(0.05(0.005)0.10) after the **power multreg** command below. Among the results, we see that an R-squared value of 0.075, with a sample size of 100, yields a power of 0.805.

```
. power multreg, r2(0.05(0.005)0.10) n(100)
```

Estimated power
Two-sided test

alpha	power	N	R-sq	K(# preds)
.05	.6224	100	.05	1
.05	.666	100	.055	1
.05	.706	100	.06	1
.05	.7424	100	.065	1
.05	.7753	100	.07	1
.05	.805	100	.075	1
.05	.8314	100	.08	1
.05	.855	100	.085	1
.05	.8757	100	.09	1
.05	.894	100	.095	1
.05	.91	100	.1	1

We can graphically depict the above result by adding the **graph** option, as shown below. This depicts the relationship between R-squared (on the x axis) and power (on the y axis) when the sample size is 100, as shown in figure 19.1.

```
. power multreg, r2(0.05(0.005)0.10) n(100) graph
```

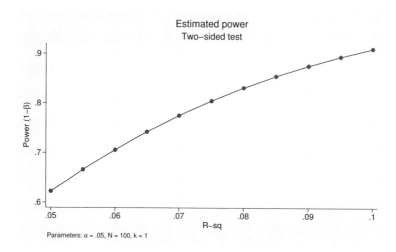

Figure 19.1: Power as a function of R-squared for $N = 100$

Let's make a graph that shows the power as a function of the R-squared values shown in the last example and includes a range of values for the sample size. The example below adds the `n(80(10)110)` option to compute the power for sample sizes ranging from 80 to 110 in increments of 10, as well as for the R-squared values ranging from 5% to 10% in 0.5% increments. This command creates the graph shown in figure 19.2.

```
. power multreg, r2(0.05(0.005)0.10) n(80(10)110) graph
```

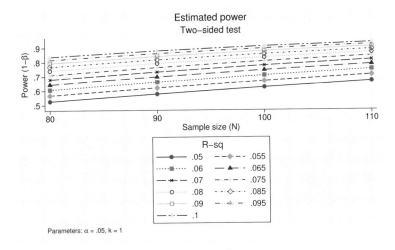

Figure 19.2: Power as a function of R-squared and sample size

---

Note: Displaying tables and graphs

Adding the `graph` option to the `power` command suppresses the display of the table of results. You can override this default behavior by adding the `table` option to display the results as a table (as well as the graph).

The graph in figure 19.2 is not very easy to read. We can make it easier to read by placing the R-squared values on the x axis. We can control the display of the graph by adding suboptions within the `graph()` option. In the example below, the `graph(xdim(r2))` option is used to graph the values specified via the `r2()` option on the x axis. The `power multreg` command below creates the graph shown in figure 19.3.

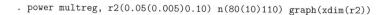

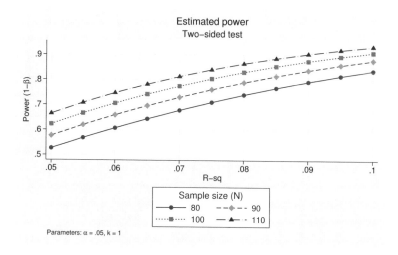

Figure 19.3: Power as a function of sample size (on x axis) and R-squared

We can further customize the graphs created by the `power` command by inserting options that we would ordinarily use with the `graph twoway` command as suboptions within the `graph()` option. (For a list of such options, see `help twoway options`.) The example below adds the `legend(rows(1))` suboption within the `graph()` option to display the legend as a single row (leaving more room for the actual graph). The example also adds the `yline(.8)` suboption within the `graph()` option to add a reference line on the y axis. We can then use this reference line to identify combinations of R-squared (on the x axis) and sample size (as separate lines) that correspond to a power of 0.80. Figure 19.4 shows the graph created by the `power multreg` command.

```
. power multreg, r2(0.05(0.005)0.10) n(80(10)110)
>     graph(xdim(r2) legend(rows(1)) yline(.8))
```

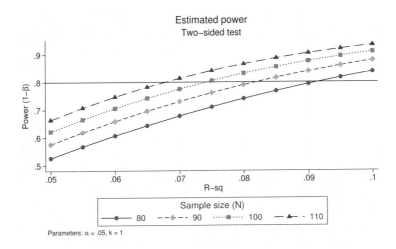

Figure 19.4: Figure of power with a reference line on the y axis

Inspecting figure 19.4, we can see the combinations of sample size and R-squared that lead to 80% power. For example, the power is 0.80 for a sample size of $N = 110$ combined with an R-squared of about 0.0068. Or the power would be 0.80 with a sample size of $N = 90$ combined with an R-squared of approximately 0.082. The `power multreg` command could be used as illustrated in this section to identify the exact sample size and R-squared values that would lead to the desired level of power.

In this section, we have seen how to use the `power multreg` command to compute power for a simple regression model. In the next section, we will see how to compute power for a multiple regression model using the `power multreg` command.

Note: Power for correlation

Although I have not shown it in this book, the `power` command includes support for computing the power for one correlation. You can see `help power onecorrelation` for more information. You can also use the `power` command for comparing the correlation between two independent groups, as described in `help power twocorrelations`.

19.3   Power for multiple regression

Let's consider a power analysis for a multiple regression model in which our desire is to test an entire group of predictors. To give this some context, let's consider the example from section 17.3.2 in which we focused on the role of four different kinds of socializing behaviors in predicting happiness. In that example, the four socializing variables were added after considering the contribution of demographics and health (we will consider such a model in the following section). For now, imagine a model in which the four socializing variables are the only predictors with happiness as the outcome. Our goal is to test the joint contribution of the four socializing variables on happiness. We can use the `power multreg` command to compute the power for this type of model by specifying the `k()` option to indicate the number of predictors.

Say that our desired sample size is 100 and that we expect the four socializing variables to explain 10% of the variance in happiness. We can compute the power for this scenario using the `power multreg` command below. Note the addition of the `k(4)` option to indicate the test of four predictor variables.

```
. power multreg, r2(0.10) n(100) k(4)
Estimated power
Two-sided test
```

alpha	power	N	R-sq	K(# preds)
.05	.7432	100	.1	4

The power for this test is 0.7432. We can use the `r2()` option as we did in the previous section to evaluate the power across a variety of R-squared values. We can also use the `n()` option to investigate different combinations of sample size. The key difference in the use of the `power multreg` command for multiple regression models is the addition of the `k()` option to specify the number of predictors.

19.4   Power for a nested multiple regression

Let's now consider an example that is like the example from section 17.3.2 in which we focused on the overall contribution of four different kinds of socializing behaviors in predicting happiness after accounting for the contribution of demographics and health factors. The focus of this test is the contribution of the four socializing variables after we account for the demographic and health factors. Say that we want to replicate the results from section 17.3.2, assessing the contribution of the four socializing variables in step 3 after accounting for the five demographic and health variables from steps 1 and 2. Looking back at that nested regression model, we see that the R-squared value for the model at step 2 was 0.1353 and that the R-squared value for the model in step 3 was 0.1526.

Let's say that our desire is to replicate this result using a sample size of $N = 300$. We can compute the power for such a test using the `power nestreg` command shown below. The `r2f(.1526)` option is used to specify the R-squared in step 3, the full model. The `kf(9)` option is used to specify that there are 9 predictors in the full model. The `r2r(0.1353)` option is used to specify the R-squared for the reduced model, the R-squared value from step 2. The `kr(5)` option is used to specify that there are 5 predictors in the reduced model. Finally, the `n(300)` option indicates the sample size is 300. The power for this test is 0.4719.

```
. power nestreg, r2f(.1526) kf(9) r2r(0.1353) kr(5) n(300)
Estimated power
Two-sided test
```

alpha	power	N	R2-Full	K-Full	R2-Reduced	K-Reduced
.05	.4719	300	0.1526	9	0.1353	5

Let's consider a range of sample-size values from 300 to 700 in increments of 50. Among the results, we find that a sample size of 550 yields a power of 0.7678, a sample size of 600 yields a power of 0.808, and a sample size of 650 yields a power of 0.8424.

```
. power nestreg, r2f(.1526) kf(9) r2r(0.1353) kr(5) n(300(50)700)
Estimated power
Two-sided test
```

alpha	power	N	R2-Full	K-Full	R2-Reduced	K-Reduced
.05	.4719	300	0.1526	9	0.1353	5
.05	.5432	350	0.1526	9	0.1353	5
.05	.6089	400	0.1526	9	0.1353	5
.05	.6684	450	0.1526	9	0.1353	5
.05	.7213	500	0.1526	9	0.1353	5
.05	.7678	550	0.1526	9	0.1353	5
.05	.808	600	0.1526	9	0.1353	5
.05	.8424	650	0.1526	9	0.1353	5
.05	.8715	700	0.1526	9	0.1353	5

As we did with the `power multreg` command, we can add the `graph` option to create a graphical display of the previous power analysis, as shown in figure 19.5.

. `power nestreg, r2r(0.1353) kr(5) r2f(.1526) kf(9) n(300(50)700) graph`

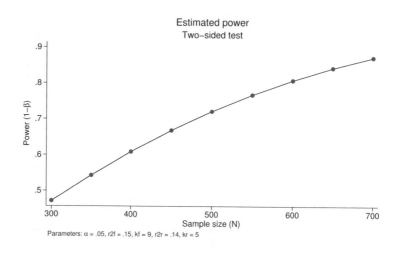

Figure 19.5: Power for nested regression as a function of sample size

Let's suppose that our future study will include three additional socializing variables, but we are unsure about how much variance the total of seven socializing variables would explain. We can use the `r2f()` values to explore a range of values for the explained variance in the full model. Let's consider R-squared values for the full model ranging from 14.5% to 16.5% in 0.5% increments by specifying the `r2f(0.1450(0.0050)0.1650)` option. The `kf(12)` option specifies that the full model includes a total of 12 predictors (5 demographics + 7 socializing). For simplicity, let's round the explained variance for the reduced model to 13% by specifying `r2r(0.13)`. The `kr(5)` option specified that there are 5 predictors in the reduced model (5 demographics). Let's use the `n(500)` option to compute the power for this scenario with a sample size of 500. As we see in the table below, we achieve a power of 0.8075 when the R-squared for the full model is 15.5%.

```
. power nestreg, r2f(0.1450(0.0050)0.1650) kf(12) r2r(0.13) kr(5) n(500)
Estimated power
Two-sided test
```

alpha	power	N	R2-Full	K-Full	R2-Reduced	K-Reduced
.05	.5397	500	0.1450	12	0.1300	5
.05	.6923	500	0.1500	12	0.1300	5
.05	.8075	500	0.1550	12	0.1300	5
.05	.8866	500	0.1600	12	0.1300	5
.05	.9367	500	0.1650	12	0.1300	5

Let's augment the previous power analysis to consider sample sizes ranging from 300 to 700 in increments of 50 by using the n(300(50)700) option. We can display this as a graph with 1 row for the legend and a reference line at 0.80 by adding the graph(legend(rows(1)) yline(0.80)) option. This creates the graph shown in figure 19.6.

```
. power nestreg, r2r(0.13) kr(5) r2f(0.1450(0.0050)0.1650) kf(12)
>     n(300(50)700) graph(legend(rows(1)) yline(0.80))
```

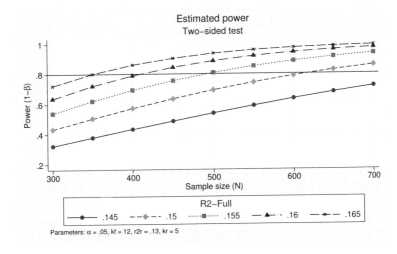

Figure 19.6: Power as a function of sample size and *R*-squared for full model

Looking at the reference line, we can see the combinations of sample size and explained variance for the full model that yield a power of 0.80. Let's focus on the combination where the *R*-squared for the full model is 0.16. This achieves 80% power just beyond a sample size of 400. The **power nestreg** command below computes the power for sample sizes ranging from 400 to 410 in 1-unit increments; finding a sample size of 409 would provide 80% power when the *R*-squared for the full model is 0.16.

```
. power nestreg, r2r(0.13) kr(5) r2f(.16) kf(12) n(400(1)410)
Estimated power
Two-sided test
```

alpha	power	N	R2-Full	K-Full	R2-Reduced	K-Reduced
.05	.7892	400	0.1600	12	0.1300	5
.05	.7904	401	0.1600	12	0.1300	5
.05	.7916	402	0.1600	12	0.1300	5
.05	.7928	403	0.1600	12	0.1300	5
.05	.794	404	0.1600	12	0.1300	5
.05	.7953	405	0.1600	12	0.1300	5
.05	.7964	406	0.1600	12	0.1300	5
.05	.7976	407	0.1600	12	0.1300	5
.05	.7988	408	0.1600	12	0.1300	5
.05	.8	409	0.1600	12	0.1300	5
.05	.8012	410	0.1600	12	0.1300	5

Note: Varying the R-squared reduced

We saw how we could examine varying values of the R-squared for the full model via the `r2f()` option. Although not shown, the `power nestreg` command allows you to examine different values of the R-squared for the reduced model via the `r2r()` option. You could specify `r2r(0.10(0.01)0.15`, and the power would be computed given that the R-squared for the reduced model ranges from 0.10 to 0.15 in increments of 0.01. This seems less intuitive than varying the R-squared for the full model, but this functionality is present if you should find a use for it.

19.5   Closing thoughts

This chapter has illustrated how you can perform power analyses for simple regression and multiple regression using the `power multreg` command. It has also illustrated how you can compute power for a nested regression model using the `power nestreg` command. Both of these are user-written commands (written by the author). Because of the way that the `power` command is written, it easily accommodates new methods for computing power, and then those methods inherit the functionality of the other built-in `power` commands. You can see `help power multreg` and `help power nestreg` for help using the `power multreg` and `power nestreg` commands. You can also see `help power` for more information about the `power` command.

Note: Writing your own power method

The Stata documentation includes information describing how you can extend the `power` command to include your own power methods. You can see `help power userwritten` for information about how you can create your own power method and integrate it within the `power` command.

Part V

Stata overview

The previous parts of this book focused on how to perform certain analyses and illustrated the commands used for performing those analyses. To use an analogy, compare the previous parts of the book to building a car, and the relevant tools were pulled out of the toolbox for performing the task of building the car, but I did not spend much time talking about the particular merits of each tool. This part of the book focuses on the use of the tools themselves. That is, the chapters in this part focus on the commands themselves, describing and illustrating additional features of the commands.

The first chapter in this part (chapter 20) shows common features of estimation commands. Even though Stata has a very large number of estimation commands, these commands share a number of features in common. This is by design, not by accident. Because estimation commands work in such similar ways, what you learn about the common behavior of one estimation command translates over to the use of other estimation commands. This chapter is about the features (behaviors) that estimation commands share in common.

Chapter 21 discusses postestimation commands (commands you can issue after estimation commands such as the `anova` or `regress` command). In particular, this chapter provides additional details about the `contrast` command, the `margins` command, the `marginsplot` command, and the `pwcompare` command.

Chapter 22 provides a brief overview and examples of common data management tasks using Stata. It illustrates reading data into Stata, saving datasets, labeling data, creating and recoding variables, keeping and dropping variables, keeping and dropping observations, combining datasets, and reshaping datasets.

The final chapter, chapter 23, recognizes that many readers of this book might be familiar with IBM® SPSS®. If you are such a reader, you might find yourself asking how to translate SPSS commands into Stata. This chapter lists commonly used SPSS commands (in alphabetical order) and shows the equivalent Stata command with an example (or examples) of how to use the command.

20 Common features of estimation commands

20.1 Chapter overview

In this book, I have mainly focused on showing you how to analyze data using the `anova` command, the `mixed` command, and the `regress` command. The `anova`, `mixed`, and `regress` commands are a special kind of command that Stata calls an "estimation command". These are called estimation commands because they fit a statistical model. Stata has a wide variety of estimation commands,[1] yet they all work similarly. In fact, if you type `help estcom`, you will see a help file that provides an extensive list of commonalities among estimation commands. This section focuses on six commonalities among estimation commands.

Section 20.2 describes how estimation commands follow a common syntax. Estimation commands also permit you to use the `if` or `in` specification to analyze data from subsamples, as described in section 20.3. It is common for estimation commands to support estimation via robust variances and sometimes via bootstrap or jackknife estimation of variances, as described in section 20.4. There are many prefix commands that

---

1. You can see `help estimation commands` for a list of all the different Stata estimation commands.

can be applied to estimation commands, five of which are illustrated in section 20.5. Section 20.6 illustrates how you can specify the confidence level used for displaying confidence intervals when using estimation commands. The chapter concludes by noting that following an estimation command, you can use a common set of postestimation commands to perform further analyses. The following chapter (chapter 21) discusses postestimation commands in greater detail.

20.2   Common syntax

Estimation commands follow a common syntax. Consider the example below that predicts happy7 from socfrend and socrel. Note how the outcome variable (happy7) is followed by the predictor variables.

```
. use gss2012_sbs

. regress happy7 socfrend socrel
```

Source	SS	df	MS	Number of obs	=	1,282
				F(2, 1279)	=	12.81
Model	24.9524995	2	12.4762497	Prob > F	=	0.0000
Residual	1245.54048	1,279	.973839312	R-squared	=	0.0196
				Adj R-squared	=	0.0181
Total	1270.49298	1,281	.991797798	Root MSE	=	.98683

happy7	Coef.	Std. Err.	t	P>|t|	[95% Conf. Interval]	
socfrend	.060522	.0171593	3.53	0.000	.0268585	.0941854
socrel	.0446249	.0168234	2.65	0.008	.0116203	.0776295
_cons	5.055143	.0938776	53.85	0.000	4.870972	5.239313

If you wanted to instead run a quantile regression, you would simply use the qreg command in lieu of the regress command. The structure of the command would be the same.

```
. qreg happy7 socfrend socrel
  (output omitted )
```

If Stata created a new command called xyzreg, it would probably work as shown below for predicting happy7 from socfrend and socrel.

```
. xyzreg happy7 socfrend socrel
```

The common syntax allows the inclusion of a weight specification. Nearly all estimation commands permit you to include a weight specification. The example below incorporates the probability weight variable wtss.

```
. regress happy7 socfrend socrel [pweight=wtss]
(sum of wgt is   1.2935e+03)
```

Linear regression				Number of obs	=	1,282
				F(2, 1279)	=	6.78
				Prob > F	=	0.0012
				R-squared	=	0.0156
				Root MSE	=	.96579

happy7	Coef.	Robust Std. Err.	t	P>|t|	[95% Conf. Interval]	
socfrend	.0369513	.0227083	1.63	0.104	-.0075983	.0815009
socrel	.054919	.0254532	2.16	0.031	.0049844	.1048537
_cons	5.121282	.1210573	42.30	0.000	4.88379	5.358775

Probability weights are a common form of weighting, but Stata also supports frequency weights (via `fweight`), analytic weights (via `aweight`), and importance weights (via `iweight`). The help file for each command (for example, `help regress`) shows whether the command supports weights and the types of weights supported by the command. See `help weight` for more information.

20.3   Analysis using subsamples

As part of this general syntax is the ability to add the `if` specification or the `in` specification to analyze just a subsample of the data. For example, the analysis below is restricted to just the females in the sample by specifying `if female==1`.

```
. regress happy7 socfrend socrel if female==1
```

Source	SS	df	MS		Number of obs	=	697
					F(2, 694)	=	6.89
Model	12.2949725	2	6.14748626		Prob > F	=	0.0011
Residual	619.653378	694	.892872302		R-squared	=	0.0195
					Adj R-squared	=	0.0166
Total	631.94835	696	.907971767		Root MSE	=	.94492

happy7	Coef.	Std. Err.	t	P>|t|	[95% Conf. Interval]	
socfrend	.0742293	.0222383	3.34	0.001	.0305669	.1178918
socrel	.0178566	.0223367	0.80	0.424	-.025999	.0617123
_cons	5.140234	.126095	40.76	0.000	4.89266	5.387807

The `regress` command below uses the `in` specification, restricting the analysis to the first 100 observations in the dataset. This can be useful for testing out the syntax of a command (by using fewer observations) if you have an enormous dataset and are performing a computationally intense analysis command.

```
. regress happy7 socfrend socrel in 1/100
```

Source	SS	df	MS			
Model	.252357191	2	.126178595			
Residual	57.3570178	61	.94027898			
Total	57.609375	63	.914434524			

Number of obs	=	64
F(2, 61)	=	0.13
Prob > F	=	0.8747
R-squared	=	0.0044
Adj R-squared	=	-0.0283
Root MSE	=	.96968

happy7	Coef.	Std. Err.	t	P>|t|	[95% Conf. Interval]	
socfrend	.0308276	.0741077	0.42	0.679	-.1173599	.1790151
socrel	.0143442	.0732103	0.20	0.845	-.1320489	.1607372
_cons	5.248647	.3768031	13.93	0.000	4.495183	6.002112

20.4 Robust standard errors

Homoskedasticity (or sometimes called homogeneity of variance) is one of the key assumptions of regression models. Violating this assumption can lead to p-values that are not trustworthy. Stata offers a robust variance estimator (called the Huber/White/sandwich estimator of variance) that is robust to violations of this assumption. You can request this method for the calculation of robust standard errors via the vce(robust) option, as illustrated below.

```
. regress happy7 socfrend socrel, vce(robust)
```

Linear regression

Number of obs	=	1,282
F(2, 1279)	=	10.38
Prob > F	=	0.0000
R-squared	=	0.0196
Root MSE	=	.98683

happy7	Coef.	Robust Std. Err.	t	P>|t|	[95% Conf. Interval]	
socfrend	.060522	.0194235	3.12	0.002	.0224166	.0986274
socrel	.0446249	.0193757	2.30	0.021	.0066132	.0826366
_cons	5.055143	.109352	46.23	0.000	4.840613	5.269672

In addition, most commands also permit you to specify the vce(bootstrap) option to obtain bootstrapped standard errors or vce(jackknife) to obtain standard errors via jackknife estimation.

20.5   Prefix commands

There are a number of `prefix` commands that you can specify in front of an estimation command. This section describes five commonly used prefix commands, namely, the `by:`, `nestreg:`, `stepwise:`, `svy:`, and `mi estimate:` prefix commands. You can see the entire list of prefix commands by typing `help prefix`.

20.5.1   The by: prefix

I think that the best way to explain the `by:` prefix is by showing you an example. In the example below, the `by:` prefix is used to perform the specified `regress` command separately for males and for females.

```
. sort female
. by female: regress happy7 socfrend socrel

-> female = Male
```

Source	SS	df	MS		Number of obs	=	585
					F(2, 582)	=	7.10
Model	15.1920507	2	7.59602533		Prob > F	=	0.0009
Residual	622.831881	582	1.07015787		R-squared	=	0.0238
					Adj R-squared	=	0.0205
Total	638.023932	584	1.09250673		Root MSE	=	1.0345

happy7	Coef.	Std. Err.	t	P>|t|	[95% Conf.	Interval]
socfrend	.044255	.0267689	1.65	0.099	-.0083203	.0968304
socrel	.0726928	.0257954	2.82	0.005	.0220293	.1233563
_cons	4.982311	.1408143	35.38	0.000	4.705745	5.258877

```
-> female = Female
```

Source	SS	df	MS		Number of obs	=	697
					F(2, 694)	=	6.89
Model	12.2949725	2	6.14748626		Prob > F	=	0.0011
Residual	619.653378	694	.892872302		R-squared	=	0.0195
					Adj R-squared	=	0.0166
Total	631.94835	696	.907971767		Root MSE	=	.94492

happy7	Coef.	Std. Err.	t	P>|t|	[95% Conf.	Interval]
socfrend	.0742293	.0222383	3.34	0.001	.0305669	.1178918
socrel	.0178566	.0223367	0.80	0.424	-.025999	.0617123
_cons	5.140234	.126095	40.76	0.000	4.89266	5.387807

As you can see, the `regress` command was estimated once for males and then again for females. In other words, it was estimated once for every level of the variable specified after the `by` prefix. The above two commands can be shortened to.

```
. bysort female: regress happy7 socfrend socrel
  (output omitted)
```

Had we specified `bysort marital:`, this would have run five separate regression analyses, one for every level of marital status. For more information, you can type `help by`.

20.5.2 The nestreg: prefix

The `nestreg:` prefix permits you to formulate nested models in which variables are entered in blocks that you specify. The significance of each block is tested after accounting for the contribution of previous blocks.

An example is shown below in which the `nestreg` prefix is used in front of the `regress` command.[2] The `regress` command predicts the variable `happy7` from nine variables that are grouped into three blocks (as indicated by the parentheses). As indicated by the comment, the first block of variables is demographics, the second block is one variable (self-reported health), and the third block is composed of four variables regarding social behaviors.

```
. * Blocks: (1) demogs & (2) health & (3) social behaviors
. nestreg, quietly : regress happy7 (age educ children class)
>       (health)
>       (socrel socommun socfrend socbar)
Block  1: age educ children class
Block  2: health
Block  3: socrel socommun socfrend socbar
```

Block	F	Block df	Residual df	Pr > F	R2	Change in R2
1	16.67	4	1257	0.0000	0.0504	
2	123.40	1	1256	0.0000	0.1353	0.0850
3	6.37	4	1252	0.0000	0.1526	0.0172

The output shows the contribution of the first block, the net contribution of the second block (after accounting for the first block), and the net contribution of the third block (after accounting for the first two blocks). You can see section 17.3 for more details about nested regression models (including more details about this example). You can also see `help nestreg` for more information about the `nestreg` command.

---

2. I added the `quietly` option to the `nestreg` command so it will omit the output of each `regress` command for the sake of saving printed space in the book. In practice, you would omit this option.

20.5.3   The stepwise: prefix

As implied by the name, the `stepwise` prefix is used for performing stepwise selection of variables. Although I am showing you how to use this command, I want to say that I am not a fan of this method because I think it can easily be misapplied.

In the example below, I use stepwise regression with forward selection in predicting `happy7` from a group of nine predictors.

```
. stepwise, pe(.05): regress happy7 age educ children class
>    health socrel soccommun socfrend socbar
                     begin with empty model
p = 0.0000 <  0.0500  adding  health
p = 0.0000 <  0.0500  adding  class
p = 0.0000 <  0.0500  adding  children
p = 0.0002 <  0.0500  adding  socfrend
p = 0.0130 <  0.0500  adding  socommun
p = 0.0430 <  0.0500  adding  socrel
```

Source	SS	df	MS		Number of obs	=	1,262
					F(6, 1255)	=	37.31
Model	188.507779	6	31.4179632		Prob > F	=	0.0000
Residual	1056.91298	1,255	.842161738		R-squared	=	0.1514
					Adj R-squared	=	0.1473
Total	1245.42076	1,261	.98764533		Root MSE	=	.91769

happy7	Coef.	Std. Err.	t	P>|t|	[95% Conf.	Interval]
health	.2592685	.0239306	10.83	0.000	.2123201	.3062169
class	.1939354	.0368361	5.26	0.000	.1216684	.2662025
children	.0826923	.0160225	5.16	0.000	.0512585	.1141261
socfrend	.0441151	.0172725	2.55	0.011	.0102289	.0780013
socommun	.0289651	.0132581	2.18	0.029	.0029545	.0549757
socrel	.0326699	.0161264	2.03	0.043	.0010322	.0643076
_cons	3.585914	.1441992	24.87	0.000	3.303016	3.868812

Section 17.4 contains more details about stepwise models. You can also see `help stepwise` for more information about the `stepwise` command.

20.5.4   The svy: prefix

The `svy:` prefix command allows you to analyze the data while accounting for the sampling design. The sampling design is first specified via the `svyset` command. Then, estimation commands that are prefaced by the `svy:` prefix will provide estimates that account for the survey design.

The `svyset` command below informs Stata that the sampling design that used a primary sampling unit is indicated by the variable `sampcode`, that the sampling weight is indicated by the variable `wtss`, and that the variable `vstrat` indicates the stratification.

```
. svyset vpsu [weight=wtss], strata(vstrat)
(sampling weights assumed)

      pweight: wtss
          VCE: linearized
  Single unit: missing
     Strata 1: vstrat
        SU 1: vpsu
       FPC 1: <zero>
```

Subsequent estimation commands that begin with the `svy:` prefix will perform computations accounting for this sampling design. For example, the `svy` prefix is used to perform a regression predicting `happy7` from nine predictors while accounting for the sampling design (but the output is omitted to save space).

```
svy : regress happy7 age educ children class health ///
   socrel socommun socfrend socbar
```

You can learn more about the analysis of data from complex sampling designs as well as more about the use of the `svy` prefix by typing `help svy`.

Note: Saving datasets after svyset

If you save a Stata dataset after specifying the survey design via the `svyset` command, that survey design specification will be saved as part of the Stata dataset. That means that you could subsequently `use` the dataset and then simply issue estimation commands using the `svy` prefix and that those analyses would account for the survey design.

20.5.5   The mi estimate: prefix

The `mi estimate:` command permits you to analyze multiply imputed datasets. This is a complex topic for which Stata has written an entire manual devoted to multiple imputation as well as the Stata commands for performing multiple imputation. You can see `help mi` for more information.

20.6   Setting confidence levels

All estimation commands show confidence intervals for point estimates. By default, the confidence level for these intervals is 95%. In the example below, I add the `level(99)` option to specify that I want 99% confidence intervals to be displayed.

```
. regress happy7 socfrend socrel, level(99)

      Source |       SS           df       MS      Number of obs   =     1,282
-------------+----------------------------------   F(2, 1279)      =     12.81
       Model |  24.9524995         2  12.4762497   Prob > F        =    0.0000
    Residual |  1245.54048     1,279  .973839312   R-squared       =    0.0196
-------------+----------------------------------   Adj R-squared   =    0.0181
       Total |  1270.49298     1,281  .991797798   Root MSE        =    .98683

      happy7 |      Coef.   Std. Err.      t    P>|t|     [99% Conf. Interval]
-------------+----------------------------------------------------------------
    socfrend |    .060522   .0171593     3.53   0.000     .0162565    .1047875
      socrel |   .0446249   .0168234     2.65   0.008     .0012258     .088024
       _cons |   5.055143   .0938776    53.85   0.000     4.812969    5.297316
```

For more information, you can see `help level`.

20.7   Postestimation commands

You can issue postestimation commands after issuing an estimation command. Such commands include `contrast`, `margins`, `marginsplot`, and `pwcompare`. You can see chapter 21 for more information about these four commonly used postestimation commands. You can type `help postestimation` for information regarding all postestimation commands. To focus on the postestimation commands relevant after a particular command (for example, `regress`), you can type `help regress postestimation`, and this will show all the postestimation commands that you can use after the `regress` command.

20.8   Closing thoughts

One of the features that I think makes Stata easy to use is the fact that estimation commands behave in such similar ways. When I learn about the general behavior of one command, that helps me learn about the general behavior of other commands. For even more details about the common features of estimation commands, you can see `help estcom`.

21 Postestimation commands

21.1 Chapter overview

In chapter 20, I focused on features that are common among Stata estimation commands. One of the final features mentioned, in section 20.7, is that there are a shared set of commands that you can use following an estimation command. These special commands are called postestimation commands because they can be typed after an estimation command. This chapter provides additional details about commonly used postestimation commands, namely, the `contrast` command (section 21.2), the `margins` command (section 21.3), the `marginsplot` command (section 21.4), and the `pwcompare` command (section 21.5).

21.2 The contrast command

This section discusses additional details about the `contrast` command, focusing on options that control the display of the output. Let's begin using an example of a one-way analysis of variance (ANOVA) from section 5.4 that used happiness as the dependent variable (DV) and marital status as the independent variable (IV). The dataset for this example is used below. Then, the `anova` command is used to perform an ANOVA with happiness as the DV and marital status as the IV.

```
. use gss2012_sbs, clear
. anova happy7 marital
```

| | Number of obs = | 1,284 | R-squared | = | 0.0408 |
| | Root MSE   = | .977102 | Adj R-squared | = | 0.0378 |

Source	Partial SS	df	MS	F	Prob>F
Model	51.89968	4	12.97492	13.59	0.0000
marital	51.89968	4	12.97492	13.59	0.0000
Residual	1221.0972	1,279	.95472807		
Total	1272.9969	1,283	.99220334		

Note! Factor variables

In previous chapters, I would have described the variable `marital` as the IV, focusing on the role the variable played in the model. In this chapter, I will often call such a variable a "factor variable". This is a Stata-specific term that refers to a categorical variable that has been entered into an ANOVA model or to using a regression model by using the `i.` prefix (for example, `i.race`). Stata treats factor variables differently, knowing that they are categorical variables, and these Stata postestimation commands understand how to treat such variables differently from continuous variables (for example, age).

The `contrast` command can be used to test the overall effect of an IV. For example, the `contrast` command below tests the overall effect of `marital`.[1]

```
. contrast marital
Contrasts of marginal linear predictions
Margins      : asbalanced
```

	df	F	P>F
marital	4	13.59	0.0000
Denominator	1279		

The `contrast` command can also be used to apply contrast operators to an IV to dissect main effects or interactions. Consider the output we obtain from the `contrast` command when we apply a contrast coefficient, such as `r.marital`.

---

1. This mirrors the results from the original `anova` command.

```
. contrast r.marital
Contrasts of marginal linear predictions
Margins        : asbalanced
```

	df	F	P>F
marital			
(widowed vs married)	1	7.63	0.0058
(divorced vs married)	1	33.39	0.0000
(separated vs married)	1	14.52	0.0001
(never married vs married)	1	28.07	0.0000
Joint	4	13.59	0.0000
Denominator	1279		

	Contrast	Std. Err.	[95% Conf. Interval]	
marital				
(widowed vs married)	-.2841316	.1028462	-.4858973	-.0823659
(divorced vs married)	-.4615629	.0798813	-.6182756	-.3048502
(separated vs married)	-.5949905	.156161	-.9013504	-.2886306
(never married vs married)	-.3488696	.0658518	-.478059	-.2196801

The output is divided into two portions. The top portion shows F tests of the specified contrast, while the second portion of the output shows the estimate of the contrast, standard error, and confidence interval (CI).

If we wanted to suppress the second portion of the output, we could use the option `noeffects`, as shown below.

```
. contrast r.marital, noeffects
Contrasts of marginal linear predictions
Margins        : asbalanced
```

	df	F	P>F
marital			
(widowed vs married)	1	7.63	0.0058
(divorced vs married)	1	33.39	0.0000
(separated vs married)	1	14.52	0.0001
(never married vs married)	1	28.07	0.0000
Joint	4	13.59	0.0000
Denominator	1279		

We can suppress the first portion of the output with the `nowald` option, as shown in the following example.

```
. contrast r.marital, nowald
Contrasts of marginal linear predictions
Margins       : asbalanced
```

	Contrast	Std. Err.	[95% Conf.	Interval]
marital				
(widowed vs married)	-.2841316	.1028462	-.4858973	-.0823659
(divorced vs married)	-.4615629	.0798813	-.6182756	-.3048502
(separated vs married)	-.5949905	.156161	-.9013504	-.2886306
(never married vs married)	-.3488696	.0658518	-.478059	-.2196801

By default, 95% confidence intervals are displayed. You can use the level() option to specify a different confidence level. The example below illustrates a 99% confidence interval.

```
. contrast r.marital, nowald level(99)
Contrasts of marginal linear predictions
Margins       : asbalanced
```

	Contrast	Std. Err.	[99% Conf.	Interval]
marital				
(widowed vs married)	-.2841316	.1028462	-.5494417	-.0188215
(divorced vs married)	-.4615629	.0798813	-.6676309	-.2554949
(separated vs married)	-.5949905	.156161	-.9978358	-.1921453
(never married vs married)	-.3488696	.0658518	-.5187461	-.1789931

By specifying the effects option, we see both the confidence interval and the *p*-values associated with each contrast. On your computer screen, this likely produces a very compact and efficient output that shows the size of the contrast, its standard error, significance, and a confidence interval. However, because of the narrowness of the printed page for this book, multiple lines are needed to display the first column that describes the contrast.

```
. contrast r.marital, nowald effects
Contrasts of marginal linear predictions
Margins        : asbalanced
```

	Contrast	Std. Err.	t	P>|t|	[95% Conf. Interval]	
marital (widowed vs married)	-.2841316	.1028462	-2.76	0.006	-.4858973	-.0823659
(divorced vs married)	-.4615629	.0798813	-5.78	0.000	-.6182756	-.3048502
(separated vs married)	-.5949905	.156161	-3.81	0.000	-.9013504	-.2886306
(never ma.. vs married)	-.3488696	.0658518	-5.30	0.000	-.478059	-.2196801

In lieu of using the **effects** option, consider the **pveffects** option. This offers an output style that is compact and displays the t-values and p-values associated with each contrast. This output style displays each contrast on one line (on the printed page) but at the compromise of omitting the confidence interval from the output. This output style is often used in this book.

```
. contrast r.marital, nowald pveffects
Contrasts of marginal linear predictions
Margins        : asbalanced
```

	Contrast	Std. Err.	t	P>|t|
marital				
(widowed vs married)	-.2841316	.1028462	-2.76	0.006
(divorced vs married)	-.4615629	.0798813	-5.78	0.000
(separated vs married)	-.5949905	.156161	-3.81	0.000
(never married vs married)	-.3488696	.0658518	-5.30	0.000

The **contrast** command can adjust the significance levels to account for multiple comparisons using the **mcompare()** option. You can specify the **bonferroni**, **sidak**, or **scheffe** options with **mcompare()** to obtain Bonferroni's, Šidák's, or Scheffé's method of adjustment (respectively). The example below illustrates using the **scheffe** option.

```
. contrast r.marital, nowald pveffects mcompare(scheffe)
Contrasts of marginal linear predictions
Margins        : asbalanced
```

	Number of Comparisons
marital	4

	Contrast	Std. Err.	Scheffe t	P>|t|
marital				
(widowed vs married)	-.2841316	.1028462	-2.76	0.107
(divorced vs married)	-.4615629	.0798813	-5.78	0.000
(separated vs married)	-.5949905	.156161	-3.81	0.006
(never married vs married)	-.3488696	.0658518	-5.30	0.000

This section has certainly not covered all the features of the `contrast` command. For additional information about the `contrast` command, see `help contrast`.

21.3   The margins command

This section illustrates some of the options that you can use with the `margins` command. In particular, this section illustrates the `at()` option (section 21.3.1), computing margins with factor variables (section 21.3.2), computing margins with factor variables and the `at()` option (section 21.3.3), and the `dydx()` option (section 21.3.4). For complete details about the `margins` command, see `help margins`.

Consider the following analysis of covariance (ANCOVA) in which the DV is `happy7`, the primary predictor of interest (that is, the IV) is `marital`, and the remaining variables in the model are considered covariates. The analysis is run below using the `anova` command.

```
. use gss2012_sbs, clear
. anova  happy7 i.marital i.female c.age c.educ c.children
>             c.socrel c.socommun c.socfrend c.socbar c.health
```

| | | Number of obs = | 1,272 | R-squared | = | 0.1595 |
| | | Root MSE   = | .916765 | Adj R-squared = | | 0.1508 |

Source	Partial SS	df	MS	F	Prob>F
Model	200.66547	13	15.435805	18.37	0.0000
marital	33.054678	4	8.2636696	9.83	0.0000
female	.63782803	1	.63782803	0.76	0.3838
age	.6741216	1	.6741216	0.80	0.3706
educ	.01093839	1	.01093839	0.01	0.9092
children	7.5524785	1	7.5524785	8.99	0.0028
socrel	2.8942581	1	2.8942581	3.44	0.0637
socommun	5.538802	1	5.538802	6.59	0.0104
socfrend	9.910579	1	9.910579	11.79	0.0006
socbar	.00952823	1	.00952823	0.01	0.9152
health	97.119216	1	97.119216	115.56	0.0000
Residual	1057.296	1,258	.84045788		
Total	1257.9615	1,271	.98974152		

Based on this analysis, let's explore how the `at()` option can be used to compute adjusted means at specific levels of the covariates.

21.3.1   The at() option

The `at()` option allows us to specify the values of covariates when computing adjusted means. For example, we can estimate the adjusted mean of happiness for a 40-year-old female with 12 years of education by using the `at()` option shown below. For people with these characteristics, their adjusted mean of happiness is 5.51 (95% CI = $[5.43, 5.59]$).

```
. margins, at(age=40 female=1 educ=12)
Predictive margins                          Number of obs   =        1,272
Expression  : Linear prediction, predict()
at          : female          =            1
              age             =           40
              educ            =           12
```

	Margin	Delta-method Std. Err.	t	P>|t|	[95% Conf. Interval]
_cons	5.511734	.040264	136.89	0.000	5.432742 5.590726

In addition to showing the adjusted mean and the 95% confidence interval for the adjusted mean, the output also includes a t-value and a p-value ($t = 136.89$, $p < 0.001$). We always get excited when we see a p-value that is below 0.05, but in this case, the excitement is not really warranted because this is merely testing the null hypothesis that the adjusted mean of happiness equals 0. That is a pretty meaningless hypothesis. I will frequently include the `nopvalues` option with the `margins` command to suppress the display of the *t*-value and *p*-value because these statistics are really not interesting. This option is illustrated below.

```
. margins, at(age=40 female=1 educ=12) nopvalues
Predictive margins                               Number of obs   =      1,272
Expression    : Linear prediction, predict()
at            : female          =         1
                age             =        40
                educ            =        12

                            Delta-method
                 Margin     Std. Err.      [95% Conf. Interval]

         _cons  5.511734    .040264       5.432742    5.590726
```

Suppose we want to compute the adjusted mean of happiness for people who are 20 to 80 years of age in 10-year increments while adjusting for all other covariates. We can compute this using the `at()` option shown in the following `margins` command. (Note the inclusion of the `vsquish` option to suppress the display of empty lines in the output.)

```
. margins, at(age=(20 30 40 50 60 70 80)) vsquish nopvalues
Predictive margins                               Number of obs   =      1,272
Expression    : Linear prediction, predict()
1._at         : age             =        20
2._at         : age             =        30
3._at         : age             =        40
4._at         : age             =        50
5._at         : age             =        60
6._at         : age             =        70
7._at         : age             =        80

                            Delta-method
                 Margin     Std. Err.      [95% Conf. Interval]

           _at
            1    5.457677    .0592664      5.341405    5.573949
            2    5.474989    .0426798      5.391258    5.558721
            3    5.492302    .0296311       5.43417    5.550434
            4    5.509615    .0261116      5.458388    5.560842
            5    5.526928    .0351142      5.458039    5.595817
            6     5.54424    .0503148       5.44553    5.642951
            7    5.561553    .0676574      5.428819    5.694287
```

Rather than typing the values 20 30 40 50 60 70 80, we can specify 20(10)80, as shown in the example below.

```
. margins, at(age=(20(10)80)) vsquish nopvalues
(output omitted)
```

Rather than specifying an exact age, we can specify that we want `age` held constant at the 25th percentile using the `at()` option shown below.

```
. margins, at((p25) age) vsquish nopvalues
Predictive margins                              Number of obs    =     1,272
Expression    : Linear prediction, predict()
at            : age              =         33 (p25)
```

	Margin	Delta-method Std. Err.	[95% Conf. Interval]	
_cons	5.480183	.0382103	5.40522	5.555146

In place of `p25`, we could specify any value ranging from `p1` (the 1st percentile) to `p99` (the 99th percentile). You can also specify `min`, `max`, or `mean`.

The `at()` option below holds the variables `age` and `educ` constant at the 25th percentile and `children` constant at 0.

```
. margins, at((p25) age educ children=0) vsquish nopvalues
Predictive margins                              Number of obs    =     1,272
Expression    : Linear prediction, predict()
at            : age              =         33 (p25)
                educ             =         12 (p25)
                children         =          0
```

	Margin	Delta-method Std. Err.	[95% Conf. Interval]	
_cons	5.377557	.0507477	5.277998	5.477116

We can specify the `at()` option multiple times to compute adjusted means for different combinations of covariate values. For example, the `margins` command below computes the adjusted mean once holding age constant at 20 and education constant at 10 and then again holding age constant at 50 and education constant at 15.

```
. margins, at(age=20 educ=10) at(age=50 educ=15) vsquish nopvalues
Predictive margins                              Number of obs    =      1,272
Expression   : Linear prediction, predict()
1._at        : age              =     20
               educ             =     10
2._at        : age              =     50
               educ             =     15
```

	Margin	Delta-method Std. Err.	[95% Conf. Interval]	
_at				
1	5.453899	.0648174	5.326737	5.581061
2	5.511084	.0288804	5.454425	5.567743

The legend describing the `at()` values can take up a lot of space. We can suppress the display of this with the `noatlegend` option.

```
. margins, at(age=20 educ=10) at(age=50 educ=15) vsquish noatlegend nopvalues
Predictive margins                              Number of obs    =      1,272
Expression   : Linear prediction, predict()
```

	Margin	Delta-method Std. Err.	[95% Conf. Interval]	
_at				
1	5.453899	.0648174	5.326737	5.581061
2	5.511084	.0288804	5.454425	5.567743

The example below computes adjusted means holding the variables educ and age constant at the 25th percentile, then again at the 50th percentile, then again at the 75th percentile.

```
. margins, at((p25) educ age)
>           at((p50) educ age)
>           at((p75) educ age) noatlegend vsquish nopvalues
Predictive margins                              Number of obs    =      1,272
Expression   : Linear prediction, predict()
```

	Margin	Delta-method Std. Err.	[95% Conf. Interval]	
_at				
1	5.478504	.0397575	5.400506	5.556503
2	5.50206	.0264018	5.450264	5.553857
3	5.531178	.0410497	5.450644	5.611711

21.3.2 Margins with factor variables

Suppose we want to compute the adjusted mean of happiness for the five levels of marital status. We could compute this using the following `margins` command.

```
. margins, at(marital=(1 2 3 4 5)) vsquish nopvalues
Predictive margins                              Number of obs    =      1,272
Expression    : Linear prediction, predict()
1._at         : marital         =            1
2._at         : marital         =            2
3._at         : marital         =            3
4._at         : marital         =            4
5._at         : marital         =            5
```

	Margin	Delta-method Std. Err.	[95% Conf. Interval]	
_at				
1	5.675395	.0396093	5.597688	5.753103
2	5.488235	.1009073	5.29027	5.686201
3	5.319401	.066439	5.189057	5.449744
4	5.066517	.1430949	4.785786	5.347248
5	5.393548	.0572373	5.281257	5.505839

Because `marital` is a factor variable, we can more easily compute this as shown below. This relieves us of the need to manually specify the levels of `marital`.

```
. margins marital, nopvalues
Predictive margins                              Number of obs    =      1,272
Expression    : Linear prediction, predict()
```

	Margin	Delta-method Std. Err.	[95% Conf. Interval]	
marital				
married	5.675395	.0396093	5.597688	5.753103
widowed	5.488235	.1009073	5.29027	5.686201
divorced	5.319401	.066439	5.189057	5.449744
separated	5.066517	.1430949	4.785786	5.347248
never married	5.393548	.0572373	5.281257	5.505839

We can apply contrast operators to factor variables. For example, the `ar.` contrast operator computes reverse adjacent group contrasts. In this case, the p-values of these tests are very interesting, so I do not include the `nopvalues` option with this (and related) commands.

```
. margins ar.marital
```

Contrasts of predictive margins

Expression   : Linear prediction, predict()

	df	F	P>F
marital			
(widowed vs married)	1	3.05	0.0809
(divorced vs widowed)	1	2.12	0.1454
(separated vs divorced)	1	2.58	0.1086
(never married vs separated)	1	4.47	0.0347
Joint	4	9.83	0.0000
Denominator	1258		

	Contrast	Delta-method Std. Err.	[95% Conf. Interval]	
marital				
(widowed vs married)	-.18716	.1071532	-.3973787	.0230587
(divorced vs widowed)	-.1688347	.1158975	-.3962085	.0585391
(separated vs divorced)	-.2528838	.1575109	-.5618967	.0561291
(never married vs separated)	.3270313	.154727	.02348	.6305826

We can include the `contrast(effects)` option to produce a display of both the significance values and the confidence intervals in the bottom portion of the output. The first column of the output wraps across lines (because of the restrictions of the width of a printed page). The output will probably look nicer on your computer display.

```
. margins ar.marital, contrast(effects)
```

Contrasts of predictive margins

Expression   : Linear prediction, predict()

	df	F	P>F
marital			
(widowed vs married)	1	3.05	0.0809
(divorced vs widowed)	1	2.12	0.1454
(separated vs divorced)	1	2.58	0.1086
(never married vs separated)	1	4.47	0.0347
Joint	4	9.83	0.0000
Denominator	1258		

	Contrast	Delta-method Std. Err.	t	P>|t|	[95% Conf. Interval]	
marital (widowed vs married)	-.18716	.1071532	-1.75	0.081	-.3973787	.0230587
(divorced vs widowed)	-.1688347	.1158975	-1.46	0.145	-.3962085	.0585391
(separated vs divorced)	-.2528838	.1575109	-1.61	0.109	-.5618967	.0561291
(never ma.. vs separated)	.3270313	.154727	2.11	0.035	.02348	.6305826

Tip! Customizing the margins output

Using the contrast(pveffects) option, you can obtain output that shows t-values and p-values but omits the confidence interval. Or you could instead specify contrast(cieffects) to include the confidence interval, but no significance test, for each contrast.

The upper portion of the output now is unnecessary because we can see the significance tests in the lower portion of the output. We can suppress the upper portion of the output by adding the nowald option.

```
. margins ar.marital, contrast(nowald effects)
Contrasts of predictive margins
Expression    : Linear prediction, predict()
```

	Contrast	Delta-method Std. Err.	t	P>|t|	[95% Conf. Interval]	
marital (widowed vs married)	-.18716	.1071532	-1.75	0.081	-.3973787	.0230587
(divorced vs widowed)	-.1688347	.1158975	-1.46	0.145	-.3962085	.0585391
(separated vs divorced)	-.2528838	.1575109	-1.61	0.109	-.5618967	.0561291
(never ma.. vs separated)	.3270313	.154727	2.11	0.035	.02348	.6305826

The output above does not look terrific because of the need to wrap the labels associated with marital to make the output fit on the printed page. If the printed page was wider, I would use this style of output because the output is both compact and fairly comprehensive (providing both significance tests and confidence intervals for each contrast).

As a compromise, I will often specify the contrast(nowald pveffects) option to display compact output that includes significance tests for each contrast, as shown below.

```
. margins ar.marital, contrast(nowald pveffects)
Contrasts of predictive margins
Expression    : Linear prediction, predict()
```

	Contrast	Delta-method Std. Err.	t	P>|t|
marital (widowed vs married)	-.18716	.1071532	-1.75	0.081
(divorced vs widowed)	-.1688347	.1158975	-1.46	0.145
(separated vs divorced)	-.2528838	.1575109	-1.61	0.109
(never married vs separated)	.3270313	.154727	2.11	0.035

Using this style of output is not meant to imply that confidence intervals are unimportant but is a compromise that saves space while providing the minimally essential details associated with each contrast.

The `mcompare` option can be used to adjust the displayed significance values to account for multiple comparisons. You can specify `bonferroni`, `sidak`, or `scheffe` with `mcompare()` to obtain Bonferroni's, Šidák's, or Scheffé's method of adjustment (respectively). The example below illustrates using the `sidak` option.

```
. margins ar.marital, contrast(nowald pveffects) mcompare(sidak)
Contrasts of predictive margins
Expression    : Linear prediction, predict()
```

	Number of Comparisons
marital	4

	Contrast	Delta-method Std. Err.	Sidak t	P>|t|
marital				
(widowed vs married)	-.18716	.1071532	-1.75	0.287
(divorced vs widowed)	-.1688347	.1158975	-1.46	0.467
(separated vs divorced)	-.2528838	.1575109	-1.61	0.369
(never married vs separated)	.3270313	.154727	2.11	0.132

21.3.3 Margins with factor variables and the at() option

We can specify both factor variables and `at()` options with the `margins` command. The example below estimates the adjusted mean of happiness by marital status holding age constant at 30 and then again at 50, adjusting for all other covariates.

```
. margins marital, at(age=(30 50)) vsquish nopvalues
Predictive margins                                    Number of obs    =     1,272
Expression  : Linear prediction, predict()
1._at       : age                =          30
2._at       : age                =          50
```

	Margin	Delta-method Std. Err.	[95% Conf. Interval]	
_at#marital				
1#married	5.644882	.0531434	5.540622	5.749141
1#widowed	5.457722	.1189978	5.224266	5.691178
1#divorced	5.288887	.0795758	5.132771	5.445003
1#separated	5.036003	.1463321	4.748921	5.323085
1#never married	5.363034	.0550966	5.254943	5.471126
2#married	5.679507	.0397145	5.601593	5.757421
2#widowed	5.492347	.0991147	5.297899	5.686795
2#divorced	5.323512	.0658268	5.19437	5.452655
2#separated	5.070629	.1432739	4.789547	5.351711
2#never married	5.39766	.0590425	5.281827	5.513492

The first 5 lines of output correspond to the adjusted means for the 5 levels of marital status when age is held constant at 30. The second 5 lines (6 to 10) of the output show the adjusted means for the 4 levels of marital status when age is held constant at 50.

21.3.4   The dydx() option

Let's consider the ANCOVA model from section 8.3, where the DV was optimism, the covariate was depression measured before the start of the study, and the IV had three levels (1 = control, 2 = traditional therapy, 3 = optimism therapy). Using the `anova` command below, we fit an ANCOVA model where the IV and covariate are entered into the model but no interaction.

```
. use opt-ancova4.dta, clear
. anova opt i.treat c.depscore
                          Number of obs =        300    R-squared       =  0.4388
                          Root MSE      =    7.16382    Adj R-squared   =  0.4332
```

Source	Partial SS	df	MS	F	Prob>F
Model	11879.704	3	3959.9012	77.16	0.0000
treat	3034.9292	2	1517.4646	29.57	0.0000
depscore	9403.7837	1	9403.7837	183.24	0.0000
Residual	15190.816	296	51.320325		
Total	27070.52	299	90.536856		

Say that we wanted to focus on the relationship between the covariate (`depscore`) and the DV (`opt`). Let's first visualize this relationship by using the `margins` command followed by the `marginsplot` command. The `margins` command is used below to estimate the adjusted mean of optimism when depression scores are 0, 20, 40, and 60. Then, the `marginsplot` command is used to graph the results of the `margins` command (see figure 21.1), showing the relationship between the covariate (`depscore`) and the DV (`opt`).[2]

```
. margins, at(depscore=(0(20)60)) vsquish
Predictive margins                          Number of obs    =        300
Expression    : Linear prediction, predict()
1._at         : depscore        =           0
2._at         : depscore        =          20
3._at         : depscore        =          40
4._at         : depscore        =          60
```

	Margin	Delta-method Std. Err.	t	P>|t|	[95% Conf. Interval]	
_at						
1	55.92426	.7497045	74.60	0.000	54.44883	57.39969
2	47.39174	.4136341	114.57	0.000	46.5777	48.20578
3	38.85922	.7581366	51.26	0.000	37.3672	40.35124
4	30.3267	1.331575	22.78	0.000	27.70615	32.94725

```
. marginsplot, noci
  Variables that uniquely identify margins: depscore
```

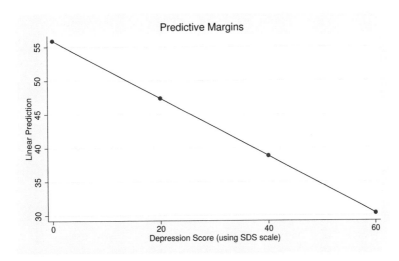

Figure 21.1: Adjusted mean of optimism as a function of depression score

As shown in figure 21.1, the relationship between depression and optimism is negative. Greater depression is associated with less optimism. We can quantify this rela-

---

2. After accounting for other predictors in the model.

tionship by computing the slope of the line depicted in figure 21.1 using the `margins` command, shown below. When we include the `dydx(depscore)` option, the `margins` command computes the slope with respect to `depscore`.

```
. margins, dydx(depscore)
Average marginal effects                         Number of obs    =       300
Expression    : Linear prediction, predict()
dy/dx w.r.t.  : depscore
```

	dy/dx	Delta-method Std. Err.	t	P>|t|	[95% Conf. Interval]	
depscore	-.426626	.0315167	-13.54	0.000	-.4886512	-.3646008

The results of the `margins` command tell us that the regression line depicted in figure 21.1 has a slope of -0.43. In other words, we expect a -0.43 unit decrease in optimism for every unit increase in depression, after accounting for the other predictors in the model.

Now, let's consider a second ANCOVA model in which we include an interaction between the IV and the covariate, as shown in the `anova` command below.

```
. anova opt i.treat##c.depscore
```

		Number of obs = 300	R-squared = 0.4971		
		Root MSE = 6.80489	Adj R-squared = 0.4885		
Source	Partial SS	df	MS	F	Prob>F
Model	13456.382	5	2691.2763	58.12	0.0000
treat	3266.27	2	1633.135	35.27	0.0000
depscore	8903.3987	1	8903.3987	192.27	0.0000
treat#depscore	1576.678	2	788.33902	17.02	0.0000
Residual	13614.138	294	46.306593		
Total	27070.52	299	90.536856		

We see that the `treat` by `depscore` interaction is significant [$F(2, 294) = 17.02$, $p < 0.001$]. We can visualize this interaction by first using the `margins` command to estimate the adjusted mean of the DV as a function of both the IV (treatment) and the covariate (depression score). The `margins` command below estimates the adjusted mean of optimism at each level of treatment crossed by 4 levels of depression (0, 20, 40, and 60). This is immediately followed by the `marginsplot` command, which creates the graph shown in figure 21.2, providing a visualization of the `treat#depscore` interaction.

```
. margins treat, at(depscore=(0(20)60)) vsquish
Adjusted predictions                           Number of obs      =        300
Expression    : Linear prediction, predict()
1._at         : depscore       =          0
2._at         : depscore       =         20
3._at         : depscore       =         40
4._at         : depscore       =         60
```

	Margin	Delta-method Std. Err.	t	P>|t|	[95% Conf. Interval]	
_at#treat						
1#Con	48.52259	1.229369	39.47	0.000	46.10311	50.94206
1#TT	55.78129	1.24348	44.86	0.000	53.33404	58.22854
1#OT	63.12268	1.22912	51.36	0.000	60.70369	65.54167
2#Con	43.07592	.6835653	63.02	0.000	41.73062	44.42123
2#TT	49.42358	.6805675	72.62	0.000	48.08418	50.76298
2#OT	49.95576	.6809653	73.36	0.000	48.61557	51.29594
3#Con	37.62926	1.339181	28.10	0.000	34.99367	40.26486
3#TT	43.06586	1.226283	35.12	0.000	40.65246	45.47927
3#OT	36.78884	1.187065	30.99	0.000	34.45262	39.12506
4#Con	32.1826	2.343028	13.74	0.000	27.57137	36.79383
4#TT	36.70814	2.160564	16.99	0.000	32.45601	40.96028
4#OT	23.62192	2.084937	11.33	0.000	19.51862	27.72521

```
. marginsplot, noci
Variables that uniquely identify margins: depscore treat
```

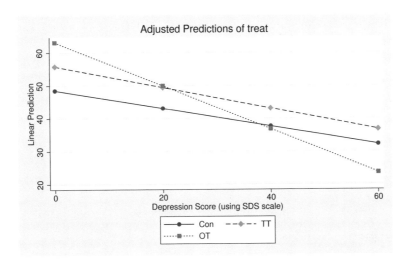

Figure 21.2: Visualizing the treatment by depression interaction (DV = Optimism)

The graph in figure 21.2 shows three regression lines (associated with the three levels of treatment). Each of these regression lines has its own individual slope. We can estimate the slope of each line using the `margins` command below. By specifying `treat` and the `dydx(depscore)` option with the `margins` command, we see that the output shows the slope with respect to `depscore` separately for each level of `treat`, as visualized in figure 21.2.

```
. margins treat, dydx(depscore)
Average marginal effects                          Number of obs    =        300
Expression    : Linear prediction, predict()
dy/dx w.r.t.  : depscore

                         Delta-method
              dy/dx    Std. Err.        t     P>|t|     [95% Conf. Interval]

depscore
     treat
       Con   -.2723331   .0544315    -5.00    0.000    -.3794579   -.1652083
        TT   -.3178858   .0515227    -6.17    0.000    -.4192858   -.2164858
        OT    -.658346   .0499053   -13.19    0.000    -.7565629    -.560129
```

The output shows three slopes corresponding to the slopes of the three regression lines depicted in figure 21.2. The slope for the control group is -0.27, compared with -0.32 for the traditional therapy group and -0.66 for the optimism therapy group.

Say that we wanted to test a specific hypothesis regarding the equality of these slopes. We can do so by applying contrast operators to `treat` to make specific comparisons among the slopes. In the example below, the `r3b1` contrast operator is used, which compares the slope for the optimism therapy group with the slope for the control group (group 3 versus group 1). The difference in the slopes (optimism therapy versus control) is -0.39 (95% CI $= [-0.53, -0.24]$) and is significant [$F(1, 294) = 27.32$, $p < 0.001$].

```
. margins r3b1. treat, dydx(depscore)
Contrasts of average marginal effects
Expression    : Linear prediction, predict()
dy/dx w.r.t.  : depscore

                     df          F        P>F

depscore
       treat          1      27.32     0.0000

Denominator          294


                  Contrast Delta-method
                     dy/dx    Std. Err.    [95% Conf. Interval]

depscore
       treat
(OT vs Con)      -.3860129    .0738467     -.531348   -.2406778
```

21.4   The marginsplot command

The body of the book has shown numerous examples using the `marginsplot` command, but those examples have focused on creating utilitarian graphs without much regard for customizing the look of the graphs. This section illustrates some of the options you can use to customize the appearance of graphs created by the `marginsplot` command. Some of the options are specific to the `marginsplot` command, while others are options supported by the `graph twoway` command. You can see `help graph twoway` for more information about such options as well as Mitchell (2012b).

Let's consider a multiple regression model that predicts the education of the respondent from the father's education, mother's education, and age of the respondent. The `margins` command is then used to compute the adjusted means when the father's education ranges from 0 to 20. The `marginsplot` command can then be used to graph the adjusted means computed by the `margins` command.

```
. use gss2012_sbs
. regress educ paeduc maeduc age
  (output omitted )
. margins, at(paeduc=(0(1)20))
  (output omitted )
```

`marginsplot`

This shows the adjusted means as a function of the father's education. The following examples illustrate how you can use options to customize this graph.

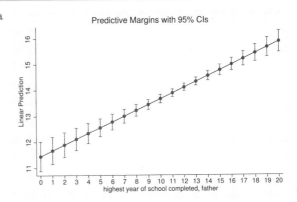

`marginsplot, `**`title(Title) subtitle(Subtitle) xtitle(X title) ///`**
    `ytitle(Y title) note(Note) caption(Caption)`

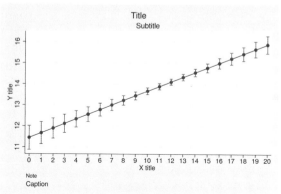

You can add titles to the graph using the `title()`, `subtitle()`, `xtitle()`, and `ytitle()` options. The `note()` and `caption()` options can also be used to annotate the graph.

`marginsplot, `**`xlabel(0(5)20) ylabel(10(1)18, angle(0))`**

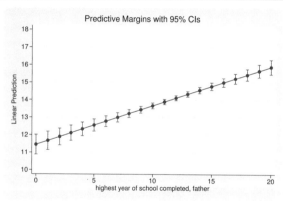

The `xlabel()` and `ylabel()` options can be used to change the labeling of the x and y axis. In addition, the `angle(0)` option is used to specify the angle of the display of the labels on the y axis. In this example, the labels are displayed using an angle of zero degrees (that is, horizontal display of the labels).

`marginsplot, `**`xscale(range(-1 21)) yscale(range(8 20))`**

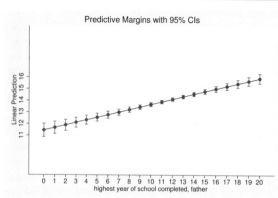

The `xscale()` and `yscale()` options can be used to expand the scale of the x and y axes.

marginsplot, **plotopts(clwidth(thick) msymbol(Oh) msize(large))**

The plotopts() option allows you
to include options that control the
look of the line and markers. The
clwidth() option is used to make
the fitted line thick, and the
msymbol() and msize() options are
used to draw the markers as large
hollow circles.

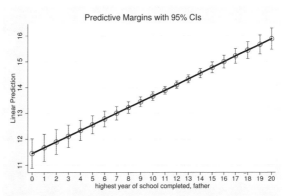

marginsplot, **ciopts(lwidth(vthick) msize(huge))**

The ciopts() option allows you to
include options that control the
look of the confidence interval. The
lwidth() option makes the lines for
the CIs thick, and the msize()
options makes the cap of each
confidence interval huge.

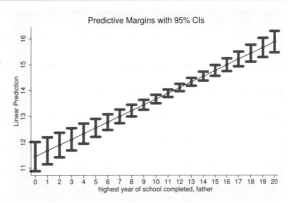

marginsplot, **recast(line) recastci(rarea)**

The recast(line) option specifies
that the fitted line be drawn like a
twoway line graph. The
recastci(rarea) option specifies
that the confidence interval be
drawn like a twoway rarea graph.

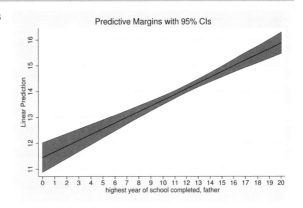

`marginsplot, `**`noci`**

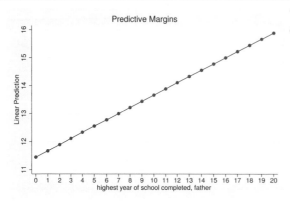

The `noci` option suppresses display of the confidence interval.

`marginsplot, noci `**`addplot(scatter educ paeduc, msymbol(o))`**

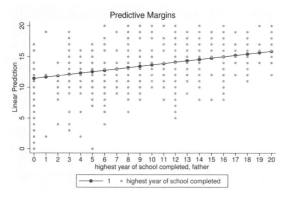

The `addplot()` option can be used to overlay a new graph onto the graph created by the `margins` command. In this case, it overlays a scatterplot of `educ` and `paeduc`.

`marginsplot, `**`scheme(economist)`**

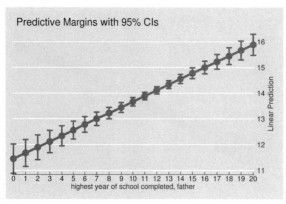

The `scheme()` option can be used to change the overall look of the graph. In this example, the `economist` scheme is used.

Let's now consider another example, this one based on the three-by-three design illustrated in section 7.4 of chapter 7. The commands below use the dataset, run the `anova` command, and use the `margins` command to obtain the mean of the DV by `treat` and `depstat`. The output of these commands is suppressed to save space. The `marginsplot` can be used to graph the results of the interaction as computed by the `margins` command.

```
. use opt-3by3
. anova opt treat##depstat
  (output omitted)
. margins treat##depstat
  (output omitted)
```

marginsplot

This shows the graph created by the `marginsplot` command. The following examples illustrate how to customize this graph, focusing on the plot dimension and the legend associated with the plot dimension.

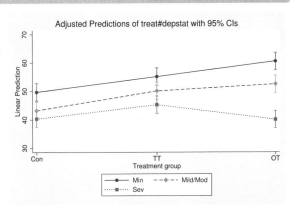

marginsplot, **xdim(depstat)**

The `xdimension()` option controls which variable is placed on the x axis. This option is used to graph `depstat` on the x axis, and hence, `treat` is graphed as separate lines.

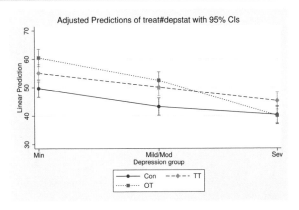

marginsplot, **plotdim(depstat)**

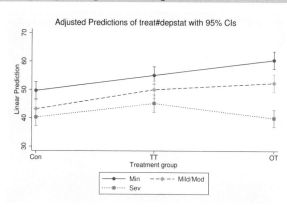

The plotdim() option controls which variable is graphed using the plot dimension, that is, graphed using separate lines. In this example, depstat is graphed using separate lines, and thus treat is placed on the x axis.

marginsplot, **plotdim(depstat, labels("Non-dep" "Mild dep" "Sev. dep"))**

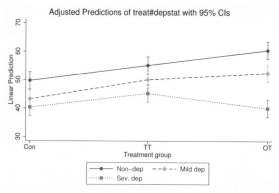

The labels() suboption within the plotdim() option changes the labels used for the plot dimension.

marginsplot, ///
    plotdim(depstat, elabels(1 "Non-depressed" 3 "Severely depressed"))

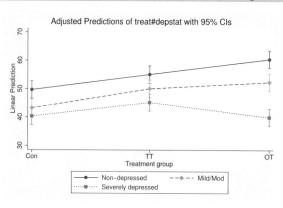

Using the elabels() suboption, you can selectively modify the labels of your choice. This example modifies the labels for the first and third groups, leaving the label for the second group unchanged.

`marginsplot, plotdim(depstat, nosimplelabels)`

Adding the `nosimplelabels` option changes the plot label to the variable name, an equal sign, and the value label for the group.

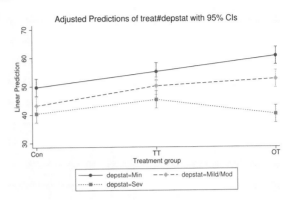

`marginsplot, plotdim(depstat, nolabels)`

Adding the `nolabels` option changes the plot label to the variable name, an equal sign, and the numeric value for the group.

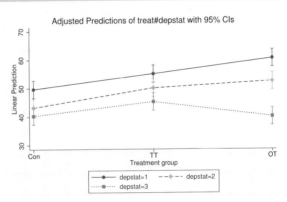

`marginsplot, plotdim(depstat, allsimplelabels)`

Using the `allsimplelabels` option yields a label that is solely composed of the value label for each group.

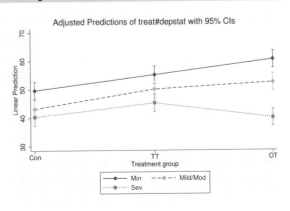

`marginsplot, ` **`plotdim(depstat, allsimplelabels nolabels)`**

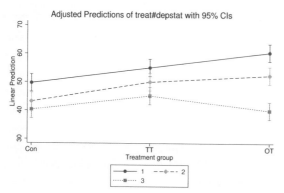

Using the `allsimplelabels` and `nolabels` options displays a label that is solely composed of the numeric value for each group.

`marginsplot, ` **`legend(subtitle("Treatment") rows(1))`**

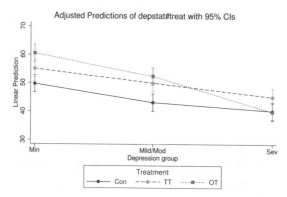

This example includes the `legend()` option to customize the display of the legend, adding a subtitle and displaying the legend keys in one row.

`marginsplot, legend(subtitle("Treatment") rows(1) ` **`ring(0) pos(11))`**

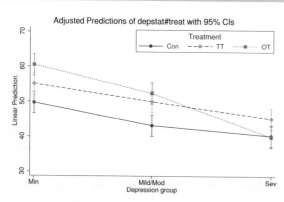

The `ring()` and `pos()` suboptions are added to the `legend()` option to display the legend within the graph in the one o'clock position.

`marginsplot, xlabel(, labsize(large))`

The `xlabel(, labsize())` option
is used to control the size of the
labels on the x axis, in this case
making the labels large.

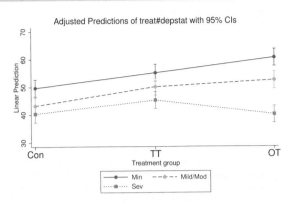

`marginsplot, xlabel(1 "Control" 2 "Traditional" 3 "Optimism Therapy")`

The `xlabel()` option is used to
control the labeling of the x axis, as
shown in this example. The next
example illustrates how to address
the issue of the optimism therapy
label being cut off.

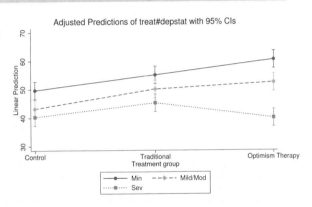

`marginsplot, xlabel(1 "Control" 2 "Traditional" 3 "Optimism Therapy") ///`
    `xscale(range(.75 3.25))`

The `xscale(range())` option is
used to expand the range of the x
axis to make additional room for
the longer x-axis labels.

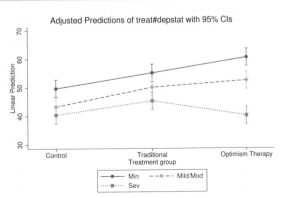

Let's now consider an example where the `marginsplot` command involves by-groups as an additional dimension. This example comes from section 9.2, in which a two-by-two-by-two ANOVA is used to predict the DV happiness from three IVs: depression status, treatment, and season. The `anova` command is run below, which produces a significant three-way interaction. The `margins` command is then used to compute the mean of happiness by depression status, treatment, and season. The output of these commands are omitted to save space.

```
. use hap-2by2by2, clear
. anova hap depstat##treat##season
  (output omitted)
. margins depstat#treat#season, nopvalues
  (output omitted)
```

marginsplot, noci

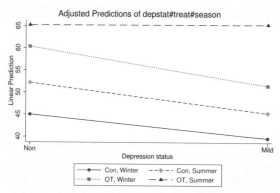

This shows the graph created by the `marginsplot` command, graphing the average happiness computed by the `margins` command. The `noci` option is used to suppress the confidence intervals. The following examples customize the graph, focusing on issues related to the by dimension.

marginsplot, noci bydim(season)

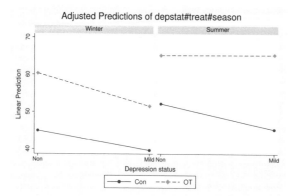

The `bydim()` option is used to specify that separate graphs should be made by `season`.

`marginsplot, noci bydim(season) byopts(cols(1))`

The `byopts()` option allows you to specify suboptions that control the way that the separate graphs are combined. In this example, the `cols(1)` option causes the graphs to be displayed in one column.

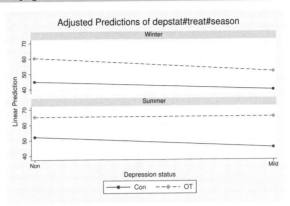

`marginsplot, noci bydim(season) ///`
`    byopts(title(Title) subtitle(Subtitle) note(Note) caption(Caption))`

The `byopts()` option can be used to control the overall title, subtitle, note, and caption for the graph. By placing such options within the `byopts()` option, you can impact the overall title, subtitle, note, and caption. Contrast this with the next example.

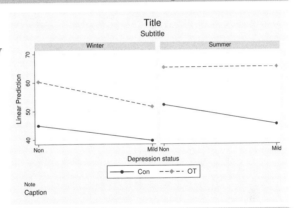

`marginsplot, noci bydim(season) ///`
`    title(Title) subtitle(Subtitle) note(Note) caption(Caption)`

By placing these options outside the `byopts()` option, you can control the title, subtitle, note, and caption for each of the graphs.

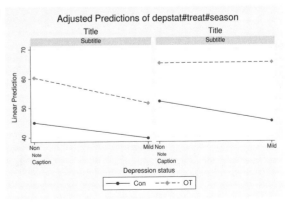

This section has illustrated some of the ways that you can customize graphs created by the `marginsplot` command. For more information, you can see `help marginsplot`, `help graph twoway`, and Mitchell (2012b).

21.5   The pwcompare command

This section covers additional details about the `pwcompare` command, which allows you to make pairwise comparisons among means from a factor variable.

The `mcompare()` option permits you to select a method for adjusting for multiple comparisons. The default method is to perform no adjustment for multiple comparisons. Four methods are provided that can be used with balanced or unbalanced data: Tukey's method, Bonferroni's method, Šidák's method, and Scheffé's method. These can be selected by specifying, respectively, `tukey`, `bonferroni`, `sidak`, or `scheffe` within the `mcompare()` option. Three additional methods are provided but require balanced data and can be used only after a linear modeling command (that is, `anova`, `manova`, `regress`, or `mvreg`). These are the Student–Newman–Keuls, Duncan, and Dunnett methods. These can be specified using, respectively, the `snk`, `duncan`, or `dunnett` option.

Let's consider an example of a one-way ANCOVA using unbalanced data, where `happy7` is the DV, `marital` is the IV, and `age` is a covariate.

```
. use gss2012_sbs, clear

. regress happy7 i.marital age
```

Source	SS	df	MS		Number of obs	=	1,282
					F(5, 1276)	=	11.16
Model	53.3217777	5	10.6643555		Prob > F	=	0.0000
Residual	1219.1751	1,276	.955466381		R-squared	=	0.0419
					Adj R-squared	=	0.0381
Total	1272.49688	1,281	.993362123		Root MSE	=	.97748

happy7	Coef.	Std. Err.	t	P>|t|	[95% Conf. Interval]	
marital						
widowed	-.2438361	.1123112	-2.17	0.030	-.4641709	-.0235013
divorced	-.4528681	.0804728	-5.63	0.000	-.6107417	-.2949945
separated	-.6027107	.1563491	-3.85	0.000	-.9094402	-.2959812
never mar..	-.3824663	.0725681	-5.27	0.000	-.5248322	-.2401004
age	-.0020545	.0019295	-1.06	0.287	-.0058399	.0017308
_cons	5.816756	.1036413	56.12	0.000	5.61343	6.020082

The `pwcompare` command (shown below) performs pairwise comparisons of the adjusted means (after adjusting for `age`). The default output includes the difference in the adjusted means (in the `Contrast` column) and a confidence interval for the difference. When the confidence interval excludes zero, the difference is significant.

```
. pwcompare marital
Pairwise comparisons of marginal linear predictions
Margins      : asbalanced
```

	Contrast	Std. Err.	Unadjusted [95% Conf. Interval]	
marital				
widowed vs married	-.2438361	.1123112	-.4641709	-.0235013
divorced vs married	-.4528681	.0804728	-.6107417	-.2949945
separated vs married	-.6027107	.1563491	-.9094402	-.2959812
never married vs married	-.3824663	.0725681	-.5248322	-.2401004
divorced vs widowed	-.209032	.12227	-.4489043	.0308403
separated vs widowed	-.3588746	.1851419	-.7220907	.0043415
never married vs widowed	-.1386302	.1313014	-.3962206	.1189601
separated vs divorced	-.1498426	.1664839	-.4764549	.1767696
never married vs divorced	.0704018	.0948233	-.1156248	.2564284
never married vs separated	.2202444	.1613186	-.0962344	.5367231

We can use the `pveffects` option to show the significance test of the difference in the means instead of the confidence interval. (We could instead specify `effects` to show both the significance tests and the confidence intervals.) The results show significant differences for the first four comparisons, while the remaining differences are all not significant. Note how the significant comparisons all involve comparisons with those who are married.

```
. pwcompare marital, pveffects
Pairwise comparisons of marginal linear predictions
Margins      : asbalanced
```

	Contrast	Std. Err.	Unadjusted t	P>|t|
marital				
widowed vs married	-.2438361	.1123112	-2.17	0.030
divorced vs married	-.4528681	.0804728	-5.63	0.000
separated vs married	-.6027107	.1563491	-3.85	0.000
never married vs married	-.3824663	.0725681	-5.27	0.000
divorced vs widowed	-.209032	.12227	-1.71	0.088
separated vs widowed	-.3588746	.1851419	-1.94	0.053
never married vs widowed	-.1386302	.1313014	-1.06	0.291
separated vs divorced	-.1498426	.1664839	-0.90	0.368
never married vs divorced	.0704018	.0948233	0.74	0.458
never married vs separated	.2202444	.1613186	1.37	0.172

By including the `cimargins` option as well, you get not only the pairwise comparisons of the means but also an estimate of each of the means with a confidence interval.

```
. pwcompare marital, pveffects cimargins
Pairwise comparisons of marginal linear predictions
Margins      : asbalanced
```

	Margin	Std. Err.	Unadjusted [95% Conf. Interval]	
marital				
married	5.816756	.1036413	5.61343	6.020082
widowed	5.57292	.1686286	5.2421	5.90374
divorced	5.363888	.1252124	5.118243	5.609533
separated	5.214045	.1752395	4.870256	5.557835
never married	5.43429	.0830535	5.271353	5.597226

	Contrast	Std. Err.	Unadjusted t	P>|t|
marital				
widowed vs married	-.2438361	.1123112	-2.17	0.030
divorced vs married	-.4528681	.0804728	-5.63	0.000
separated vs married	-.6027107	.1563491	-3.85	0.000
never married vs married	-.3824663	.0725681	-5.27	0.000
divorced vs widowed	-.209032	.12227	-1.71	0.088
separated vs widowed	-.3588746	.1851419	-1.94	0.053
never married vs widowed	-.1386302	.1313014	-1.06	0.291
separated vs divorced	-.1498426	.1664839	-0.90	0.368
never married vs divorced	.0704018	.0948233	0.74	0.458
never married vs separated	.2202444	.1613186	1.37	0.172

Specifying the **groups** option displays group codes that signify groups that are not significantly different from each other. This output indicates that groups 1, 2, 3, and 4 are all not significantly different from each other because they all belong to group A. Group 1 is significantly different from groups 2, 3, and 4 because that group does not belong to group A. This provides a concise summary of which pairwise differences are significant and nonsignificant.

```
. pwcompare marital, groups
Pairwise comparisons of marginal linear predictions
Margins       : asbalanced
```

	Margin	Std. Err.	Unadjusted Groups
marital			
married	5.816756	.1036413	
widowed	5.57292	.1686286	A
divorced	5.363888	.1252124	A
separated	5.214045	.1752395	A
never married	5.43429	.0830535	A

```
Note: Margins sharing a letter in the group label
      are not significantly different at the 5%
      level.
```

So far, these pairwise comparisons have not adjusted for multiple comparisons. Let's adjust for multiple comparisons by adding the mcompare(tukey) option.[3] After making this adjustment, we see that the comparison of the widowed with the married group is no longer significant ($p = 0.191$). The other three comparisons of the married group with the divorced, separated, and never married groups are all significant ($ps \leq 0.001$).

```
. pwcompare marital, pveffects mcompare(tukey)
Pairwise comparisons of marginal linear predictions
Margins       : asbalanced
```

	Number of Comparisons
marital	10

	Contrast	Std. Err.	Tukey t	P>|t|
marital				
widowed vs married	-.2438361	.1123112	-2.17	0.191
divorced vs married	-.4528681	.0804728	-5.63	0.000
separated vs married	-.6027107	.1563491	-3.85	0.001
never married vs married	-.3824663	.0725681	-5.27	0.000
divorced vs widowed	-.209032	.12227	-1.71	0.428
separated vs widowed	-.3588746	.1851419	-1.94	0.297
never married vs widowed	-.1386302	.1313014	-1.06	0.829
separated vs divorced	-.1498426	.1664839	-0.90	0.897
never married vs divorced	.0704018	.0948233	0.74	0.946
never married vs separated	.2202444	.1613186	1.37	0.650

---

3. If we preferred, we could have used the mcompare(bonferroni), mcompare(sidak), or the mcompare(scheffe) options to obtain Bonferroni's, Šidák's, or Scheffé's method of adjustment (respectively).

Finally, let's consider an example of a one-way ANOVA using balanced data, that is, in which all groups have equal cell sizes. Let's use the example from section 5.6 that looked at pain ratings (the DV) as a function of medication dosage (the IV). The dosage group is stored in the variable `dosegrp` and is coded as follows: 1 = 0 mg, 2 = 50 mg, 3 = 100 mg, 4 = 150 mg, 5 = 200 mg, and 6 = 250 mg. The `anova` command below performs a one-way ANOVA with `pain` as the DV and `dosegrp` as the IV.

```
. use pain
. anova pain dosegrp
                              Number of obs =        180    R-squared     =  0.4602
                              Root MSE      =   10.4724     Adj R-squared =  0.4447

          Source |  Partial SS          df          MS           F      Prob>F

           Model |   16271.694           5      3254.3389      29.67    0.0000

         dosegrp |   16271.694           5      3254.3389      29.67    0.0000

        Residual |   19082.633         174      109.67031

           Total |   35354.328         179      197.51021
```

Let's use the `pwcompare` command to form pairwise comparisons of all dosage groups using Tukey's method of adjustment for multiple comparisons. Further, let's create a table of groups that are not significantly different by using the `group` option.

```
. pwcompare dosegrp, mcompare(tukey) group
Pairwise comparisons of marginal linear predictions
Margins        : asbalanced
```

	Number of Comparisons
dosegrp	15

	Margin	Std. Err.	Tukey Groups
dosegrp			
Zero (control)	71.83333	1.911982	A
50 mg	70.6	1.911982	A
100 mg	72.13333	1.911982	A
150 mg	70.4	1.911982	A
200 mg	54.7	1.911982	B
250 mg	48.3	1.911982	B

```
Note: Margins sharing a letter in the group label
      are not significantly different at the 5%
      level.
```

The output shows that groups 5 and 6 both belong to group B and are not significantly different. Also, groups 1, 2, 3, and 4 all belong to group A and are not significantly different. For this example with balanced data, the Student–Newman–Keuls, Duncan, or Dunnett method could have been selected by specifying, respectively, the `snk`, `duncan`, or, `dunnett` suboption instead of the `tukey` suboption. We also could have selected the `bonferroni`, `sidak`, or `scheffe` option with `mcompare()` to obtain, respectively, Bonferroni's, Šidák's, or Scheffé's method of adjustment. These methods can be used with either balanced or unbalanced data.

You can find more information about the `pwcompare` command by seeing `help pwcompare`.

21.6   Closing thoughts

In this book, I have frequently shown how to apply `contrast`, `margins`, `marginsplot`, and `pwcompare`. This chapter focused on the commands themselves, illustrating options when using these commands in one central place. In this book, I have covered other postestimation commands as well. For example, I illustrated the `test` command in section 14.4 and the `lincom` command in section 14.4.3 and the `estimates table` command in chapter 16.

There are many other postestimation commands that I have not addressed. For information about all the postestimation commands supported by Stata, you can type

```
. help postestimation commands
```

This will show all the different postestimation commands included in Stata. This list may include commands that are not appropriate following the estimation command that you just used. Say that you are interested in seeing the postestimation commands that can be used following the `regress` command. In that case, you can type the following command.

```
. help regress postestimation
```

This will show the postestimation commands that are appropriate after using the `regress` command.

You can insert any estimation command in place of `regress`, for example,

```
. help rreg postestimation
```

The above command will show you the postestimation commands that can be used after the `rreg` command (for robust regression).

22 Stata data management commands

22.1 Chapter overview

This chapter provides a brief overview and examples of common data management tasks using Stata. It illustrates reading data into Stata (section 22.2), saving datasets (section 22.3), labeling data (section 22.4), creating and recoding variables (section 22.5), keeping and dropping variables (section 22.6), keeping and dropping observations (section 22.7), combining datasets (section 22.8), and reshaping datasets (section 22.9).

22.2   Reading data into Stata

22.2.1   Reading Stata datasets

Obviously, the easiest kind of file that you can read into Stata is a Stata dataset. Such datasets have a .dta extension and can be read into Stata via the use command. As shown below, the use command reads the Stata dataset named cardio1.dta into memory. The list command is then used to show the values for all the variables in the entire dataset.

```
. use cardio1
. list
```

	id	age	bp1	bp2	bp3	bp4	bp5	pl1	pl2	pl3	pl4	pl5
1.	1	40	115	86	129	105	127	54	87	93	81	92
2.	2	30	123	136	107	111	120	92	88	125	87	58
3.	3	16	124	122	101	109	112	105	97	128	57	68
4.	4	23	105	115	121	129	137	52	79	71	106	39
5.	5	18	116	128	112	125	111	70	64	52	68	59

Sometimes, when you go to use a dataset, Stata might refuse to obey your request, as illustrated below.

```
. use cardio1
no; data in memory would be lost
r(4);
```

This error message is trying to tell you that you have unsaved changes in your current dataset. Before you can read a new dataset into memory, you need to either save the current dataset or clear out the current dataset. If you want to save the current dataset, you can do so with the save command. On the other hand, if you do not care to save the current dataset, you can use the clear command to clear the current dataset from memory so that you can then use another dataset, as illustrated below.

```
. clear
. use cardio1
```

Optionally, you can combine the above two steps into a single step by specifying the use command with the clear option.

```
. use cardio1, clear
```

22.2.2   Reading Excel workbooks

You can read Excel workbooks into Stata using the `import excel` command. The example below reads the Excel workbook named `cardio.xls` into Stata. When we include the `firstrow` option, the first row of the worksheet is used to name the variables in the Stata dataset.

```
. import excel using cardio.xls, firstrow
```

The `list` command below shows that this file was successfully read into Stata.

```
. list
```

	id	age	bp1	bp2	bp3	bp4	bp5	pl1	pl2	pl3	pl4	pl5
1.	1	40	115	86	129	105	127	54	87	93	81	92
2.	2	30	123	136	107	111	120	92	88	125	87	58
3.	3	16	124	122	101	109	112	105	97	128	57	68
4.	4	23	105	115	121	129	137	52	79	71	106	39
5.	5	18	116	128	112	125	111	70	64	52	68	59

Video tutorial: Import Excel workbooks to Stata

See a video demonstrating how to import Excel workbooks into Stata at http://www.stata.com/sbs/import-excel. The *Stata Blog* also illustrates a real-world example of reading Excel workbooks at http://blog.stata.com/2012/06/25/using-import-excel-with-real-world-data/.

The `import excel` option includes additional options to control how the workbook is read into Stata. For example, the `sheet()` option allows you to specify the worksheet to be loaded, and the `cellrange()` option allows you to specify the range of cells to be loaded. You can see `help import excel` for additional details.

> **Video tutorial: Paste Excel workbooks to Stata**
> You can copy data from Excel and paste them directly into the Stata Data Editor
> as shown in the video at http://www.stata.com/sbs/paste-excel.

22.2.3 Reading comma-separated files

Another common file format is a comma-separated file. Such files usually have a `.csv`
extension and often have variable names in the first row. Consider the file `cardio.csv`
shown below using the `type` command.

```
. type cardio.csv
id,mf,ag,bp1,bp2,bp3,bp4,bp5,pl1,pl2,pl3,pl4,pl5
1,1,40,115,86,129,105,127,54,87,93,81,92
2,1,30,123,136,107,111,120,92,88,125,87,58
3,2,16,124,122,101,109,112,105,97,128,57,68
4,2,23,105,115,121,129,137,52,79,71,106,39
5,1,18,116,128,112,125,111,70,64,52,68,59
```

We can see that this file contains variable names on the first row and then five rows
of data in which each variable is separated by a comma. We can read this file into Stata
using the `import delimited` command.

```
. import delimited using cardio.csv
(13 vars, 5 obs)
```

Using the `list` command, we can see that this file was successfully read into Stata.[1]

```
. list, noobs
```

id	mf	ag	bp1	bp2	bp3	bp4	bp5	pl1	pl2	pl3	pl4	pl5
1	1	40	115	86	129	105	127	54	87	93	81	92
2	1	30	123	136	107	111	120	92	88	125	87	58
3	2	16	124	122	101	109	112	105	97	128	57	68
4	2	23	105	115	121	129	137	52	79	71	106	39
5	1	18	116	128	112	125	111	70	64	52	68	59

22.2.4 Reading other file formats

If you type `help import`, you can see other kinds of files that Stata can read, including
SAS XPORT Transport format files as well as various raw-data file formats.

---

1. I added the `noobs` option to suppress the observation number. Otherwise, the observations would
be so wide they would not each fit onto one line.

However, there are some datasets that Stata cannot directly read, including native SAS datasets or native SPSS datasets. For such cases, I suggest that you consider purchasing Stat/Transfer, which you can obtain directly from StataCorp. Stat/Transfer can convert virtually any kind of dataset from one format to another format (for example, from Stata to SAS, from SPSS to Stata, from SAS to SPSS, from Access to Excel, and so forth). I have used Stat/Transfer for many years and have been consistently impressed with its ease of use and how well it works. Speaking for myself, I cannot imagine doing my daily work without Stat/Transfer.

Video tutorial: Stat/Transfer

See a video demonstrating how you can convert data to Stata with Stat/Transfer at http://www.stata.com/sbs/stat-transfer.

22.3   Saving data

The way that you save a Stata dataset is through the `save` command. This command saves the dataset that is currently in memory using the name that you specify. In the example below, the dataset in memory is saved as `dissertation.dta`. Note that we did not need to specify the `.dta` extension.

```
. save dissertation
file dissertation.dta saved
```

However, if the file `dissertation.dta` already exists, Stata will give the following error message.

```
. save dissertation
file dissertation.dta already exists
r(602);
```

Stata is protecting you from overwriting an existing dataset by warning you that the dataset already exists. If you want to overwrite `dissertation.dta` with the dataset currently in memory, then you can include the `replace` option to indicate that you want to replace the existing file, as illustrated below.

```
. save dissertation, replace
file dissertation.dta saved
```

Say that you use the `save` command to create the dataset `dissertation.dta` and share it with a friend of yours who had an older version of Stata than you do. When your friend tries to use the file, he or she will likely receive an error like this.

```
. use dissertation
dissertation.dta not Stata format
(610);
```

This is because the older version of Stata does not understand how to read your newer Stata dataset. To share a dataset with someone who is using an older version of Stata, you can use the `saveold` command instead of the `save` command. The help file for the `saveold` command will specify the versions of Stata that will be able to read a `.dta` file created by the `saveold` command. See `help saveold` for more details.

Note: Exporting data to other formats

In section 22.2, we saw that Stata can read data in a variety of formats, including Excel workbooks and comma-separated files. Likewise, Stata can save data in these file formats as well. You can see `help export excel` for information about exporting data to an Excel workbook and `help export delimited` for information about exporting comma-separated files.

22.4   Labeling data

Let's return back to the `cardio.csv` comma-separated file, which we read in section 22.2.3. Let's read this below and then use the `describe` command. This shows that the variables in the dataset do not have variable labels or value labels.

```
. import delimited using cardio.csv
(13 vars, 5 obs)

. describe

Contains data
  obs:             5
  vars:           13
  size:          105

              storage   display    value
variable name   type     format    label      variable label

id            byte      %8.0g
mf            byte      %8.0g
ag            byte      %8.0g
bp1           int       %8.0g
bp2           int       %8.0g
bp3           int       %8.0g
bp4           int       %8.0g
bp5           int       %8.0g
pl1           int       %8.0g
pl2           byte      %8.0g
pl3           int       %8.0g
pl4           int       %8.0g
pl5           byte      %8.0g

Sorted by:
      Note: Dataset has changed since last saved.
```

The following sections illustrate how you can create variable labels and value labels.

22.4.1 Variable labels

The `label variable` command can be used to apply labels to variables. It is used below to label the variables `id`, `mf`, and `ag`.

```
. label variable id "ID for person"
. label variable mf "Gender of person"
. label variable ag "Age of person"
```

The `describe` command shows that these variables are now labeled with the labels we supplied.

```
. describe
Contains data
  obs:             5
  vars:           13
  size:          105
```

variable name	storage type	display format	value label	variable label
id	byte	%8.0g		ID for person
mf	byte	%8.0g		Gender of person
ag	byte	%8.0g		Age of person
bp1	int	%8.0g		
bp2	int	%8.0g		
bp3	int	%8.0g		
bp4	int	%8.0g		
bp5	int	%8.0g		
pl1	int	%8.0g		
pl2	byte	%8.0g		
pl3	int	%8.0g		
pl4	int	%8.0g		
pl5	byte	%8.0g		

```
Sorted by:
     Note: Dataset has changed since last saved.
```

22.4.2 A looping trick

Say we also wanted to assign labels to the variables `bp1` `to` `bp5` as well as labels to the variables `pl1` `to` `pl5`. We could manually do this by typing five `label variable` commands to label `bp1` `to` `bp5` and five `label variable` commands to label the variables `pl1` `to` `pl5`.

Instead, I am going to show you a trick by using the `forvalues` command.

```
. forvalues trialnum=1/5 {
  2.   label variable bp`trialnum´ "Systolic BP: Trial `trialnum´"
  3.   label variable pl`trialnum´ "Pulse: Trial `trialnum´"
  4. }
```

Before I explain how this works, let's look at the results of the `describe` command.

```
. describe

Contains data
  obs:             5
  vars:           13
  size:          105
```

variable name	storage type	display format	value label	variable label
id	byte	%8.0g		ID for person
mf	byte	%8.0g		Gender of person
ag	byte	%8.0g		Age of person
bp1	int	%8.0g		Systolic BP: Trial 1
bp2	int	%8.0g		Systolic BP: Trial 2
bp3	int	%8.0g		Systolic BP: Trial 3
bp4	int	%8.0g		Systolic BP: Trial 4
bp5	int	%8.0g		Systolic BP: Trial 5
pl1	int	%8.0g		Pulse: Trial 1
pl2	byte	%8.0g		Pulse: Trial 2
pl3	int	%8.0g		Pulse: Trial 3
pl4	int	%8.0g		Pulse: Trial 4
pl5	byte	%8.0g		Pulse: Trial 5

```
Sorted by:
     Note: Dataset has changed since last saved.
```

The `forvalues` command says that it is going to run the commands between the open curly brace and the closed curly brace, each time assigning a different value to the local macro variable called `trialnum`. The first time these commands are run, `trialnum` has a value of 1, and then the second time these commands are run, `trialnum` has a value of 2, and so forth. The fifth (and final) time these commands are run, `trialnum` has a value of 5

Note that within each of the `label variable` commands, I specified 'trialnum'. The left and right quotes that surround `trialnum` tell Stata to replace 'trialnum' with the contents of the local macro named `trialnum`. So the first time this group of commands (within the `foreach` loop) is executed, the value of 'trialnum' is replaced with the value of 1, and the next time this group of commands is executed, 'trialnum' is replaced with the value of 2, until the fifth time, in which 'trialnum' is replaced with the value of 5.

In this case, where we only had five trials, this loop may not have saved us that much time. But if we had 100 trials, it would have been a great time saver, saving us typing 200 `label variable` commands!

Note! Where are these quotes?

It can be tricky to use the correct quotation marks when you want to type `'trialnum'`. First, I call ` a left quote. On U.S. keyboards, it is usually on the same key along with the tilde (~), often positioned above the **Tab** key. Second, I call ' a right quote, and it is located below the double quote (") on your keyboard. The left and right quotes hug the name of the macro, making it clear where the macro begins and ends.

22.4.3   Value labels

As the name implies, assigning value labels assigns labels to the values of variables. Such labels are most commonly used for categorical variables, especially when the meaning of the categories can be unclear. For example, consider the variable `mf`, which represents the gender of the person. Using the `fre` command, we see that this variable is coded 1 and 2, but we do not know which code represents a male and which represents a female.

```
. fre mf
mf — Gender of person
```

		Freq.	Percent	Valid	Cum.
Valid	1	3	60.00	60.00	60.00
	2	2	40.00	40.00	100.00
	Total	5	100.00	100.00	

In this case, the value of 1 represents male and 2 represents female. Let's assign value labels to this variable so that we can clearly know that 1 represents male and 2 represents female. Assigning value labels is a two-step process in Stata. The first step is to use the `label define` command to define a label. The second step is to use the `label values` command to attach the label to a variable.

In the example below, the `label define` command creates the label named `gender`, and then the `label values` command attaches the label `gender` to the variable `mf`.

```
. label define gender 1 "Male" 2 "Female"
. label values mf gender
```

We can see the values and labels for the variable `mf` with the `codebook` command below.

```
. codebook mf
```

mf				Gender of person

```
                 type:  numeric (byte)
                label:  gender
                range:  [1,2]                          units:  1
        unique values:  2                          missing .:  0/5
           tabulation:  Freq.    Numeric  Label
                            3          1  Male
                            2          2  Female
```

> **Tip! Coding for dummies**
>
> In the above example, the coding of the variable `gender` was very ambiguous. Suppose I had called the variable `female` and used the values 1 and 0. Even in the absence of value labels, you would be able to surmise that a value of 1 indicates being a female and a value of 0 indicates being a male. That does not obviate the utility of value labels to be positive that the coding is correct, but it is a coding style that can convey the meaning of the values even if value labels are absent.

22.5   Creating and recoding variables

Stata offers a variety of commands for creating and modifying the contents of variables. This section illustrates the use of the `generate` command to create variables, and the `replace` command to modify the contents of existing variables. This section also illustrates the `egen` command, a powerful cousin of the `generate` command. This section concludes with an example of the `recode` command for recoding the values of a categorical variable.

22.5.1   Creating new variables with generate

To illustrate how to create new variables, I will start by using `cardio1.dta`.

```
. use cardio1
```

As we have seen before, this dataset has 5 measures of pulse rate for trials 1 to 5 stored in the variables named `pl1`, `pl2`, `pl3`, `pl4`, and `pl5`. Say we wanted to create an average pulse rate across all five trials. Below we use the `generate` command to create a variable named `plavg`, which is the average pulse of the five trials.

```
. generate plavg = (pl1 + pl2 + pl3 + pl4 + pl5) / 5
```

The `list` command below lists the variable `id` followed by all the variables that begin with `pl` (by specifying `pl*`).

```
. list id pl*
```

	id	pl1	pl2	pl3	pl4	pl5	plavg
1.	1	54	87	93	81	92	81.4
2.	2	92	88	125	87	58	90
3.	3	105	97	128	57	68	91
4.	4	52	79	71	106	39	69.4
5.	5	70	64	52	68	59	62.6

Tip! Naming variables

Note how I called the average pulse rate `plavg` instead of, for example, `avgpl`. By choosing to name the average `plavg`, I knew that I could specify `pl*` to refer to the individual pulse rates and the average pulse rate. I often plan the names of my Stata variable names strategically to allow myself to be able to use the wildcards of `*` and `?` to quickly and easily refer to groups of variables. For more information, see `help varlist`.

22.5.2   Modifying existing variables with replace

After further consideration, I decide that it would be better to compute this average omitting the pulse rate from the first trial (because the participants are probably not warmed up during the first trial). To compute the average pulse rate for the second through fifth trials, I think it would seem logical to use the `generate` command again. But as we see below, this leads to an error message. We receive this error because the `generate` command is for creating a *new* variable, but the variable `plavg` already exists. Stata does this to protect us from accidentally overwriting the contents of an existing variable.

```
. generate plavg = (pl2 + pl3 + pl4 + pl5) / 4
plavg already defined
r(110);
```

Instead of using the `generate` command (from above), we instead use the `replace` command (shown below). These commands work in the same way, except that the `generate` command creates a new variable, while the `replace` command changes the contents of an existing variable. The `replace` command is used below to replace the existing `plavg` variable with the average pulse rate for the second through fifth trials.

```
. replace plavg = (pl2 + pl3 + pl4 + pl5) / 4
(5 real changes made)
```

We can confirm that this change worked via the `list` command below.

```
. list id pl*
```

	id	pl1	pl2	pl3	pl4	pl5	plavg
1.	1	54	87	93	81	92	88.25
2.	2	92	88	125	87	58	89.5
3.	3	105	97	128	57	68	87.5
4.	4	52	79	71	106	39	73.75
5.	5	70	64	52	68	59	60.75

22.5.3   Extensions to generate egen

Say that we wanted to compute the minimum pulse across the second through fifth trials. This would be tricky, except that Stata offers a command called `egen` (think, extensions to generate) that allows you to create variables drawing from a wide variety of functions. In this case, we can use the `rowmin()` function to obtain the minimum pulse for the second through fifth trial, as illustrated below.

```
. egen plmin = rowmin(pl2 pl3 pl4 pl5)
```

We can then see these four pulse measures and the minimum value with the `list` command.

```
. list id pl2 pl3 pl4 pl5 plmin
```

	id	pl2	pl3	pl4	pl5	plmin
1.	1	87	93	81	92	81
2.	2	88	125	87	58	58
3.	3	97	128	57	68	57
4.	4	79	71	106	39	39
5.	5	64	52	68	59	52

We could have computed the maximum of the pulse values using the `rowmax()` function, or we could have computed the standard deviation of the pulse values using the `rowsd()` function. You can see `help egen` for a list of all the different functions that you can use.

22.5.4   Recode

The recode command is very useful when you want to create a new variable that will have discrete values based on the values of one other variable. To illustrate the recode command, I will use the General Social Survey (GSS) dataset.

```
. use gss2012_sbs
```

Let's look at the distribution of the variable race using the fre command. As you can see, this variable is coded 1 for white, 2 for black, and 3 for other.

```
. fre race
```

race — race of respondent

		Freq.	Percent	Valid	Cum.
Valid	1 white	1477	74.82	74.82	74.82
	2 black	301	15.25	15.25	90.07
	3 other	196	9.93	9.93	100.00
	Total	1974	100.00	100.00	

Say that we wanted to create a dummy variable that is coded 1 for white and 0 for everyone else. This is easily done with the recode command shown below, in which the newly recoded variable is called white.

```
. recode race (1=1 "White") (2 3=0 "Non-White"), gen(white)
(497 differences between race and white)
```

The recode command not only allowed us to create the new variable white with the desired coding scheme but also allowed us to assign value labels for each of the newly coded values. We can see this using the fre command below, seeing that 0 is labeled as Non-White and that 1 is labeled White.

```
. fre white
```

white — RECODE of race (race of respondent)

		Freq.	Percent	Valid	Cum.
Valid	0 Non-White	497	25.18	25.18	25.18
	1 White	1477	74.82	74.82	100.00
	Total	1974	100.00	100.00	

When I recode a variable in this way, I like to double check my results by cross-tabulating the original variable against the newly recoded variable, as shown below.

```
. tabulate race white, missing
```

| | RECODE of race (race of respondent) | | |
race of respondent	Non-White	White	Total
white	0	1,477	1,477
black	301	0	301
other	196	0	196
Total	497	1,477	1,974

By cross comparing the values of the new variable (in the columns) with the original variable (in the rows), we can confirm that the recoding was performed correctly.

22.6   Keeping and dropping variables

Consider cardio1.dta, which contains an ID variable, an age variable, five measurements of blood pressure (named bp1 to bp5), and five measurements of pulse (named pl1 to pl5). The use command (below) reads the dataset into memory, and then the describe command shows the variables in the dataset.

```
. use cardio1

. describe

Contains data from cardio1.dta
  obs:              5
  vars:            12                          22 Dec 2009 19:50
  size:           100
```

variable name	storage type	display format	value label	variable label
id	byte	%3.0f		Identification variable
age	byte	%3.0f		Age of person
bp1	int	%3.0f		Systolic BP: Trial 1
bp2	int	%3.0f		Systolic BP: Trial 2
bp3	int	%3.0f		Systolic BP: Trial 3
bp4	int	%3.0f		Systolic BP: Trial 4
bp5	int	%3.0f		Systolic BP: Trial 5
pl1	int	%3.0f		Pulse: Trial 1
pl2	byte	%3.0f		Pulse: Trial 2
pl3	int	%3.0f		Pulse: Trial 3
pl4	int	%3.0f		Pulse: Trial 4
pl5	byte	%3.0f		Pulse: Trial 5

```
Sorted by:
```

Say that we wanted to get rid of (that is, drop) the variables measuring the pulse rate. The drop command below drops the variables pl1, pl2, pl3, pl4, and pl5. The describe command below shows that these variables are no longer part of the dataset in memory.

```
. drop pl1 pl2 pl3 pl4 pl5

. describe

Contains data from cardio1.dta
  obs:            5
  vars:           7                             22 Dec 2009 19:50
  size:          60

              storage   display   value
variable name   type    format    label    variable label

id             byte     %3.0f              Identification variable
age            byte     %3.0f              Age of person
bp1            int      %3.0f              Systolic BP: Trial 1
bp2            int      %3.0f              Systolic BP: Trial 2
bp3            int      %3.0f              Systolic BP: Trial 3
bp4            int      %3.0f              Systolic BP: Trial 4
bp5            int      %3.0f              Systolic BP: Trial 5

Sorted by:
      Note: Dataset has changed since last saved.
```

Stata also allows you to specify groups of variables using the * or ? as a wildcard. Instead of manually specifying pl1 pl2 pl3 pl4 pl5 in the previous drop command, we could have specified pl*, and that would have had the same effect, dropping all the variables that started with pl.

Note: Using the ? wildcard

Imagine that cardio1.dta had also included variables named player and plan but that we wanted to drop only the pulse-related variables. In this case, had we specified pl* with the drop command, we would have dropped not only the pulse-related variables but would also player and plan. In that case, we could have specified pl?. This drops only variables that start with pl and are exactly three characters wide. ? indicates that the third character can be any *single* arbitrary character.

Let's use cardio1.dta again, but imagine that we want to keep just a handful of variables, namely, id, age, and bp1. Rather than using the drop command, we can use the keep command to keep just these variables.

```
. use cardio1

. keep id age bp1
```

We can see that the keep command had the desired effect via the describe command below.

```
. describe
Contains data from cardio1.dta
  obs:            5
 vars:            3                              22 Dec 2009 19:50
 size:           20

              storage   display    value
variable name  type     format     label      variable label

id             byte     %3.0f                  Identification variable
age            byte     %3.0f                  Age of person
bp1            int      %3.0f                  Systolic BP: Trial 1

Sorted by:
     Note: Dataset has changed since last saved.
```

22.7 Keeping and dropping observations

You can eliminate observations from the dataset in memory by using either the `keep if` command or the `drop if` command. As you probably can infer, the `keep if` command works by specifying a condition that observations must meet to remain in the dataset, whereas the `drop if` command works by specifying a condition that, if met, results in the elimination of observations from the dataset. In both cases, this impacts only the dataset in memory.

Let's use the GSS dataset and the `fre` command to show the frequency distribution of the variable `marital`.

```
. use gss2012_sbs
. fre marital
marital — marital status
```

		Freq.	Percent	Valid	Cum.
Valid	1 married	900	45.59	45.59	45.59
	2 widowed	163	8.26	8.26	53.85
	3 divorced	317	16.06	16.06	69.91
	4 separated	68	3.44	3.44	73.35
	5 never married	526	26.65	26.65	100.00
	Total	1974	100.00	100.00	

Say that we wanted to perform a series of analyses that excluded everyone who was separated but included everyone else. We can drop the observations for those who are separated via the `drop if` command.

```
. drop if marital==4
(68 observations deleted)
```

Note: The double equal sign

Note the use of the double equal sign when specifying `if marital==4`. This is used to test the equality of two values, so `marital==4` asks whether the variable `marital` equals 4. To Stata, a single equal sign means variable assignment, as in the statement `generate marital = 4`, in which the equal sign says to *assign* the value of 4 to the variable `marital`.

We can now use the `fre` command again and see that the 68 observations for those who were separated have been dropped from the dataset in memory.

```
. fre marital
marital — marital status
```

		Freq.	Percent	Valid	Cum.
Valid	1 married	900	47.22	47.22	47.22
	2 widowed	163	8.55	8.55	55.77
	3 divorced	317	16.63	16.63	72.40
	5 never married	526	27.60	27.60	100.00
	Total	1906	100.00	100.00	

Now, let's consider an example of using the `keep if` command. To start this example, we will use the GSS dataset again.

```
. use gss2012_sbs
```

Say that we wanted to keep just those observations for people who are married. In this case, we can use the `keep if` command specifying `if married==1` as the condition for keeping observations in the dataset in memory.

```
. keep if marital==1
(1,074 observations deleted)
```

The `fre` command below shows that this command was successful.

```
. fre married
married — marital: married=1, unmarried=0 (recoded)
```

		Freq.	Percent	Valid	Cum.
Valid	1 Married	900	100.00	100.00	100.00

22.8   Combining datasets

Sometimes, you have two or more datasets that you want to combine. This section illustrates two different ways that you can combine datasets, either through the `append` command or through the `merge` command.

The `append` command is used when you have two or more datasets that contain the same variables but represent different observations. When you use the `append` command, the datasets are appended together, as if stacked one atop the other.

I also illustrate a simple example of performing a one-to-one merge using the `merge` command (there are other ways you can merge files, and these are described in `help merge`). In a one-to-one merge, there are two datasets (with different variables), but they contain information about the same observations. Each dataset contains a *key* variable that identifies observations within each dataset and connects the observations between the two datasets.

22.8.1   Appending datasets

This section will provide a very simple example of appending datasets. Let's begin by looking at the contents of two datasets that we will append, one called `moms.dta` and one called `dads.dta`.

Consider the file `moms.dta`, which contains a family ID variable and the mother's age, race, and whether she graduated from high school.

```
. use moms
. list

     +---------------------------+
     | famid   age   race    hs  |
     |---------------------------|
  1. |     3    24      2     1  |
  2. |     2    28      1     1  |
  3. |     4    21      1     0  |
  4. |     1    33      2     1  |
     +---------------------------+
```

Next, consider the file `dads.dta`, which also contains a family ID variable and the father's age, race, and whether he graduated from high school.

```
. use dads
. list
```

	famid	age	race	hs
1.	1	21	1	0
2.	4	25	2	1
3.	2	25	1	1
4.	3	31	2	1

Note how moms.dta and dads.dta both contain the same variables (with the same variable names). Using the append command, we can combine these two datasets by stacking the files one atop the other. In the example below, the clear command is used to first clear out any data from working memory. Then, after the append command, we list the datasets we want to combine (as many as we like). Below we append moms.dta with dads.dta.

```
. clear
. append using moms dads
```

The list command shows us the combined dataset, in which we see the four observations from moms.dta followed by the four observations from dads.dta.

```
. list
```

	famid	age	race	hs
1.	3	24	2	1
2.	2	28	1	1
3.	4	21	1	0
4.	1	33	2	1
5.	1	21	1	0
6.	4	25	2	1
7.	2	25	1	1
8.	3	31	2	1

Let's repeat the append command from above but include the generate() option. In the example below, I specify generate(dtasrc), which creates a variable called dtasrc that contains a value of 1 if the observation comes from the first (that is, moms.dta) dataset and a value of 2 if the observation comes from the second (that is, dads.dta) dataset.

```
. clear
. append using moms dads, generate(dtasrc)
```

The `list` command shows us the combined dataset, including the variable `dtasrc`, which contains 1 to indicate that the observation is from `moms.dta` and a 2 to indicate that the observation is from `dads.dta`.

```
. list
```

	dtasrc	famid	age	race	hs
1.	1	3	24	2	1
2.	1	2	28	1	1
3.	1	4	21	1	0
4.	1	1	33	2	1
5.	2	1	21	1	0
6.	2	4	25	2	1
7.	2	2	25	1	1
8.	2	3	31	2	1

Note: Naming of variables for appending

Note how the variables were named in the `moms.dta` and `dads.dta` datasets. The variable `age` was used in both datasets to indicate the age of the person (for both moms and dads), and the variable `race` was used to indicate the race of the person (for both moms and dads). If our goal is to append datasets, it is important to use the same variable names for variables we want to be appended together into the same variable.

22.8.2 Merging datasets

Let's consider two datasets that are very similar to `moms.dta` and `dads.dta`. These two datasets are named `moms1.dta` and `dads1.dta`. Let's first look at `moms1.dta`, which contains a family ID variable and the mother's age, race, and whether she graduated from high school.

```
. use moms1
. list
```

	famid	mage	mrace	mhs
1.	1	33	2	1
2.	2	28	1	1
3.	3	24	2	1
4.	4	21	1	0

Also consider the file `dads1.dta`, which contains a family ID variable and the father's age, race, and whether he graduated from high school.

```
. use dads1
. list
```

	famid	dage	drace	dhs
1.	1	21	1	0
2.	2	25	1	1
3.	3	31	2	1
4.	4	25	2	1

Note that both files contain a variable named `famid` that can be used to link the mom with the dad.

Let's merge `moms1.dta` with `dads1.dta`. To do so, we start by using `moms1.dta`. We then use the `merge` command to perform a one-to-one merge in which the variable `famid` is used to merge the data in memory with `dads1.dta`. The `merge` command shows that four observations were matched.

```
. use moms1
. merge 1:1 famid using dads1
```

Result	# of obs.	
not matched	0	
matched	4	(_merge==3)

The `list` command shows the resulting dataset, which contains the family ID, the variables from `moms1.dta`, the variables from `dads1.dta`, and the variable `_merge`, which shows the matching status for each observation.

```
. list
```

	famid	mage	mrace	mhs	dage	drace	dhs	_merge
1.	1	33	2	1	21	1	0	matched (3)
2.	2	28	1	1	25	1	1	matched (3)
3.	3	24	2	1	31	2	1	matched (3)
4.	4	21	1	0	25	2	1	matched (3)

Note: Naming of variables for merging

Note how the variables were named in the `moms1.dta` and `dads1.dta` datasets. For example, the variable `mage` was used in `moms1.dta` to indicate the age of the mom, and the variable `dage` was used in `dads1.dta` to indicate the age of the dad. If our goal is to merge datasets, it is important to avoid using the same variable names for variables (except for the *key* variable).

22.9   Reshaping datasets

This section illustrates how you can convert datasets from a wide format to a long format and how you can convert datasets from a long format to a wide format.

22.9.1   Reshaping datasets wide to long

`cardio_wide.dta` contains information about six people in which the five blood pressure measurements are stored in the variables `bp1` to `bp5` and the five pulse measurements are stored in the variables `pl1` to `pl5`.

```
. use cardio_wide
. list
```

	id	age	bp1	bp2	bp3	bp4	bp5	pl1	pl2	pl3	pl4	pl5
1.	1	40	115	86	129	105	127	54	87	93	81	92
2.	2	30	123	136	107	111	120	92	88	125	87	58
3.	3	16	124	122	101	109	112	105	97	128	57	68
4.	4	23	105	115	121	129	137	52	79	71	106	39
5.	5	18	116	128	112	125	111	70	64	52	68	59
6.	6	27	108	126	124	131	107	74	78	92	99	80

This wide format is useful if we want to analyze the variables `bp1` to `bp5` and `pl1` to `pl5` with commands like `correlate`, `alpha`, or `factor`. However, if we want to analyze the data using the techniques illustrated in chapters 12 and 13, we would need to convert the dataset into a long format. This is easily performed using the `reshape long` command.

The `reshape long` command below converts this wide dataset into a long dataset. We specify the stem of the wide variables that we want to be converted into a long form, namely, `bp` and `pl`. The `i(id)` option indicates that the observations in the wide dataset are identified by the variable `id`. The `j()` option indicates that the suffix of `bp` and `pl` should be stored in a new variable that will be named `trial`.

```
. reshape long bp pl, i(id) j(trial)
(note: j = 1 2 3 4 5)
Data                              wide   ->   long

Number of obs.                       6   ->     30
Number of variables                 12   ->      5
j variable (5 values)                    ->   trial
xij variables:
                       bp1 bp2 ... bp5   ->   bp
                       pl1 pl2 ... pl5   ->   pl
```

The `list` command shows that the wide dataset, which had six observations and five measurements per observation, now has been converted into a long dataset with 30 observations in which the variable `id` identifies the person and the variable `trial` identifies the trial from which the blood pressure (`bp`) and pulse (`pl`) measurement originated.

```
. list, sepby(id)
```

	id	trial	age	bp	pl
1.	1	1	40	115	54
2.	1	2	40	86	87
3.	1	3	40	129	93
4.	1	4	40	105	81
5.	1	5	40	127	92
6.	2	1	30	123	92
7.	2	2	30	136	88
8.	2	3	30	107	125
9.	2	4	30	111	87
10.	2	5	30	120	58
11.	3	1	16	124	105
12.	3	2	16	122	97
13.	3	3	16	101	128
14.	3	4	16	109	57
15.	3	5	16	112	68
16.	4	1	23	105	52
17.	4	2	23	115	79
18.	4	3	23	121	71
19.	4	4	23	129	106
20.	4	5	23	137	39
21.	5	1	18	116	70
22.	5	2	18	128	64
23.	5	3	18	112	52
24.	5	4	18	125	68
25.	5	5	18	111	59
26.	6	1	27	108	74
27.	6	2	27	126	78
28.	6	3	27	124	92
29.	6	4	27	131	99
30.	6	5	27	107	80

For more information about reshaping datasets from wide format to long format, see `help reshape`.

22.9.2 Reshaping datasets long to wide

The dataset `cardio_long.dta` contains information about six people, each containing five observations (representing five trials) in which their blood pressure (`bp`) and pulse (`pl`) were observed. This dataset is very useful for analyzing the data using the techniques illustrated in chapters 12 and 13. But if we wanted to examine the correlations among the blood pressure and pulse rate observations across time, we would need to convert the dataset into a wide format. The wide format would also be required to use the `alpha` command or `factor` command with respect to the blood pressure or pulse measures.

```
. use cardio_long

. list, sepby(id)
```

	id	trial	age	bp	pl
1.	1	1	40	115	54
2.	1	2	40	86	87
3.	1	3	40	129	93
4.	1	4	40	105	81
5.	1	5	40	127	92
6.	2	1	30	123	92
7.	2	2	30	136	88
8.	2	3	30	107	125
9.	2	4	30	111	87
10.	2	5	30	120	58
11.	3	1	16	124	105
12.	3	2	16	122	97
13.	3	3	16	101	128
14.	3	4	16	109	57
15.	3	5	16	112	68
16.	4	1	23	105	52
17.	4	2	23	115	79
18.	4	3	23	121	71
19.	4	4	23	129	106
20.	4	5	23	137	39
21.	5	1	18	116	70
22.	5	2	18	128	64
23.	5	3	18	112	52
24.	5	4	18	125	68
25.	5	5	18	111	59
26.	6	1	27	108	74
27.	6	2	27	126	78
28.	6	3	27	124	92
29.	6	4	27	131	99
30.	6	5	27	107	80

The `reshape wide` command below converts the long dataset into a wide format. Specifying the variables `bp` and `pl` indicates that those variables are to be reshaped from long to wide. The `i(id)` option is used to indicate that the observations in the wide format are uniquely identified by the variable `id`, and the `j(trial)` option indicates that the values of `trial` should be used to form the suffix of `bp` and `pl` as the data are reshaped from long to wide format.

```
. reshape wide bp pl, i(id) j(trial)
(note: j = 1 2 3 4 5)
```

Data	long	->	wide
Number of obs.	30	->	6
Number of variables	5	->	12
j variable (5 values)	trial	->	(dropped)
xij variables:			
	bp	->	bp1 bp2 ... bp5
	pl	->	pl1 pl2 ... pl5

The `list` command shows that the dataset now has six observations with blood pressure stored in `bp1` to `bp5` and the five pulse measurements stored in the variables `pl1` to `pl5`.

```
. list
```

	id	bp1	pl1	bp2	pl2	bp3	pl3	bp4	pl4	bp5	pl5	age
1.	1	115	54	86	87	129	93	105	81	127	92	40
2.	2	123	92	136	88	107	125	111	87	120	58	30
3.	3	124	105	122	97	101	128	109	57	112	68	16
4.	4	105	52	115	79	121	71	129	106	137	39	23
5.	5	116	70	128	64	112	52	125	68	111	59	18
6.	6	108	74	126	78	124	92	131	99	107	80	27

You can see `help reshape` for more information about reshaping datasets from long to wide format.

22.10   Closing thoughts

This chapter has provided a very brief overview of some common data management commands using simple examples to illustrate basic data management tasks. For more detailed information about data management commands using Stata, you can see `help contents_data`. For a topic-oriented introduction to data management using Stata, I would recommend my book *Data Management Using Stata: A Practical Handbook*.

23 Stata equivalents of common IBM SPSS Commands

23.1   Chapter overview

When taking a language class, you often would have a book that would quickly translate English words into another language (say, Spanish). In some cases, there would be one exact word that was the equivalent of the English word, and in other cases, the translation would be a bit more murky. There might be two or more equivalent words, or the words would have similar but not exact meanings. Well, think of this chapter as an IBM® SPSS® to Stata translation, where sometimes, the SPSS command has an exact equivalent, and sometimes, there is a little bit more in the translation.

This chapter lists commonly used SPSS commands (in alphabetical order) and shows the equivalent (or near equivalent) Stata command along with one or more examples of the Stata command. The aim of this chapter is to help you learn, for example, that the Stata equivalent of the SPSS AGGREGATE command is the collapse command and to show a simple example of the collapse command. To that end, the examples are intentionally simple (and sometimes might be overly simplistic). But once you know that the command you want is, for example, the collapse command, you can then type help collapse and access the Stata documentation for all the details about the use of the collapse command.

23.2   ADD FILES

Stata equivalent: append
Example: Consider the file moms.dta, which contains a family ID variable, the mother's age, race, and whether she graduated from high school.

```
. use moms
. list
```

	famid	age	race	hs
1.	3	24	2	1
2.	2	28	1	1
3.	4	21	1	0
4.	1	33	2	1

Also consider the file `dads.dta`, which contains a family ID variable, the father's age, race, and whether he graduated from high school.

```
. use dads
. list
```

	famid	age	race	hs
1.	1	21	1	0
2.	4	25	2	1
3.	2	25	1	1
4.	3	31	2	1

The `append` command can be used to stack these files together. In the example below, the `clear` command is used to first clear out any data from working memory. Then, after the `append` command, we can list the datasets we want to combine (as many as we like). Below we append `moms.dta` with `dads.dta`.

```
. clear
. append using moms dads
```

The `list` command shows the resulting dataset, showing the observations for the moms followed by the observations for the dads.

```
. list
```

	famid	age	race	hs
1.	3	24	2	1
2.	2	28	1	1
3.	4	21	1	0
4.	1	33	2	1
5.	1	21	1	0
6.	4	25	2	1
7.	2	25	1	1
8.	3	31	2	1

An alternative to the above process is to first use `moms.dta` and then use the `append` command to append `dads.dta`.

```
. use moms
. append using dads
```

23.3 AGGREGATE

Stata equivalent: `collapse`

Example: `cardio_long.dta` contains information about six people, each containing five observations (representing five trials) in which their blood pressure (`bp`) and pulse (`pl`) were observed.

```
. use cardio_long
. list, sepby(id)
```

	id	trial	age	bp	pl
1.	1	1	40	115	54
2.	1	2	40	86	87
3.	1	3	40	129	93
4.	1	4	40	105	81
5.	1	5	40	127	92
6.	2	1	30	123	92
7.	2	2	30	136	88
8.	2	3	30	107	125
9.	2	4	30	111	87
10.	2	5	30	120	58
11.	3	1	16	124	105
12.	3	2	16	122	97
13.	3	3	16	101	128
14.	3	4	16	109	57
15.	3	5	16	112	68
16.	4	1	23	105	52
17.	4	2	23	115	79
18.	4	3	23	121	71
19.	4	4	23	129	106
20.	4	5	23	137	39
21.	5	1	18	116	70
22.	5	2	18	128	64
23.	5	3	18	112	52
24.	5	4	18	125	68
25.	5	5	18	111	59
26.	6	1	27	108	74
27.	6	2	27	126	78
28.	6	3	27	124	92
29.	6	4	27	131	99
30.	6	5	27	107	80

The `collapse` command below creates one observation per person (via the `by(id)` option) that contains the average of `bp` and `pl`.

```
. collapse bp pl, by(id)
. list
```

	id	bp	pl
1.	1	112	81
2.	2	119	90
3.	3	114	91
4.	4	121	69
5.	5	118	63
6.	6	119	85

23.4 ANOVA

Stata equivalent: `anova`

Example: Participants were randomly assigned to one of three treatment groups (1 = Control, 2 = Traditional therapy, 3 = Optimism therapy), and their optimism was measured. The `anova` command found significant differences among the three groups $[F(2, 299) = 8.19, p = 0.0003]$.

```
. use opt-3
. anova opt treat
```

		Number of obs =	300	R-squared	=	0.0523
		Root MSE =	10.1165	Adj R-squared	=	0.0459
Source	Partial SS		df	MS	F	Prob>F
Model	1676.3467		2	838.17333	8.19	0.0003
treat	1676.3467		2	838.17333	8.19	0.0003
Residual	30396.09		297	102.34374		
Total	32072.437		299	107.26567		

The `contrast` command is used below to compare each group with the control group by applying the **r** contrast operator to `treat`. The traditional therapy group had significantly greater optimism than the control group $[t(297) = 4.78, p = 0.001]$, and the optimism therapy group had significantly greater optimism than the control group $[t(297) = 5.22, p < 0.001]$.

```
. contrast r.treat, pveffects nowald
Contrasts of marginal linear predictions
Margins        : asbalanced
```

	Contrast	Std. Err.	t	P>|t|
treat				
(TT vs Con)	4.78	1.43069	3.34	0.001
(OT vs Con)	5.22	1.43069	3.65	0.000

The `margins` command shows that the average optimism for the control, traditional therapy, and optimism therapy groups is 44.71, 49.49, and 49.93 (respectively).

```
. margins treat
Adjusted predictions                    Number of obs     =        300
Expression    : Linear prediction, predict()
```

	Margin	Delta-method Std. Err.	t	P>|t|	[95% Conf. Interval]	
treat						
Con	44.71	1.011651	44.20	0.000	42.71909	46.70091
TT	49.49	1.011651	48.92	0.000	47.49909	51.48091
OT	49.93	1.011651	49.35	0.000	47.93909	51.92091

23.5   AUTORECODE

Stata equivalents: `destring`, `encode`
Example: `cardio1str.dta` contains a number of variables regarding the cardio fitness of five people. However, the variables were all stored as string variables. Two of the variables, `wt` and `gender`, are described below.

```
. use cardio1str
. describe wt gender
```

variable name	storage type	display format	value label	variable label
wt	str5	%5s		Weight of person
gender	str6	%6s		Gender of person

The `list` command shows the listing of these two variables.

```
. list wt gender
```

	wt	gender
1.	150.7	male
2.	186.3	male
3.	109.9	male
4.	183.4	male
5.	159.1	female

To convert `wt` from a string variable to a numeric variable, we can use the `destring` command. The `describe` command shows the variable is now a numeric variable.

```
. destring wt, replace
wt has all characters numeric; replaced as double
. describe wt

              storage   display   value
variable name   type    format    label    variable label

wt            double    %10.0g             Weight of person
```

The `list` command shows the values for these variables.

```
. list wt
```

	wt
1.	150.7
2.	186.3
3.	109.9
4.	183.4
5.	159.1

The variable `gender`, which contains the words "male" and "female", can be converted into a numeric variable with the `encode` command to create the variable `gendern`, which is a numeric version of `gender`.

```
. encode gender, generate(gendern)
```

The `codebook` command shows that `gendern` is a numeric variable that is coded 1 for `female` and 2 for `male` and that these values are labeled using value labels.

```
. codebook gendern
```

---

gendern   Gender of person

```
                 type:  numeric (long)
                label:  gendern
                range:  [1,2]                         units:  1
        unique values:  2                           missing .:  0/5
           tabulation:  Freq.   Numeric  Label
                            1         1  female
                            4         2  male
```

23.6   CASESTOVARS

Stata equivalent: `reshape wide`

Example: `cardio_long.dta` contains information about six people, each containing five observations (representing five trials) in which their blood pressure (`bp`) and pulse (`pl`) were observed.

```
. use cardio_long
. list, sepby(id)
```

	id	trial	age	bp	pl
1.	1	1	40	115	54
2.	1	2	40	86	87
3.	1	3	40	129	93
4.	1	4	40	105	81
5.	1	5	40	127	92
6.	2	1	30	123	92
7.	2	2	30	136	88
8.	2	3	30	107	125
9.	2	4	30	111	87
10.	2	5	30	120	58
11.	3	1	16	124	105
12.	3	2	16	122	97
13.	3	3	16	101	128
14.	3	4	16	109	57
15.	3	5	16	112	68
16.	4	1	23	105	52
17.	4	2	23	115	79
18.	4	3	23	121	71
19.	4	4	23	129	106
20.	4	5	23	137	39
21.	5	1	18	116	70
22.	5	2	18	128	64
23.	5	3	18	112	52
24.	5	4	18	125	68
25.	5	5	18	111	59
26.	6	1	27	108	74
27.	6	2	27	126	78
28.	6	3	27	124	92
29.	6	4	27	131	99
30.	6	5	27	107	80

The reshape wide command converts this into a wide dataset in which the five blood pressure measurements are stored in the variables bp1 to bp5 and the five pulse measurements are stored in the variables pl1 to pl5. The i(id) option indicates that id identifies each observation in the wide dataset. The j(trial) option indicates that the suffix of bp and pl in the wide dataset will be drawn from the variable trial.

```
. reshape wide bp pl, i(id) j(trial)
(note: j = 1 2 3 4 5)
Data                              long   ->    wide
```

Data	long	->	wide
Number of obs.	30	->	6
Number of variables	5	->	12
j variable (5 values)	trial	->	(dropped)
xij variables:			
	bp	->	bp1 bp2 ... bp5
	pl	->	pl1 pl2 ... pl5

The `list` command shows that the dataset now has six observations with blood pressure stored in `bp1` to `bp5` and the five pulse measurements stored in the variables `pl1` to `pl5`.

```
. list
```

	id	bp1	pl1	bp2	pl2	bp3	pl3	bp4	pl4	bp5	pl5	age
1.	1	115	54	86	87	129	93	105	81	127	92	40
2.	2	123	92	136	88	107	125	111	87	120	58	30
3.	3	124	105	122	97	101	128	109	57	112	68	16
4.	4	105	52	115	79	121	71	129	106	137	39	23
5.	5	116	70	128	64	112	52	125	68	111	59	18
6.	6	108	74	126	78	124	92	131	99	107	80	27

23.7 COMPUTE

Stata equivalents: `generate`, `replace`
Example: `wws2.dta` contains information from the women's work survey. The variable `wage` contains the woman's hourly wage. The `generate` command computes a new variable `wageweek` that contains the wage multiplied by 40 to represent her weekly wage.

```
. use wws2
(Working Women Survey w/fixes)
. generate wageweek = wage*40
(2 missing values generated)
. summarize wage wageweek
```

Variable	Obs	Mean	Std. Dev.	Min	Max
wage	2,244	7.796781	5.82459	0	40.74659
wageweek	2,244	311.8712	232.9836	0	1629.864

After further consideration, I decide that it would be better to compute a weekly wage as the hourly wage multiplied by the hours the woman works per week. The `replace` command is used to change the existing `wageweek` variable to contain `wage` multiplied by `hours`.

```
. replace wageweek = wage*hours
(1,152 real changes made, 4 to missing)
```

23.8   CORRELATIONS

Stata equivalents: `correlate, pwcorr`
Examples: Using the dataset `gss2012_sbs.dta`, we use the `correlate` command to compute the correlations among `educ`, `happy7`, and `health` using listwise deletion.

```
. use gss2012_sbs

. correlate educ happy7 health
(obs=1,279)
                 educ    happy7    health

       educ    1.0000
     happy7    0.0737    1.0000
     health    0.2163    0.3182    1.0000
```

The `pwcorr` command is used to obtain correlations using pairwise deletion. The `obs` and `sig` options are included to show the number of observations and the p-value associated with each correlation.

```
. pwcorr educ happy7 health, obs sig
                 educ    happy7    health

       educ    1.0000

                 1972

     happy7    0.0688    1.0000
               0.0137
                 1284      1284

     health    0.2273    0.3182    1.0000
               0.0000    0.0000
                 1290      1279      1290
```

23.9   CROSSTABS

Stata equivalent: `tabulate`
Example: Using the General Social Survey dataset from 2012, we use the `tabulate` command to form a cross-tabulation of `marital` by `female`.

```
. use gss2012_sbs, clear

. tabulate marital female
```

marital status	Is R female (yes=1 no=0)? Male	Female	Total
married	411	489	900
widowed	45	118	163
divorced	139	178	317
separated	27	41	68
never married	264	262	526
Total	886	1,088	1,974

In the example below, the `column` option is used to show column percentages, and the `chi2` and `exact` options are used to show a chi-squared test and Fisher's exact test (respectively).

```
. tabulate marital female, column chi2 exact
```

Key
frequency
column percentage

```
Enumerating sample-space combinations:
stage 5:   enumerations = 1
stage 4:   enumerations = 41
stage 3:   enumerations = 2061
stage 2:   enumerations = 114517
stage 1:   enumerations = 0
```

marital status	Is R female (yes=1 no=0)? Male	Female	Total
married	411	489	900
	46.39	44.94	45.59
widowed	45	118	163
	5.08	10.85	8.26
divorced	139	178	317
	15.69	16.36	16.06
separated	27	41	68
	3.05	3.77	3.44
never married	264	262	526
	29.80	24.08	26.65
Total	886	1,088	1,974
	100.00	100.00	100.00

```
          Pearson chi2(4) =  26.7507   Pr = 0.000
           Fisher's exact =              0.000
```

Other common options include the `cell` option (for cell percentages) and the `row` option (for row percentages).

23.10   DATA LIST

Stata equivalents: `infile`, `infix`, `import delimited`
Example: Consider the raw data file `cardio2miss.txt`, shown below using the `type` command.

```
. type cardio2miss.txt
1 40 54 115 87 86 93 129 81 105 -2 -2
2 30 92 123 88 136 125 107 87 111 58 120
3 16 105 -1 97 122 128 101 57 109 68 112
4 23 52 105 79 115 71 121 106 129 39 137
5 18 70 116 -1 128 52 112 68 125 59 111
```

As you can see, this is a space-separated raw data file. The variables contained in the file are id, age, five blood pressure measurements, and five pulse measurements. This raw data file can be read into Stata using the `infile` command, shown below.

```
. clear
. infile id age bp1 bp2 bp3 bp4 bp5 pl1 pl2 pl3 pl4 pl5 using cardio2miss.txt
(5 observations read)
```

The `list` command shows contents of the variables after you read the data into Stata.

```
. list
```

	id	age	bp1	bp2	bp3	bp4	bp5	pl1	pl2	pl3	pl4	pl5
1.	1	40	54	115	87	86	93	129	81	105	-2	-2
2.	2	30	92	123	88	136	125	107	87	111	58	120
3.	3	16	105	-1	97	122	128	101	57	109	68	112
4.	4	23	52	105	79	115	71	121	106	129	39	137
5.	5	18	70	116	-1	128	52	112	68	125	59	111

The `import delimited` command can be used to read comma-separated or tab-separated raw datasets. The `infix` command reads fixed-column datasets.

Tip: Importing and exporting Excel workbooks

In Stata, you can use the `import excel` command to read Excel workbooks and you can use the `export excel` command to export the variables in memory to an Excel workbook.

23.11   DELETE VARIABLES

Stata equivalent: `keep`, `drop`
Example: The dataset `cardio1.dta` contains an ID variable, an age variable, five measurements of blood pressure (named **bp1** to **bp5**), and five measurements of pulse (named **pl1** to **pl5**).

```
. use cardio1

. describe

Contains data from cardio1.dta
  obs:             5
  vars:           12                          22 Dec 2009 19:50
  size:          100

              storage   display   value
variable name   type    format    label    variable label

id             byte    %3.0f              Identification variable
age            byte    %3.0f              Age of person
bp1            int     %3.0f              Systolic BP: Trial 1
bp2            int     %3.0f              Systolic BP: Trial 2
bp3            int     %3.0f              Systolic BP: Trial 3
bp4            int     %3.0f              Systolic BP: Trial 4
bp5            int     %3.0f              Systolic BP: Trial 5
pl1            int     %3.0f              Pulse: Trial 1
pl2            byte    %3.0f              Pulse: Trial 2
pl3            int     %3.0f              Pulse: Trial 3
pl4            int     %3.0f              Pulse: Trial 4
pl5            byte    %3.0f              Pulse: Trial 5

Sorted by:
```

The `drop` command specifies variables to be dropped from the working dataset. The `drop` command below drops the variables **pl1**, **pl2**, **pl3**, **pl4**, and **pl5**. The `describe` command shows the variables that remain in the dataset.

```
. drop pl1 pl2 pl3 pl4 pl5

. describe

Contains data from cardio1.dta
  obs:            5
  vars:           7                              22 Dec 2009 19:50
  size:          60
```

variable name	storage type	display format	value label	variable label
id	byte	%3.0f		Identification variable
age	byte	%3.0f		Age of person
bp1	int	%3.0f		Systolic BP: Trial 1
bp2	int	%3.0f		Systolic BP: Trial 2
bp3	int	%3.0f		Systolic BP: Trial 3
bp4	int	%3.0f		Systolic BP: Trial 4
bp5	int	%3.0f		Systolic BP: Trial 5

```
Sorted by:
     Note: Dataset has changed since last saved.
```

Tip: Abbreviating variables

In SPSS, you could use pl1 TO pl5 to indicate the variables from pl1 to pl5. The Stata equivalent is pl1-pl5, where the dash is the Stata equivalent of the SPSS TO. The dash specifies the variables from bp1 to bp5 based on their consecutive position in the dataset.

Stata also allows you to abbreviate variables using * or ? as a wildcard. For example, pl* would refer to all the variables that begin with the letters pl, or *1 would refer to all variables that end with 1. Specifying pl? would refer to all variables that start with pl and have any single additional character (that is, pl1 but not variables like plug or plank).

The keep command keeps just the named variables. After using cardio1.dta, we use the keep command to keep just the variables id, age, and all the variables that start with bp. The describe command shows the variables that remain in the dataset.

```
. use cardio1

. keep id age bp*

. describe
Contains data from cardio1.dta
  obs:            5
  vars:           7                              22 Dec 2009 19:50
  size:          60
```

variable name	storage type	display format	value label	variable label
id	byte	%3.0f		Identification variable
age	byte	%3.0f		Age of person
bp1	int	%3.0f		Systolic BP: Trial 1
bp2	int	%3.0f		Systolic BP: Trial 2
bp3	int	%3.0f		Systolic BP: Trial 3
bp4	int	%3.0f		Systolic BP: Trial 4
bp5	int	%3.0f		Systolic BP: Trial 5

```
Sorted by:
     Note: Dataset has changed since last saved.
```

23.12   DESCRIPTIVES

Stata equivalent: `summarize`

Example: The dataset `cardio1.dta` contains an ID variable, an age variable, five measurements of blood pressure (named `bp1` to `bp5`), and five measurements of pulse (named `pl1` to `pl5`). The `summarize` command below provides summary statistics for all the variables in the dataset.

```
. use cardio1

. summarize
```

Variable	Obs	Mean	Std. Dev.	Min	Max
id	5	3	1.581139	1	5
age	5	25.4	9.787747	16	40
bp1	5	116.6	7.635444	105	124
bp2	5	117.4	19.17811	86	136
bp3	5	114	11.13553	101	129
bp4	5	115.8	10.54514	105	129
bp5	5	121.4	10.87658	111	137
pl1	5	74.6	23.36236	52	105
pl2	5	83	12.38951	64	97
pl3	5	93.8	33.20693	52	128
pl4	5	79.8	18.70027	57	106
pl5	5	63.2	19.25357	39	92

The `summarize` command below shows summary statistics for the blood pressure measures (that is, variables that start with `bp`).

```
. summarize bp*
    Variable |        Obs        Mean    Std. Dev.        Min         Max
-------------+--------------------------------------------------------------
         bp1 |          5       116.6    7.635444        105         124
         bp2 |          5       117.4    19.17811         86         136
         bp3 |          5         114    11.13553        101         129
         bp4 |          5       115.8    10.54514        105         129
         bp5 |          5       121.4    10.87658        111         137
```

The `summarize` command below shows summary statistics for the variables that end in 1.

```
. summarize *1
    Variable |        Obs        Mean    Std. Dev.        Min         Max
-------------+--------------------------------------------------------------
         bp1 |          5       116.6    7.635444        105         124
         pl1 |          5        74.6    23.36236         52         105
```

The `detail` option can be used to obtain detailed summary statistics for the given variables.

23.13   DISPLAY

Stata equivalent: `describe`

Example: The dataset `cardio1.dta` contains an ID variable, an age variable, five measurements of blood pressure (named `bp1` to `bp5`), and five measurements of pulse (named `pl1` to `pl5`). The `describe` command below shows the variables in the dataset along with their storage type, display format, value labels, and variable labels.

```
. use cardio1

. describe

Contains data from cardio1.dta
  obs:             5
  vars:           12                       22 Dec 2009 19:50
  size:          100
```

variable name	storage type	display format	value label	variable label
id	byte	%3.0f		Identification variable
age	byte	%3.0f		Age of person
bp1	int	%3.0f		Systolic BP: Trial 1
bp2	int	%3.0f		Systolic BP: Trial 2
bp3	int	%3.0f		Systolic BP: Trial 3
bp4	int	%3.0f		Systolic BP: Trial 4
bp5	int	%3.0f		Systolic BP: Trial 5
pl1	int	%3.0f		Pulse: Trial 1
pl2	byte	%3.0f		Pulse: Trial 2
pl3	int	%3.0f		Pulse: Trial 3
pl4	int	%3.0f		Pulse: Trial 4
pl5	byte	%3.0f		Pulse: Trial 5

```
Sorted by:
```

23.14 DOCUMENT

Stata equivalent: `notes` and `note`
Examples: Consider the dataset `wws2.dta`, which is used below.

```
. use wws2, clear
(Working Women Survey w/fixes)
```

Using the `notes` command, we can see notes (documentation) that have been previously stored inside the dataset.

```
. notes
_dta:
  1. This is a hypothetical dataset and should not be used for analysis
     purposes
```

We can add our own comments using the `note` command.

```
. note : Don´t publish any results based on this hypothetical dataset.
```

You can also add notes to specific variables; for example, you can type

```
. note idcode: This is the unique ID for this dataset.
```

Using the `notes` command, we can again display all the notes in the dataset. We see the original note that was contained within the dataset and the two additional notes that were just added.

```
. notes
_dta:
    1.  This is a hypothetical dataset and should not be used for analysis
        purposes
    2.  Don't publish any results based on this hypothetical dataset.
idcode:
    1.  This is the unique ID for this dataset.
```

When the dataset is saved, all of these notes would be saved with the dataset.

Note: Make a note of it

I highly encourage you to liberally use the `note` command to add comments to your datasets. As you saw, you can add overall notes to the dataset as well as notes for specific variables. I use the `note` command frequently as part of my daily work to add documentation to my datasets. When I use such datasets in the future, the notes help me remember key pieces of information that help me understand and use the dataset. You can type `help notes` to learn more about how to create and view notes.

23.15 FACTOR

Stata equivalent: `factor`
Example: Here is a simple example taken directly from the Stata help file for `factor`.

```
. webuse bg2
(Physician-cost data)
. factor bg2cost1-bg2cost6
  (output omitted)
```

For details about the `factor` command, see `help factor`.

23.16 FILTER

Stata equivalent: `if` condition
Examples: Using the dataset `gss2012_sbs.dta`, we use the `summarize` command to show summary statistics for happiness for just the people who are married (that is, `married` is equal to 1).

```
. use gss2012_sbs, clear
. summarize happy7 if married==1
    Variable |        Obs        Mean    Std. Dev.        Min        Max
-------------+--------------------------------------------------------
      happy7 |        577    5.714038     .918385          1          7
```

The `keep if` command can be used to remove observations from the dataset in working memory (the equivalent of the SPSS `SELECT IF` command). The `keep if` command is used below to keep just those who are married.

```
. keep if married==1
(1,074 observations deleted)
```

Note: Double equal

Note that I typed `if married==1` and not `if married=1`. This is because Stata (like many programming languages) considers `married=1` (one equal sign) to signify the assignment of the value 1 to the variable `married`, whereas `married==1` (two equal signs) is a test of whether the variable `married` is equal to 1.

23.17   FORMATS

Stata equivalent: `format`
I feel this is a seldom-used command. See `help format` for more details about using this command.

23.18   FREQUENCIES

Stata equivalent: `tabulate`
Examples: Using the dataset gss2012_sbs.dta, we use the `tabulate` command to show a frequency table for the variable `happy7`.

```
. use gss2012_sbs, clear

. tabulate happy7
```

how happy R is (recoded)	Freq.	Percent	Cum.
Completely unhappy	5	0.39	0.39
Very unhappy	16	1.25	1.64
Fairly unhappy	35	2.73	4.36
Neither happy nor unhappy	77	6.00	10.36
Fairly happy	440	34.27	44.63
Very happy	563	43.85	88.47
Completely happy	148	11.53	100.00
Total	1,284	100.00	

This shows us that among the 1,284 people with a valid answer to this item, 148 said that they were Completely happy. This composed 11.53% of the valid responses. Let's repeat this command by adding the `missing` option.

```
. tabulate happy7, missing
```

how happy R is (recoded)	Freq.	Percent	Cum.
Completely unhappy	5	0.25	0.25
Very unhappy	16	0.81	1.06
Fairly unhappy	35	1.77	2.84
Neither happy nor unhappy	77	3.90	6.74
Fairly happy	440	22.29	29.03
Very happy	563	28.52	57.55
Completely happy	148	7.50	65.05
Cannot choose	11	0.56	65.60
Inapplicable	672	34.04	99.65
No Answer	7	0.35	100.00
Total	1,974	100.00	

Now we see a table that also includes those with missing responses to the variable `happy7`. We still see that 148 said that they were completely happy and that this composed 7.50% of the 1,974 people in the entire dataset.

For output that is more similar to the SPSS `FREQUENCIES` command, you can download the `fre` command, written by Ben Jann (a Stata user from the University of Bern), using the `ssc` command shown below.[1]

```
. ssc install fre
```

After downloading this command, you can execute it like any other Stata command, as shown below.

---

1. See section 1.2.7 for more information about accessing user-written programs in Stata.

```
. fre happy7
```

happy7 — how happy R is (recoded)

			Freq.	Percent	Valid	Cum.
Valid	1	Completely unhappy	5	0.25	0.39	0.39
	2	Very unhappy	16	0.81	1.25	1.64
	3	Fairly unhappy	35	1.77	2.73	4.36
	4	Neither happy nor unhappy	77	3.90	6.00	10.36
	5	Fairly happy	440	22.29	34.27	44.63
	6	Very happy	563	28.52	43.85	88.47
	7	Completely happy	148	7.50	11.53	100.00
		Total	1284	65.05	100.00	
Missing	.c	Cannot choose	11	0.56		
	.i	Inapplicable	672	34.04		
	.n	No Answer	7	0.35		
		Total	690	34.95		
Total			1974	100.00		

You can access help for the `fre` command by typing `help fre`.

23.19 GET FILE

Stata equivalent: `use`
Example: The example below uses the dataset `gss2012_sbs.dta`.

```
. use gss2012_sbs
```

If you already have an unsaved dataset in memory (that you do not care to save), you can first use the `clear` command before using the dataset.

```
. clear
. use gss2012_sbs
```

Or you can use the `clear` option with the `use` command, as shown below.

```
. use gss2012_sbs, clear
```

23.20 GET TRANSLATE

Stata equivalent: `import`
The Stata `import` command can be used to read datasets such as SAS XPORT files and Excel workbooks (see help `import`). For greater flexibility in reading a wider variety of datasets, I suggest that you consider purchasing Stat/Transfer, which you can obtain

directly from StataCorp. You can see a video demonstrating the use of Stat/Transfer at http://www.stata.com/sbs/stat-transfer.

23.21 LOGISTIC REGRESSION

Stata equivalent: `logistic`
Example: Using the dataset `gss2012_sbs.dta`, we use the `logistic` command to perform a logistic regression predicting whether one is very happy (Yes = 1) as a function of education, age, and whether one is married.

```
. logistic vhappy educ age i.married, nolog or

Logistic regression                             Number of obs   =      1,958
                                                LR chi2(3)      =     107.80
                                                Prob > chi2     =     0.0000
Log likelihood = -1146.0463                     Pseudo R2       =     0.0449
```

vhappy	Odds Ratio	Std. Err.	z	P>|z|	[95% Conf. Interval]	
educ	1.005681	.0161567	0.35	0.724	.9745082	1.037852
age	.9973204	.0029539	-0.91	0.365	.9915476	1.003127
married						
Married	2.827507	.2904122	10.12	0.000	2.311943	3.458042
_cons	.2694878	.0731764	-4.83	0.000	.1582723	.4588529

Neither education nor age was a significant predictor of whether one was very happy. However, the odds of being happy for those who were married was 2.83 times the odds of being happy for those who were not married.

23.22 MATCH FILES

Stata equivalent: `merge`
Example: Consider the dataset `moms1.dta`, which contains a family ID variable, the mother's age, race, and whether she graduated from high school.

```
. use moms1

. list
```

	famid	mage	mrace	mhs
1.	1	33	2	1
2.	2	28	1	1
3.	3	24	2	1
4.	4	21	1	0

Also consider the dataset `dads1.dta`, which contains a family ID variable, the father's age, race, and whether he graduated from high school.

```
. use dads1
. list
```

	famid	dage	drace	dhs
1.	1	21	1	0
2.	2	25	1	1
3.	3	31	2	1
4.	4	25	2	1

The following `use` command uses the dataset `moms1.dta`. The `merge` command performs a one-to-one merge in which the variable `famid` is used to merge the data in memory with the `dads1.dta` dataset. The `merge` command shows that four observations were matched.

```
. use moms1
. merge 1:1 famid using dads1
    Result                           # of obs.

    not matched                              0
    matched                                  4  (_merge==3)
```

The `list` command shows the resulting dataset, which contains the family ID, the variables from `moms1.dta`, the variables from `dads1.dta`, and the variable `_merge`, which shows the matching status for each observation.

```
. list
```

	famid	mage	mrace	mhs	dage	drace	dhs	_merge
1.	1	33	2	1	21	1	0	matched (3)
2.	2	28	1	1	25	1	1	matched (3)
3.	3	24	2	1	31	2	1	matched (3)
4.	4	21	1	0	25	2	1	matched (3)

Note: Other types of merges

In this example, we saw a one-to-one merge. Stata also supports one-to-many merges (indicated by `1:m`) and many-to-one merges (indicated by `m:1`).

23.23   MEANS

Stata equivalent: `tabulate X, summarize(Y)`
Examples: Using the dataset `gss2012_sbs.dta`, we use the `tabulate` command to show summary statistics for happiness broken down by gender.

```
. use gss2012_sbs, clear
. tabulate female, summarize(happy7)
    Is R female |
       (yes=1   |  Summary of how happy R is (recoded)
        no=0)?  |      Mean      Std. Dev.       Freq.
----------------+-----------------------------------------
           Male |  5.4812287    1.0445573         586
         Female |  5.5186246     .95393009        698
----------------+-----------------------------------------
          Total |  5.5015576     .99609404       1,284
```

The following command forms a cross-tabulation of gender by married and shows the average happiness for each cell.

```
. tabulate female married, summarize(happy7)

    Means, Standard Deviations and Frequencies of how happy R is (recoded)

       Is R  | marital: married=1,
      female |     unmarried=0
      (yes=1 |      (recoded)
      no=0)? | Unmarried    Married  |    Total
  -----------+----------------------+-----------
        Male | 5.2924528   5.7052239 | 5.4812287
             | 1.0888043    .94356223 | 1.0445573
             |      318         268   |      586
  -----------+----------------------+-----------
      Female | 5.3573265   5.7216828 | 5.5186246
             |  .96768884   .89744356 |  .95393009
             |      389         309   |      698
  -----------+----------------------+-----------
       Total | 5.3281471   5.7140381 | 5.5015576
             | 1.0237062    .91838503 |  .99609404
             |      707         577   |     1284
```

23.24   MISSING VALUES

Stata equivalent: `mvdecode`
Example: Consider the raw data file `cardio2miss.txt`, shown below using the `type` command.

```
. type cardio2miss.txt
1 40 54 115 87 86 93 129 81 105 -2 -2
2 30 92 123 88 136 125 107 87 111 58 120
3 16 105 -1 97 122 128 101 57 109 68 112
4 23 52 105 79 115 71 121 106 129 39 137
5 18 70 116 -1 128 52 112 68 125 59 111
```

In this file, a value of -1 indicates that a value is missing because of an equipment error, and a value of -2 indicates a missing value due to a subject dropping out of the study. This raw data file is read into Stata using the `infile` command below. The `list` command shows the contents of the dataset after you read it into Stata.

```
. infile id age bp1-bp5 pl1-pl5 using cardio2miss.txt
(5 observations read)
. list
```

	id	age	bp1	bp2	bp3	bp4	bp5	pl1	pl2	pl3	pl4	pl5
1.	1	40	54	115	87	86	93	129	81	105	-2	-2
2.	2	30	92	123	88	136	125	107	87	111	58	120
3.	3	16	105	-1	97	122	128	101	57	109	68	112
4.	4	23	52	105	79	115	71	121	106	129	39	137
5.	5	18	70	116	-1	128	52	112	68	125	59	111

The `mvdecode` command can be used to convert the values of -1 and -2 into missing value codes that Stata will recognize. For any variable starting with `bp` or `pl`, a value of -1 will be converted into a missing value code of `.a`, and a value of -2 will be converted into a missing value code of `.b`.

```
. mvdecode bp* pl*, mv(-1=.a \ -2=.b)
        bp2: 1 missing value generated
        bp3: 1 missing value generated
        pl4: 1 missing value generated
        pl5: 1 missing value generated
```

The results of this can be seen in the `list` command below.

```
. list
```

	id	age	bp1	bp2	bp3	bp4	bp5	pl1	pl2	pl3	pl4	pl5
1.	1	40	54	115	87	86	93	129	81	105	.b	.b
2.	2	30	92	123	88	136	125	107	87	111	58	120
3.	3	16	105	.a	97	122	128	101	57	109	68	112
4.	4	23	52	105	79	115	71	121	106	129	39	137
5.	5	18	70	116	.a	128	52	112	68	125	59	111

For more information, see `help missing`.

23.25 MIXED

Stata equivalent: `mixed`

Example: The dataset `school_read.dta` contains the variable `read`, a standardized writing test that ranges from 0 to 1,500. These scores were obtained by randomly sampling 100 schools, and students were randomly sampled from each of the schools. The dataset includes a student-level predictor, socioeconomic status (`ses`), and a school-level predictor (`schtype`), the type of school, which is coded as follows: 1 = private, 2 = public, 3 = Catholic. Of the 100 schools, 35 were private (non-Catholic), 35 were public, and 30 were Catholic schools.

The aim of this hypothetical study is to determine whether the strength of the relationship between `ses` and `read` differs by the type of school. In other words, the goal is to determine whether there is a cross-level interaction of `ses` and `schtype` in the prediction of `read`. The `mixed` command below predicts `read` from `c.ses`, `i.schtype`, and the `i.schtype#c.ses` interaction. The random-effects portion of the model indicates that `ses` is a random effect across levels of `schoolid`. The `cov(un)` option permits the random intercept and random `ses` slope to be correlated. (The `nolog` and `noheader` options are used to save space by suppressing the iteration log and the header information.)

```
. use school_read
. mixed read i.schtype##c.ses || schoolid: ses, cov(un) nolog noheader
```

read	Coef.	Std. Err.	z	P>|z|	[95% Conf. Interval]	
schtype						
Public	41.96651	23.26979	1.80	0.071	-3.641439	87.57447
Catholic	152.4775	23.78712	6.41	0.000	105.8556	199.0994
ses	4.042064	.6734925	6.00	0.000	2.722043	5.362085
schtype#						
c.ses						
Public	-1.240951	.9512141	-1.30	0.192	-3.105296	.6233947
Catholic	-3.3688	.9865231	-3.41	0.001	-5.30235	-1.435251
_cons	519.2829	16.57457	31.33	0.000	486.7974	551.7685

Random-effects Parameters	Estimate	Std. Err.	[95% Conf. Interval]	
schoolid: Unstructured				
var(ses)	12.13332	2.128858	8.602639	17.11306
var(_cons)	4.916487	73.92184	7.82e-13	3.09e+13
cov(ses,_cons)	6.75668	31.29822	-54.58671	68.10007
var(Residual)	10015.67	264.2554	9510.902	10547.23

LR test vs. linear model: chi2(3) = 4052.81   Prob > chi2 = 0.0000
Note: LR test is conservative and provided only for reference.

The results show that the coefficient for ses does not differ for public schools versus private schools ($p = 0.192$) but that there is a significant difference in the coefficient for ses for Catholic schools versus private schools ($p = 0.001$).

23.26 MULTIPLE IMPUTATION

Stata equivalent: mi
For more information about multiple imputation in Stata, see help mi. There is an entire manual of over 350 pages devoted to this topic in Stata.

23.27 NOMREG

Stata equivalent: mlogit
Example: For this example, let's use the variable happy3 as a nominal outcome variable. This variable contains the respondent's rating of his or her happiness on a three-point scale: 3 = very happy, 2 = pretty happy, and 1 = not too happy. (For the sake of illustration, this ordered outcome will be analyzed using a multinomial logistic model. In section 23.28, this outcome will be analyzed using an ordinal logistic model.)

Let's predict this happiness rating from gender, education, and age. This is performed using the mlogit command shown below. The nolog option is included to suppress the iteration log (to save space), while the rrr option is used to display relative risk ratios (instead of coefficients). The baseoutcome(1) option specifies that level 1 (not too happy) is used as the reference outcome. This example uses the dataset gss2012_sbs.dta, as shown below.

```
. use gss2012_sbs
. mlogit happy3 i.female educ age, nolog rrr baseoutcome(1)
Multinomial logistic regression              Number of obs    =      1,958
                                             LR chi2(6)       =      37.96
                                             Prob > chi2      =     0.0000
Log likelihood = -1864.2369                  Pseudo R2        =     0.0101
```

happy3	RRR	Std. Err.	z	P>|z|	[95% Conf. Interval]	
not_too_ha~y	(base outcome)					
pretty_happy						
female						
Female	.9502947	.1306342	-0.37	0.711	.7258488	1.244143
educ	1.129255	.0241014	5.70	0.000	1.082992	1.177495
age	.9950909	.0037856	-1.29	0.196	.987699	1.002538
_cons	1.061176	.3777012	0.17	0.868	.5282281	2.131832
very_happy						
female						
Female	1.051422	.1561319	0.34	0.736	.7859175	1.40662
educ	1.121509	.0259421	4.96	0.000	1.071799	1.173525
age	.9961748	.0040973	-0.93	0.351	.9881765	1.004238
_cons	.566829	.2194677	-1.47	0.143	.2653869	1.210667

In comparing outcome 2 with outcome 1, (pretty_happy) with (not_too_happy), we see that education is a significant predictor ($p < 0.001$). Likewise, in comparing outcome 3 with outcome 1, (very_happy) with (not_too_happy), we see that education is also a significant predictor ($p < 0.001$).

23.28 PLUM

Stata equivalent: ologit
Example: As in the previous example, let's use the variable happy3 as the outcome variable. This variable contains the respondent's rating of his or her happiness on a three-point scale: 3 = very happy, 2 = pretty happy, and 1 = not too happy. Let's use the ologit command to perform an ordinal logistic regression predicting this happiness rating from gender, education, and age. The nolog option is included to suppress the iteration log (to save space), while the or option is used to display odds ratios (instead of coefficients).

```
. use gss2012_sbs
. ologit happy3 i.female educ age, nolog or
Ordered logistic regression                    Number of obs   =      1,958
                                               LR chi2(3)      =      15.45
                                               Prob > chi2     =     0.0015
Log likelihood = -1875.4906                    Pseudo R2       =     0.0041
```

happy3	Odds Ratio	Std. Err.	z	P>|z|	[95% Conf. Interval]	
female						
Female	1.06208	.0938107	0.68	0.495	.8932499	1.262821
educ	1.056022	.0152414	3.78	0.000	1.026568	1.086321
age	.9983659	.0024889	-0.66	0.512	.9934997	1.003256
/cut1	-1.133597	.2446024			-1.613009	-.6541851
/cut2	1.539478	.2457125			1.05789	2.021066

Among these three predictors, education is the only significant predictor of happiness. Greater education is associated with a greater odds of being happy ($p < 0.001$).

23.29   PROBIT

Stata equivalent: `probit`

Example: Using the dataset `gss2012_sbs.dta`, we use the `probit` command to perform a probit regression predicting whether one is very happy (yes = 1) as a function of education, age, and whether one is married.

```
. probit vhappy educ age i.married, nolog
Probit regression                              Number of obs   =      1,958
                                               LR chi2(3)      =     107.82
                                               Prob > chi2     =     0.0000
Log likelihood = -1146.0329                    Pseudo R2       =     0.0449
```

vhappy	Coef.	Std. Err.	z	P>|z|	[95% Conf. Interval]	
educ	.0042721	.009713	0.44	0.660	-.014765	.0233091
age	-.0015311	.0017404	-0.88	0.379	-.0049421	.00188
married						
Married	.6252741	.0611129	10.23	0.000	.505495	.7450532
_cons	-.8126066	.1627255	-4.99	0.000	-1.131543	-.4936705

Neither education nor age was a significant predictor of being very happy. Those who were married were more likely to be happy than those who were not married ($p < 0.001$).

23.30   RECODE

Stata equivalent: `recode`

Examples: Using the dataset gss2012_sbs.dta, we use the `codebook` command to show the values and labels associated with the variable `female`.

```
. use gss2012_sbs, clear

. codebook female

────────────────────────────────────────────────────────────────────────────
female                                              Is R female (yes=1 no=0)?
────────────────────────────────────────────────────────────────────────────

                 type:  numeric (byte)
                label:  female

                range:  [0,1]                        units:  1
         unique values:  2                         missing .:  0/1,974

            tabulation:  Freq.   Numeric  Label
                          886        0    Male
                        1,088        1    Female
```

Say we wanted to create a dummy variable called `male` (based on female) that is coded 1 = male and 0 = female. We can do so as shown below.

```
. recode female (1=0) (0=1), generate(male)
(1974 differences between female and male)

. tabulate male

  RECODE of │
  female (Is │
   R female │
     (yes=1 │
     no=0)?) │      Freq.     Percent        Cum.
────────────┼───────────────────────────────────────
          0 │      1,088       55.12       55.12
          1 │        886       44.88      100.00
────────────┼───────────────────────────────────────
      Total │      1,974      100.00
```

As a second example, let's recode the years of education (stored in the variable `educ`) to create a variable named `hs` that is coded 1 if the person has 12 or more years of education and is coded 0 for people with fewer than 12 years of education. This example applies value labels to the recoded variable, labeling 0 as "0. Not HS Grad" and labeling 1 as "1. HS Grad".

```
. recode educ (min/11=0 "0. Not HS Grad")
>              (12/max=1 "1. HS Grad"), gen(hs)
(1969 differences between educ and hs)
```

We can verify that our recoding of `educ` into `hs` was correct via the `tabstat` command below. We see that `hs` properly ranges from 0 to 11 when `hs` is 0 and that `hs` properly ranges from 12 to 20 when `hs` is 1.

```
. tabstat educ, by(hs) stat(min max)
Summary for variables: educ
     by categories of: hs (RECODE of educ (highest year of school completed))
```

hs	min	max
0. Not HS Grad	0	11
1. HS Grad	12	20
Total	0	20

Using the `tabulate` command, we see the frequencies of our recoded variable `hs`. Note how this displays the labeled values of `hs` using the labels we assigned in the `recode` command.

```
. tabulate hs
```

RECODE of educ (highest year of school completed)	Freq.	Percent	Cum.
0. Not HS Grad	318	16.13	16.13
1. HS Grad	1,654	83.87	100.00
Total	1,972	100.00	

23.31 RELIABILITY

Stata equivalent: `alpha`
Example: Here is a simple example taken directly from the help file for the `alpha` command. For more details, see `help alpha`.

```
. webuse automiss, clear
(1978 Automobile Data)
. alpha price headroom rep78 trunk weight length turn displacement
  (output omitted )
```

23.32 RENAME VARIABLES

Stata equivalent: `rename`
Examples: Consider the dataset `cardio1.dta`. The `describe` command shows the names of the variables in the dataset.

```
. use cardio1

. describe

Contains data from cardio1.dta
  obs:             5
  vars:           12                          22 Dec 2009 19:50
  size:          100
```

variable name	storage type	display format	value label	variable label
id	byte	%3.0f		Identification variable
age	byte	%3.0f		Age of person
bp1	int	%3.0f		Systolic BP: Trial 1
bp2	int	%3.0f		Systolic BP: Trial 2
bp3	int	%3.0f		Systolic BP: Trial 3
bp4	int	%3.0f		Systolic BP: Trial 4
bp5	int	%3.0f		Systolic BP: Trial 5
pl1	int	%3.0f		Pulse: Trial 1
pl2	byte	%3.0f		Pulse: Trial 2
pl3	int	%3.0f		Pulse: Trial 3
pl4	int	%3.0f		Pulse: Trial 4
pl5	byte	%3.0f		Pulse: Trial 5

```
Sorted by:
```

The `rename` command is used below to rename the variable `id` to `idvar`.

```
. rename id idvar
```

The `rename` command can also be used to rename groups of variables. For example, you can combine the `rename` command with the wildcard character * to rename all the variables that begin with `pl` with `pulse`.

```
. rename pl* pulse*
```

The `describe` command shows us the results of these `rename` commands, showing the names of the variables in the dataset.

```
. describe

Contains data from cardio1.dta
  obs:             5
  vars:           12                          22 Dec 2009 19:50
  size:          100
```

variable name	storage type	display format	value label	variable label
idvar	byte	%3.0f		Identification variable
age	byte	%3.0f		Age of person
bp1	int	%3.0f		Systolic BP: Trial 1
bp2	int	%3.0f		Systolic BP: Trial 2
bp3	int	%3.0f		Systolic BP: Trial 3
bp4	int	%3.0f		Systolic BP: Trial 4
bp5	int	%3.0f		Systolic BP: Trial 5
pulse1	int	%3.0f		Pulse: Trial 1
pulse2	byte	%3.0f		Pulse: Trial 2
pulse3	int	%3.0f		Pulse: Trial 3
pulse4	int	%3.0f		Pulse: Trial 4
pulse5	byte	%3.0f		Pulse: Trial 5

```
Sorted by:
    Note: Dataset has changed since last saved.
```

Tip: Renaming groups of variables

The `rename` command is amazingly powerful for renaming groups of variables. See `help rename group` for more information.

23.33   SAVE

Stata equivalent: `save`

Example: The `save` command saves the working dataset (in memory) using the name you supply with the `save` command. The `save` command shown below saves the working dataset (in memory) using the name `mydata.dta`. Note that the `.dta` extension is not required.

```
. save mydata
```

If the file `mydata.dta` already existed on disk, the `replace` option is needed to overwrite the file on disk.

```
. save mydata, replace
```

23.34   SELECT IF

Stata equivalents: `keep if`, `drop if`
Examples: After using `gss2012_sbs.dta`, we use the `keep if` command to keep just those who are married (if `married` is equal to 1).

```
. use gss2012_sbs, clear
. keep if married==1
(1,074 observations deleted)
```

You can also use the `drop if` command to drop unwanted observations. The example below uses `gss2012_sbs.dta` and then removes observations for those who are under age 65.

```
. use gss2012_sbs, clear
. drop if age < 65
(1,575 observations deleted)
```

23.35   SAVE TRANSLATE

Stata equivalent: `export`
The Stata `export` command can be used to save datasets as SAS XPORT files and as Excel workbooks (see help `export`). For greater flexibility in converting Stata datasets to other formats, I suggest that you consider purchasing Stat/Transfer, which you can obtain directly from StataCorp. You can see a video demonstrating the use of Stat/Transfer at http://www.stata.com/sbs/stat-transfer.

23.36   SORT CASES

Stata equivalent: `sort`
Example: Consider the dataset `cardio1.dta`. This dataset is used below, and the observations for all variables are listed.

```
. use cardio1, clear
. list
```

	id	age	bp1	bp2	bp3	bp4	bp5	pl1	pl2	pl3	pl4	pl5
1.	1	40	115	86	129	105	127	54	87	93	81	92
2.	2	30	123	136	107	111	120	92	88	125	87	58
3.	3	16	124	122	101	109	112	105	97	128	57	68
4.	4	23	105	115	121	129	137	52	79	71	106	39
5.	5	18	116	128	112	125	111	70	64	52	68	59

The `sort` command below sorts the observations based on `bp1`.

```
. sort bp1
```

The `list` command shows that the observations are now sorted based on `bp1`.

```
. list
```

	id	age	bp1	bp2	bp3	bp4	bp5	pl1	pl2	pl3	pl4	pl5
1.	4	23	105	115	121	129	137	52	79	71	106	39
2.	1	40	115	86	129	105	127	54	87	93	81	92
3.	5	18	116	128	112	125	111	70	64	52	68	59
4.	2	30	123	136	107	111	120	92	88	125	87	58
5.	3	16	124	122	101	109	112	105	97	128	57	68

23.37   SORT VARIABLES

Stata equivalent: `order`
See `help order` for more details about ordering variables within a Stata dataset.

23.38   SUMMARIZE

Stata equivalent: `tabulate X, summarize(Y)`
Examples: Using the dataset gss2012_sbs.dta, we use the `tabulate` command to show summary statistics for happiness broken down by gender.

```
. use gss2012_sbs, clear
. tabulate female, summarize(happy7)
```

Is R female (yes=1 no=0)?	Summary of how happy R is (recoded)		
	Mean	Std. Dev.	Freq.
Male	5.4812287	1.0445573	586
Female	5.5186246	.95393009	698
Total	5.5015576	.99609404	1,284

23.39   T-TEST

Stata equivalent: `ttest`
Examples: Using the dataset gss2012_sbs.dta, we use the `ttest` command to compare the average happiness of respondents who are married with the average happiness of those who are not married.

```
. use gss2012_sbs, clear

. ttest happy7, by(married)
Two-sample t test with equal variances
```

Group	Obs	Mean	Std. Err.	Std. Dev.	[95% Conf. Interval]	
Unmarrie	707	5.328147	.0385004	1.023706	5.252558	5.403736
Married	577	5.714038	.0382329	.918385	5.638945	5.789131
combined	1,284	5.501558	.0277983	.996094	5.447023	5.556093
diff		-.385891	.0548568		-.49351	-.2782721

```
        diff = mean(Unmarrie) - mean(Married)                    t =   -7.0345
Ho: diff = 0                                    degrees of freedom =       1282

    Ha: diff < 0                  Ha: diff != 0                  Ha: diff > 0
 Pr(T < t) = 0.0000        Pr(|T| > |t|) = 0.0000          Pr(T > t) = 1.0000
```

The results show that the average happiness rating for those who are not married ($M = 5.33$) is significantly lower than the rating for those who are married ($M = 5.71$), $t(1282) = -7.03$, $p < 0.001$.

23.40   VALUE LABELS

Stata equivalent: `label values`
Example: Consider the dataset `survey2.dta`, which contains data from a student survey. The `describe` command shows the names of the variables and also shows there are variable labels for each of the variables. But none of the variables have value labels.

```
. use survey2
(Survey of graduate students)

. describe

Contains data from survey2.dta
  obs:             8                          Survey of graduate students
 vars:             9                          2 Feb 2010 18:48
 size:           432
```

variable name	storage type	display format	value label	variable label
id	float	%9.0g		Unique identification variable
gender	float	%9.0g		Gender of student
race	float	%9.0g		Race of student
havechild	float	%9.0g		Given birth to a child?
ksex	float	%9.0g		Sex of child
bdays	str10	%10s		Birthday of student
income	float	%9.0g		Income of student
kbdays	str10	%10s		Birthday of child
kidname	str10	%10s		Name of child

```
Sorted by:
```

Labeling values is a two-step process in Stata. The first step is to use the `label define` command to define a label. The second step is to use the `label values` command to attach the label to a variable. In the example below, the `label define` command creates the label named `mf`, and then the `label values` command attaches the label `mf` to the variable `gender`.

```
. label define mf 1 "Male" 2 "Female"
. label values gender mf
```

We can see the values and labels for the variable `gender` with the `codebook` command below.

```
. codebook gender
```

gender				Gender of student
type:	numeric (float)			
label:	mf			
range:	[1,2]		units:	1
unique values:	2		missing .:	0/8
tabulation:	Freq.	Numeric	Label	
	3	1	Male	
	5	2	Female	

23.41 VARIABLE LABELS

Stata equivalents: `label variable`
Example: Consider the dataset `survey1.dta`, which contains data from a student survey. The `describe` command shows the names of the variables and also shows us that none of the variables are labeled.

```
. use survey1

. describe
Contains data from survey1.dta
  obs:             8
  vars:            9                              1 Jan 2010 12:13
  size:          432
```

variable name	storage type	display format	value label	variable label
id	float	%9.0g		
gender	float	%9.0g		
race	float	%9.0g		
havechild	float	%9.0g		
ksex	float	%9.0g		
bdays	str10	%10s		
income	float	%9.0g		
kbdays	str10	%10s		
kidname	str10	%10s		

```
Sorted by:
```

The label variable command can be used to apply labels to variables. It is used below to label the variables gender and race.

```
. label variable gender "Gender of student"

. label variable race "Race of student"
```

The describe command shows that these variables are now labeled with the labels we supplied.

```
. describe
Contains data from survey1.dta
  obs:             8
  vars:            9                              1 Jan 2010 12:13
  size:          432
```

variable name	storage type	display format	value label	variable label
id	float	%9.0g		
gender	float	%9.0g		Gender of student
race	float	%9.0g		Race of student
havechild	float	%9.0g		
ksex	float	%9.0g		
bdays	str10	%10s		
income	float	%9.0g		
kbdays	str10	%10s		
kidname	str10	%10s		

```
Sorted by:
```

23.42   VARSTOCASES

Stata equivalent: `reshape long`
Example: The dataset `cardio_wide.dta` contains information about six people in which five blood pressure measurements are stored in the variables `bp1` to `bp5` and five pulse measurements are stored in the variables `pl1` to `pl5`.

```
. use cardio_wide

. list
```

	id	age	bp1	bp2	bp3	bp4	bp5	pl1	pl2	pl3	pl4	pl5
1.	1	40	115	86	129	105	127	54	87	93	81	92
2.	2	30	123	136	107	111	120	92	88	125	87	58
3.	3	16	124	122	101	109	112	105	97	128	57	68
4.	4	23	105	115	121	129	137	52	79	71	106	39
5.	5	18	116	128	112	125	111	70	64	52	68	59
6.	6	27	108	126	124	131	107	74	78	92	99	80

The `reshape long` command converts this into a long dataset in which the five blood pressure measurements and five pulse measurements are stored in separate observations. The `i(id)` option indicates that the observations in the wide dataset are identified by the variable `id`. The `j(trial)` option indicates that we want suffixes of `bp` and `pl` to be stored in the variable `trial`.

```
. reshape long bp pl, i(id) j(trial)
(note: j = 1 2 3 4 5)
Data                             wide   ->   long
────────────────────────────────────────────────────
Number of obs.                      6   ->     30
Number of variables                12   ->      5
j variable (5 values)                   ->   trial
xij variables:
                       bp1 bp2 ... bp5   ->   bp
                       pl1 pl2 ... pl5   ->   pl
────────────────────────────────────────────────────
```

The `list` command shows that the dataset that had 6 observations and 5 measurements per observation has been converted into a dataset with 30 observations in which the variable `id` identifies the person and the variable `trial` identifies the trial from which the blood pressure (`bp`) and pulse (`pl`) measurement originated.

```
. list, sepby(id)
```

	id	trial	age	bp	pl
1.	1	1	40	115	54
2.	1	2	40	86	87
3.	1	3	40	129	93
4.	1	4	40	105	81
5.	1	5	40	127	92
6.	2	1	30	123	92
7.	2	2	30	136	88
8.	2	3	30	107	125
9.	2	4	30	111	87
10.	2	5	30	120	58
11.	3	1	16	124	105
12.	3	2	16	122	97
13.	3	3	16	101	128
14.	3	4	16	109	57
15.	3	5	16	112	68
16.	4	1	23	105	52
17.	4	2	23	115	79
18.	4	3	23	121	71
19.	4	4	23	129	106
20.	4	5	23	137	39
21.	5	1	18	116	70
22.	5	2	18	128	64
23.	5	3	18	112	52
24.	5	4	18	125	68
25.	5	5	18	111	59
26.	6	1	27	108	74
27.	6	2	27	126	78
28.	6	3	27	124	92
29.	6	4	27	131	99
30.	6	5	27	107	80

23.43   Closing thoughts

One of the challenges of learning a new statistical package is the constant feeling that you know how to do a task in the package you know but don't know the equivalent command to do that task in the new package. To that end, this chapter is aimed at people who know SPSS and are using this book to learn Stata. Rather than trying to translate every SPSS command, I focused on the commands that I find that people use most. By seeing not only the equivalent Stata command but also a brief example, you can quickly learn some of the most commonly used Stata commands. To that end, I close by referring you back to section 1.4 of chapter 1, which describes resources that I hope you will find useful in your journey of learning and using Stata.

References

Acock, A. C. 2013. *Discovering Structural Equation Modeling Using Stata*. Revised ed. College Station, TX: Stata Press.

Baum, C. F. 2009. *An Introduction to Stata Programming*. College Station, TX: Stata Press.

Cleves, M., W. Gould, R. G. Gutierrez, and Y. V. Marchenko. 2010. *An Introduction to Survival Analysis Using Stata*. 3rd ed. College Station, TX: Stata Press.

Cohen, J. 1992. A power primer. *Psychological Bulletin* 112: 155–159.

Cox, N. J. 2003. extremes: Stata module to list extreme values of a variable. Statistical Software Components S430801, Department of Economics, Boston College. https://ideas.repec.org/c/boc/bocode/s430801.html.

Fox, J. 1991. Regression Diagnostics: An Introduction. Sage University Papers Series on Quantitative Applications in the Social Sciences, 07-079. Newbury Park, CA: Sage.

Gallup, J. L. 2012a. A new system for formatting estimation tables. *Stata Journal* 12: 3–28.

———. 2012b. A programmer's command to build formatted statistical tables. *Stata Journal* 12: 655–673.

Huber, P. J. 1967. The behavior of maximum likelihood estimates under nonstandard conditions. In Vol. 1 of *Proceedings of the Fifth Berkeley Symposium on Mathematical Statistics and Probability*, 221–233. Berkeley: University of California Press.

Jann, B. 2005. Making regression tables from stored estimates. *Stata Journal* 5: 288–308.

———. 2007a. fre: Stata module to display one-way frequency table. Statistical Software Components S456835, Department of Economics, Boston College. https://ideas.repec.org/c/boc/bocode/s456835.html.

———. 2007b. Making regression tables simplified. *Stata Journal* 7: 227–244.

———. 2014. Plotting regression coefficients and other estimates. *Stata Journal* 14: 708–737.

Lokshin, M., and Z. Sajaia. 2008. Creating print-ready tables in Stata. *Stata Journal* 8: 374–389.

Lumley, T., P. Diehr, S. Emerson, and L. Chen. 2002. The importance of the normality assumption in large public health data sets. *Annual Review of Public Health* 23: 151–169.

Mitchell, M. N. 2010. *Data Management Using Stata: A Practical Handbook*. College Station, TX: Stata Press.

———. 2012a. *Interpreting and Visualizing Regression Models Using Stata*. College Station, TX: Stata Press.

———. 2012b. *A Visual Guide to Stata Graphics*. 3rd ed. College Station, TX: Stata Press.

Murnane, R. J., and J. B. Willett. 2011. *Methods Matter: Improving Causal Inference in Educational and Social Science Research*. New York: Oxford University Press.

Rabe-Hesketh, S., and A. Skrondal. 2012a. *Multilevel and Longitudinal Modeling Using Stata. Vol. I: Continuous Responses*. 3rd ed. College Station, TX: Stata Press.

———. 2012b. *Multilevel and Longitudinal Modeling Using Stata. Vol. II: Categorical Responses, Counts, and Survival*. 3rd ed. College Station, TX: Stata Press.

Raudenbush, S. W., and A. S. Bryk. 2002. *Hierarchical Linear Models: Applications and Data Analysis Methods*, vol. 1. Thousand Oaks, CA: Sage.

Rausch, J. R., S. E. Maxwell, and K. Kelley. 2003. Analytic methods for questions pertaining to a randomized pretest, posttest, follow-up design. *Journal of Clinical Child and Adolescent Psychology* 32: 467–486.

Schilling, M. F., A. E. Watkins, and W. Watkins. 2002. Is human height bimodal? *American Statistician* 56: 223–229.

Shadish, W. R., T. D. Cook, and D. T. Campbell. 2002. *Experimental and Quasi-Experimental Designs for Generalized Causal Inference*. Boston: Houghton Mifflin.

Singer, J. D., and J. B. Willett. 2003. *Applied Longitudinal Data Analysis: Modeling Change and Event Occurrence*. New York: Oxford University Press.

Wada, R. 2005. outreg2: Stata module to arrange regression outputs into an illustrative table. Statistical Software Components S456416, Department of Economics, Boston College. https://ideas.repec.org/c/boc/bocode/s456416.html.

Weisberg, S. 2014. *Applied Linear Regression*. 4th ed. New York: Wiley.

White, H. 1980. A heteroskedasticity-consistent covariance matrix estimator and a direct test for heteroskedasticity. *Econometrica* 48: 817–838.

Wilcox, R. R. 1987. New designs in analysis of variance. *Annual Review of Psychology* 38: 29–60.

Author index

Subject index